Cyber-Physical Systems

The SEI Series in Software Engineering

Software Engineering Institute of Carnegie Mellon University and Addison-Wesley

Visit informit.com/sei for a complete list of available publications.

The SEI Series in Software Engineering is a collaborative undertaking of the Carnegie Mellon Software Engineering Institute (SEI) and Addison-Wesley to develop and publish books on software engineering and related topics. The common goal of the SEI and Addison-Wesley is to provide the most current information on these topics in a form that is easily usable by practitioners and students.

Titles in the series describe frameworks, tools, methods, and technologies designed to help organizations, teams, and individuals improve their technical or management capabilities. Some books describe processes and practices for developing higher-quality software, acquiring programs for complex systems, or delivering services more effectively. Other books focus on software and system architecture and product-line development. Still others, from the SEI's CERT Program, describe technologies and practices needed to manage software and network security risk. These and all titles in the series address critical problems in software engineering for which practical solutions are available.

Make sure to connect with us!
informit.com/socialconnect

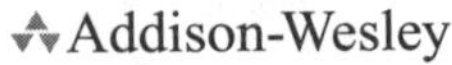

Cyber-Physical Systems

Raj Rajkumar
Dionisio de Niz
Mark Klein

Addison-Wesley

Boston • Columbus • Indianapolis • New York • San Francisco • Amsterdam • Cape Town
Dubai • London • Madrid • Milan • Munich • Paris • Montreal • Toronto • Delhi • Mexico City
São Paulo • Sydney • Hong Kong • Seoul • Singapore • Taipei • Tokyo

 Software Engineering Institute | Carnegie Mellon

The SEI Series in Software Engineering

For information about buying this title in bulk quantities, or for special sales opportunities (which may include electronic versions; custom cover designs; and content particular to your business, training goals, marketing focus, or branding interests), please contact our corporate sales department at corpsales@pearsoned.com or (800) 382-3419.

For government sales inquiries, please contact governmentsales@pearsoned.com.

For questions about sales outside the U.S., please contact intlcs@pearson.com.

Visit us on the Web: informit.com/aw

Library of Congress Control Number: 2016953412

Cover images by Xavier MARCHANT/Fotolia; Small Town Studio/Fotolia; Josemaria Toscano/Fotolia; Mario Beauregard/Fotolia; Algre/Fotolia; Antiksu/Fotolia; and James Steidl/ShutterStock.

ISBN-13: 978-0-321-92696-8
ISBN-10: 0-321-92696-X

1 16

To our families

Contents

Introduction

Ragunathan (Raj) Rajkumar, Dionisio de Niz, and Mark Klein

The National Science Foundation defines cyber-physical systems (CPS) as "engineered systems that are built from, and depend upon, the seamless integration of computational algorithms and physical components"—that is, cyber and physical components. In practical terms, this integration means that, to understand CPS behavior, we cannot focus only on the cyber part or only on the physical part. Instead, we need to consider both parts working together. For instance, if we try to verify the behavior of the airbag of a car, it is not enough to ensure that the correct instructions to inflate the airbag are executed when the system detects that a crash is occurring. We also need to verify that the execution of such instructions is completed in sync with the physical process—for example, within 20 ms—so as to ensure that the airbag is fully inflated before the driver hits the steering wheel. The seamless integration between the cyber and physical parts of a CPS involves an understanding across multiple aspects that, in this simple example, include software logic, software execution timing, and physical forces.

While the airbag example already contains important elements of a CPS, we must note that it is one of the simplest examples and does not involve the most significant challenges for CPS. In particular, both the cyber components and the physical components are very simple in this case, and their interaction can be reduced to worst-case differences between the software execution completion time and the time it takes the driver to hit the steering wheel in a crash. However, as the complexity of both the software and the physical processes grows, the complexity of their integration also increases significantly. In a large CPS—for example, one of the latest commercial aircraft—the integration of multiple physical and cyber components and the tradeoffs between their multiple aspects become very challenging. For instance, the simple use of additional lithium ion batteries in the Boeing 787 Dreamliner brings a combination of multiple constraints that must be satisfied. In particular, we not only need to be able to satisfy the power consumption requirements under different operating modes with a specific configuration of batteries (interacting with software that is run at a specific processor speed and voltage), but we also need to manage when and how battery cells are charged and discharge while preserving the required voltage and verify that the charge/discharge configurations do not overheat the battery (causing it to burst into flames, as occurred during the first few flights of the 787), which interacts with the thermal dissipation design of the system. More importantly, all of these aspects, and more, must be certified under the strict safety standards from the Federal Aviation Administration (FAA).

CPS, however, are facing even more challenges than the increased complexity coming from single systems. Notably, they are being developed to interact with other CPS without human intervention. This phenomenon resembles the way the Internet began. Specifically, the Internet started as a simple connection between two computers. However, the real revolution happened when computers all over the world were connected seamlessly and a large number of services were developed on top of this worldwide network. Such connectivity not only allowed services to be delivered all over the globe, but also enabled the collection and processing of massive amounts of information in what is now known as "Big Data." Big Data allow us to explore trends among groups of people, or even explore trends in real time when they are combined with social media such as Facebook and Twitter. In CPS, this revolution is just starting. We can already observe services that collect driving information through global positioning system (GPS) applications on smartphones and allow us to select routes with low traffic congestion, thereby taking us in the direction of a smart highway, even if it is still mediated by humans. More significant steps in this direction are occurring every day, such as in the multiple projects involving autonomous cars. Most of these cars not only know how to drive a route autonomously, but also correctly interact with other non-autonomous cars on the route.

Emergence of CPS

Before CPS emerged as a specific, yet evolving discipline, systems composed of physical and cyber components already existed. However, the interactions between these two types of components were very simple, and the supporting theoretical foundations were mostly partitioned into silos. Specifically, the theoretical foundations for computer science and physical sciences where developed independently and ignored each other. As a result, techniques to, say, verify properties in one discipline such as thermal resilience, aerodynamics, and mechanical stress were developed independently from advances in computer science such as logical clocks, model checking, type systems, and so on. These advances, in fact, abstracted away behaviors that could be important in one discipline but were not relevant in the other. This is the case, for instance, with the timeless nature of programming languages and logical verification models in which only sequences of instructions are considered. This nature contrasts with the importance of time in the evolution of physical processes such as the movement of a car and maintenance of the temperature of a room that we are trying to control.

The early realization of the interactions between computational and physical science gave birth to some simple and mostly pairwise interaction models. This is the case with real-time scheduling theory and control theory, for example. On the one hand, scheduling theory adds time to computational elements and allows us to verify the response time of computations that interact with physical processes, thereby ensuring that such a process does not deviate beyond what the computational part expects and is capable of correcting. On the other hand, control theory allows us to put together a control algorithm and a physical process and analyze whether it would be possible for the algorithm to keep the system within a desired region around a specific setpoint. While control theory uses a continuous time model in which computations happen instantaneously, it allows the addition of delays to take into account the computation time (including scheduling), making it possible to specify the periodicity of computations and provide a clean interface with scheduling.

As the complexity of the interactions between domains increases, new techniques are developed to model such interactions. This is the case, for instance, with hybrid systems—a type of state machine in which the states model computational and physical states and the transitions model computational actions and physical evolutions. While these techniques enhance the capacity to describe complex interactions, their analysis is, more often than not, intractable. In general, the complexity of these models prevents the analysis of systems of practical dimensions. As the number of scientific disciplines that need to be considered grows (e.g., functional, thermal, aerodynamics, mechanical, fault tolerance), their interactions need to be analyzed to ensure that the assumptions of each discipline and its models are not invalidated by the other disciplines' models. For instance, the speed of a processor assumed by a real-time scheduling algorithm may be invalidated if the dynamic thermal management (DTM) system of the processor decreases the speed of this processor to prevent it from overheating.

CPS Drivers

While CPS are already being built today, the challenge is to be able to understand their behavior and develop techniques to verify their reliability, security, and safety. This is, indeed, the core motivation of the scientific community around CPS. As a result, CPS has been driven by two mutually reinforcing drivers: applications and theoretical foundations.

Applications

CPS applications have allowed researchers to team up with practitioners to better understand the problems and challenges and provide solutions that can be tested in practical settings. This is the case with medical devices, for example: CPS researchers have teamed up with medical doctors to understand the origin and challenges of medical device-based errors. For instance, some errors associated with infusion pumps were caused by incorrect assumptions about how the human body processes different drugs, how to implement the safeguards to avoid overdose, and how to ensure that a nurse enters the correct information. Furthermore, the current generation of medical devices are certified only as individual devices and are not allowed to be connected to one another. As a result, health practitioners are required to coordinate the use of these devices and ensure the safety of their interactions. For instance, if an X ray of the chest is taken during an operation, it is necessary to ensure that the respirator is disabled (if one is in used). However, once the X rays are taken, the respirator needs to be re-enabled within a safe interval of time to prevent the patient from suffocating. While this invariant could be implemented in software, the current certification techniques and policies prevent this kind of integration from happening. Researchers working in this area are developing techniques to enable the certification of these interactions. This issue is discussed at length in Chapter 1, "Medical Cyber-Physical Systems."

The electric grid is another important application area due to its national importance as a critical infrastructure. A key challenge in this area is the uncoordinated nature of the actions that affect the consumption and generation of electricity by a large number of consumers and providers, respectively. In particular, each individual household is able to change its consumption with the flip of a switch; the consequences of these decisions then have an aggregate effect on the grid that needs to be balanced with the supply. Similarly, renewable power-generation sources, such as wind and solar energy, provide sporadic and unpredictable bursts of energy, making the balance of supply and demand very challenging with such systems. More importantly, the interactions between these elements are both cyber and physical in nature. On the one hand, some level of computer-mediated coordination happens between suppliers. On the other hand, the interactions with the consumer occur mostly through the physical consumption of electricity. Today, a number of techniques are already being used for the development and control of power grids to prevent damage to the infrastructure and improve reliability. However, new challenges require a new combination of cyber and physical elements that can support efficient markets, renewable energy sources, and cheaper energy prices. Chapter 2, "Energy Cyber-Physical Systems," discusses the challenges and advances in the electric grid domain.

Perhaps one of the most interesting application areas that has created its own technical innovations is sensor networks. The development and deployment of sensor networks faces challenges of space, time, energy, and reliability that are very particular to this area. The challenges facing sensor networks as well as the main technical innovations in this area are discussed in Chapter 3, "Cyber-Physical Systems Built on Wireless Sensor Networks."

Other application areas have their own momentum, and yet other emerging areas may soon appear on the horizon. This book discusses some of the most influential areas that are defining the CPS discipline.

Theoretical Foundations

The theoretical advances in CPS have largely focused on the challenges imposed by the interactions between multiple scientific domains. This is the case, for instance, with real-time scheduling. A few trends in this area are worth mentioning. First, new scheduling models to accommodate execution overloads have appeared. These models combine multiple execution budgets with a criticality classification to guarantee that during a normal operation all tasks will meet their deadlines; if an overload occurs, however, the high-criticality tasks meet their deadlines by stealing processor cycles from low-criticality tasks. The second trend comes from variations in periodicity. The so-called *rhythmic task* model allows tasks to continuously vary their periodicity following physical events of variable frequency. This is the case, for instance, with tasks triggered by the angular position of the crankshaft in the engine of a car: New scheduling analysis techniques were developed to verify the timing aspect of these systems. The real-time scheduling foundations and innovations are discussed in Chapter 9, "Real-Time Scheduling for Cyber-Physical Systems."

Cross-cutting innovations between model checking and control theory for control synthesis are among the recent noteworthy developments. In this scheme, hybrid state machine models are used to describe the behavior of the physical plant and the requirements of the computational algorithm. Then, this model is used to automatically synthesize the controller algorithm enforcing the desired specifications. This scheme is discussed at length in Chapter 4, "Symbolic Synthesis for Cyber-Physical Systems." Similarly, a number of new techniques have been developed to analyze the timing effects of a scheduling discipline in control algorithms. These issues are discussed in Chapter 5, "Software and Platform Issues in Feedback Control Systems."

Another area of interaction that has been explored is the relationship between model checking and scheduling. In this case, a new model checker called REK was developed to take advantage of the constraints that the rate-monotonic scheduler and periodic task model imposes on task inter-leavings in an effort to reduce the verification efforts. These new interactions are discussed in Chapter 6, "Logical Correctness for Hybrid Systems."

Security is another big area that is significantly affected by the presence of physical processes. In particular, the interactions between software and physical processes present new opportunities for potential attackers that make CPS security different from software-only security. This difference arises because attacks against, say, sensors to provide false sensor readings can sometimes be very difficult to distinguish from genuine readings from the physical processes. A number of innovations to prevent this kind of man-in-the-middle attack, as well as other significant techniques, are presented in Chapter 7, "Security of Cyber-Physical Systems."

For distributed systems, new techniques to enforce synchronized communication between distributed agents have proved very useful to reduce the effort needed to produce formal proofs of functional correctness in distributed real-time systems. This issue is discussed at length in Chapter 8, "Synchronization in Distributed Cyber-Physical Systems."

CPS analysis techniques rely on models, and the formal semantics of these models is a key challenge that must be addressed. Chapter 10, "Model Integration in Cyber-Physical Systems," presents the latest developments in the definition of formal semantics for models in what are called model-integration languages.

A large number of theoretical advances are discussed in this book, along with the open challenges in each area. While some advances stem from specific challenges in application areas, others expose new opportunities.

Target Audience

This book is aimed at both practitioners and researchers. For practitioners, the book profiles both the current application areas that are benefiting from CPS perspectives and the current techniques that had proved successful for the development of the current generation of CPS. For the researcher, this book provides a survey of application areas and highlights their current achievements and open challenges as well as the current scientific advances and their challenges.

The book is divided into two parts. Part I, "Cyber-Physical System Application Domains," presents the current CPS application areas that are driving the CPS revolution. Part II, "Foundations," presents the current theoretical foundations from the multiple scientific disciplines used in the development of CPS.

Part I

Cyber-Physical System Application Domains

Chapter 1

Medical Cyber-Physical Systems[1]

Insup Lee, Anaheed Ayoub, Sanjian Chen, Baekgyu Kim, Andrew King, Alexander Roederer, and Oleg Sokolsky

Medical cyber-physical systems (MCPS) are life-critical, context-aware, networked systems of medical devices that are collectively involved in treating a patient. These systems are increasingly used in hospitals to provide high-quality continuous care for patients in complex clinical scenarios. The need to design complex MCPS that are both safe and effective has presented numerous challenges, inclulding achieving high levels of assurance in system software, interoperability, context-aware decision support, autonomy, security and privacy, and certification. This chapter discusses these challenges in developing MCPS, provides case studies that illustrate these challenges and suggests ways to address them, and highlights several open research and development issues. It concludes with a discussion of the implications of MCPS for stakeholders and practitioners.

1. Research is supported in part by NSF grants CNS-1035715, IIS-1231547, and ACI-1239324, and NIH grant 1U01EB012470-01.

1.1 Introduction and Motivation

The two most significant transformations in the field of medical devices in recent times are the high degree of reliance on software-defined functionality and the wide availability of network connectivity. The former development means that software plays an ever more significant role in the overall device safety. The latter implies that, instead of stand-alone devices that can be designed, certified, and used independently of each other to treat patients, networked medical devices will work as distributed systems that simultaneously monitor and control multiple aspects of the patient's physiology. The combination of the embedded software controlling the devices, the new networking capabilities, and the complicated physical dynamics of the human body makes modern medical device systems a distinct class of cyber-physical systems (CPS).

The goal of MCPS is to improve the effectiveness of patient care by providing personalized treatment through sensing and patient model matching while ensuring safety. However, the increased scope and complexity of MCPS relative to traditional medical systems present numerous developmental challenges. These challenges need to be systematically addressed through the development of new design, composition, verification, and validation techniques. The need for these techniques presents new opportunities for researchers in MCPS and, more broadly, embedded technologies and CPS. One of the primary concerns in MCPS development is the assurance of patient safety. The new capabilities of future medical devices and the new techniques for developing MCPS with these devices will, in turn, require new regulatory procedures to approve their use for treating patients. The traditional process-based regulatory regime used by the U.S. Food and Drug Administration (FDA) to approve medical devices is becoming lengthy and prohibitively expensive owing to the increased MCPS complexity, and there is an urgent need to ease this often onerous process without compromising the level of safety it delivers.

In this chapter, we advocate a systematic approach to analysis and design of MCPS for coping with their inherent complexity. Consequently, we suggest that model-based design techniques should play a larger role in MCPS design. Models should cover not only devices and communications between them, but also, of equal importance, patients and caregivers. The use of models will allow developers to assess system properties early in the development process and build confidence in the safety and effectiveness of the system design, well before the system is built. Analysis of system safety and effectiveness performed at the modeling level needs to be complemented by generative implementation techniques that preserve properties of the model during the implementation stage. Results of model analysis, combined with the guarantees of the generation process, can form the basis for evidence-based regulatory approval. The ultimate goal is to use model-based development as the foundation for building safe and effective MCPS.

This chapter describes some of the research directions being taken to address the various challenges involved in building MCPS:

- *Stand-alone device:* A model-based high-assurance software development scheme is described for stand-alone medical devices such as patient-controlled analgesia (PCA) pumps and pacemakers.
- *Device interconnection:* A medical device interoperability framework is presented for describing, instantiating, and validating clinical interaction scenarios.
- *Adding intelligence:* A smart alarm system is presented that takes vital signs data from various interacting devices to inform caregivers of potential patient emergencies and non-operational issues about the devices.
- *Automated actuation/delivery:* A model-based closed-loop care delivery system is presented, which can autonomously deliver care to the patients based on the current state of the patient.
- *Assurance cases:* The use of assurance cases is described for organizing collections of claims, arguments, and evidence to establish the safety of a medical device system.

MCPS are viewed in a bottom-up manner in this chapter. That is, we first describe issues associated with individual devices, and then progressively increase their complexity by adding communication, intelligence, and feedback control. Preliminary discussion of some of these challenges has appeared in [Lee12].

1.2 System Description and Operational Scenarios

MCPS are safety-critical, smart systems of interconnected medical devices that are collectively involved in treating a patient within a specific *clinical scenario*. The clinical scenario determines which treatment options can be chosen and which adjustments of treatment settings need to be made in response to changes in the patient's condition.

Traditionally, decisions about the treatment options and settings have been made by the attending caregiver, who makes them by monitoring patient state using individual devices and performing manual adjustments. Thus, clinical scenarios can be viewed as closed-loop systems in which caregivers are the controllers, medical devices act as sensors and actuators, and patients are the "plants." MCPS alter this view by introducing additional computational entities that aid the caregiver in controlling the "plant." Figure 1.1 presents a conceptual overview of MCPS.

Devices used in MCPS can be categorized into two large groups based on their primary functionality:

- *Monitoring devices*, such as bedside heart rate and oxygen level monitors and sensors, which provide different kinds of clinic-relevant information about patients
- *Delivery devices*, such as infusion pumps and ventilators, which actuate therapy that is capable of changing the patient's physiological state

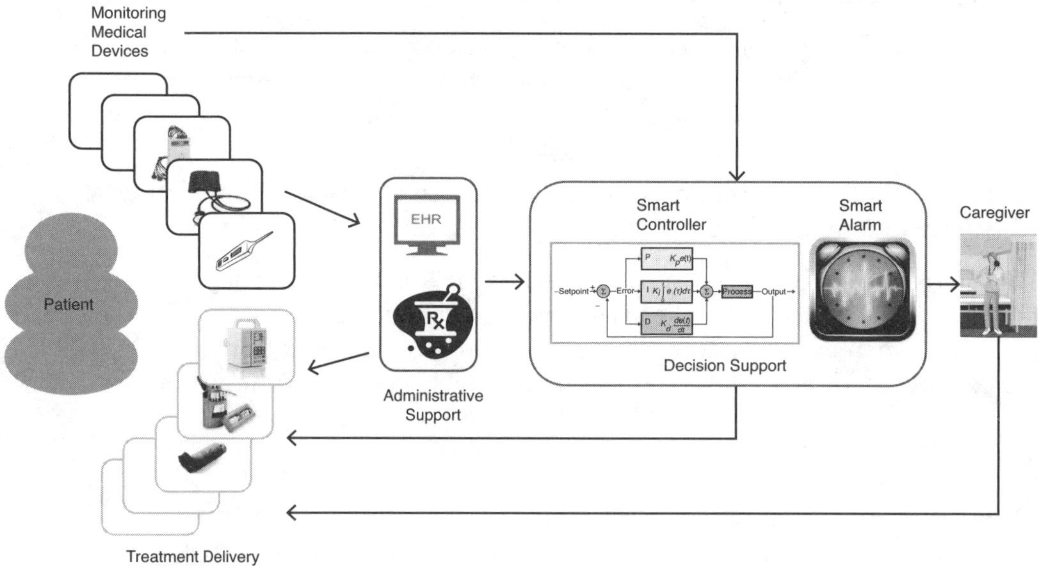

Figure 1.1: *A conceptual overview of medical cyber-physical systems*

In MCPS, interconnected monitoring devices can feed collected data to decision support or administrative support entities, each of which serves a different, albeit complementary, purpose. For example, caregivers can analyze the information provided by these devices and then use delivery devices to initiate treatment, thereby bringing the caregiver into the control loop around the patient. Alternatively, the decision support entities might utilize a smart controller to analyze the data received from the monitoring devices, estimate the state of the patient's health, and automatically initiate treatment (e.g., drug infusion) by issuing commands to delivery devices, thereby closing the loop.

Most medical devices rely on software components for carrying out their tasks. Ensuring the safety of these devices and their interoperation is crucial. One of the more effective strategies to do so is to use model-based development methods, which can ensure device safety by enabling medical device verification. This strategy also opens the door for eventually certifying the devices to meet certain safety standards.

1.2.1 Virtual Medical Devices

Given the high complexity of MCPS, any such system has to be user-centric; that is, it must be easy to set up and use, in a largely automated manner. One way to accomplish this is to develop a description of the MCPS workflow and then enforce it on physical devices. MCPS workflow can be described in terms of the number and types of devices involved, their mutual interconnections, and the clinical supervisory algorithm needed for coordination and analysis of data collected by the system. Such a description defines *virtual medical device* (VMD). VMDs are used by a *VMD app* and instantiated during the setup of actual medical devices—that is, as part of a *virtual medical device instance*.

The devices in a VMD instance are usually interconnected using some form of interoperability middleware, which is responsible for ensuring that the inter-device connections are correctly configured. The principal task of the VMD app, therefore, is to find the medical devices in a VMD instance (which may be quite large), establish network connections between them, and install the clinical algorithm into the supervisor module of the middleware for managing the interactions of the clinical workflow and the reasoning about the data produced. Basically, when the VMD app is started, the supervisor reads the VMD app specification and tries to couple all involved devices according to the specification. Once the workflow has run its course, the VMD app can perform the necessary cleanup to allow another workflow to be specified using a different combination of medical devices in the VMD instance.

1.2.2 Clinical Scenarios

Each VMD supports a specific clinical scenario with a detailed description of how devices and clinical staff work together in a clinical situation or event. Here, we describe two such scenarios: one for X ray and ventilator coordination and another for a patient-controlled analgesia (PCA) safety interlock system.

One example that illustrates how patient safety can be improved by MCPS is the development of a VMD that coordinates the interaction between an X-ray machine and a ventilator. Consider the scenario described by [Lofsky04]. X-ray images are often taken during surgical procedures. If the surgery is being performed under general anesthesia, the patient breathes with the help of a ventilator during the procedure. Because the patient on ventilator cannot hold his or her breath to let the X-ray image be taken without the blur caused by moving lungs, the ventilator has to be paused and later restarted. In some unfortunate cases, the ventilator was not restarted, leading to the death of the patient.

Interoperation of the two devices can be used in several ways to ensure that patient safety is not compromised, as discussed in [Arney09]. One possibility is to let the X-ray machine pause and restart the ventilator automatically. A safer alternative, albeit one presenting tighter timing constraints, is to let the ventilator transmit its internal state to the X-ray machine. There typically is enough time to take an X-ray image at the end of the breathing cycle, between the time when the patient has finished exhaling and the time he or she starts the next inhalation. This approach requires the X-ray machine to know precisely the instance when the air flow rate becomes close enough to zero and the time when the next inhalation starts. Then, it can decide to take a picture if enough time—taking transmission delays into account—is available.

Another clinical scenario that can easily benefit from the closed-loop approach of MCPS is patient-controlled analgesia. PCA infusion pumps are commonly used to deliver opioids for pain management—for instance, after surgery. Patients have very different reactions to the medications and require distinct dosages and delivery schedules. PCA pumps allow patients to press a button to request a dose when they decide they want it, rather than using a dosing schedule fixed by a caregiver. Some patients may decide they prefer a higher level of pain to the nausea that the drugs may cause and, therefore, press the button less often; others, who need a higher dose, can press the button more often.

A major problem with opioid medications in general is that an excessive dose can cause respiratory failure. A properly programmed PCA system should prevent an overdose by limiting how many doses it will deliver, regardless of how often the patient pushes the button. However, this safety mechanism is not sufficient to protect all patients. Some patients may still receive overdoses if the pump is misprogrammed, if the pump programmer overestimates the maximum dose that a patient can receive, if the wrong concentration of drug is loaded into the pump, or if someone other than the patient presses the button (PCA-by-proxy), among other causes. PCA infusion pumps are currently associated with a large number of adverse events, and existing safeguards such as drug libraries and programmable limits are not adequate to address all the scenarios seen in clinical practice [Nuckols08].

1.3 Key Design Drivers and Quality Attributes

While software-intensive medical devices such as infusion pumps, ventilators, and patient monitors have been used for a long time, the field of medical devices is currently undergoing a rapid transformation. The changes under way are raising new challenges in the development of high-confidence medical devices, yet are simultaneously opening up new opportunities for the research community [Lee06]. This section begins by reviewing the main trends that have emerged recently, then identifies quality attributes and challenges, and finally provides a detailed discussion of several MCPS-specific topics.

1.3.1 Trends

Four trends in MCPS are critical in the evolution of the field: software as the main driver of new features, device interconnection, closed loops that automatically adjust to physiological response, and a new focus on continuous monitoring and care. The following subsections discuss each of these trends.

1.3.1.1 New Software-Enabled Functionality

Following the general trend in the field of embedded systems, and more broadly in cyber-physical systems, introduction of new functionality is largely driven by the new possibilities that software-based development of medical device systems is offering. A prime example of the new functionality is seen in the area of robotic surgery, which requires real-time processing of high-resolution images and haptic feedback.

Another example is proton therapy treatment. One of the most technology-intensive medical procedures, it requires one of the largest-scale medical device systems. To deliver its precise doses of radiation to patients with cancer, the treatment requires precise guiding of a proton beam from a cyclotron to patients, but must be able to adapt to even minor shifts in the patient's position. Higher precision of the treatment, compared to conventional radiation therapy, allows higher radiation doses to be applied. This, in turn, places more stringent requirements on patient safety. Control of proton beams is subject to very tight timing constraints, with much less tolerance than for most medical devices. To further complicate the problem, the same beam is applied to multiple locations in the patient's body and needs to be switched from location to location, opening up the possibility of interference between beam scheduling and application. In addition to controlling the proton beam, a highly critical function of software in a proton treatment system is real-time image processing to determine the precise position of the patient and detect any patient movement. In [Rae03], the authors analyzed the safety of proton therapy machines, but their analysis concentrated on a single system, the emergency shutdown. In general, proper analysis and validation of such large and complex systems remains one of the biggest challenges facing the medical device industry.

As further evidence of the software-enabled functionality trend, even in simpler devices, such as pacemakers and infusion pumps, more and more software-based features are being added, making their device software more complex and error prone [Jeroeno4]. Rigorous approaches are required to make sure that the software in these devices operates correctly. Because these devices are relatively simple, they are good candidates for case studies of challenges and experimental development techniques. Some of these devices, such as pacemakers, are being used as challenge problems in the formal methods research community [McMaster13].

1.3.1.2 Increased Connectivity of Medical Devices

In addition to relying on software to a greater extent, medical devices are increasingly being equipped with network interfaces. In essence, interconnected medical devices form a distributed medical device system of a larger scale and complexity that must be properly designed and validated to ensure effectiveness and patient safety. Today, the networking capabilities of medical devices are primarily exploited for patient monitoring purposes (through local connection of individual devices to integrated patient monitors or for remote monitoring in a tele-ICU [Sapirstein09] setting) and for interaction with electronic health records to store patient data.

The networking capabilities of most medical devices today are limited in functionality and tend to rely on proprietary communication protocols offered by major vendors. There is, however, a growing realization among clinical professionals that open interoperability between different medical devices will lead to improved patient safety and new treatment procedures. The Medical Device Plug-and-Play (MD PnP) Interoperability initiative [Goldman05, MDPNP] is a relatively recent effort that aims to provide an open standards framework for safe and flexible interconnectivity of medical devices, with the ultimate goal of improving patient safety and health care efficiency. In addition to developing interoperability standards, the MD PnP initiative collects and demonstrates clinical scenarios in which interoperability leads to improvement over the existing practice.

1.3.1.3 Physiological Closed-Loop Systems

Traditionally, most clinical scenarios have a caregiver—and often more than one—controlling the process. For example, an anesthesiologist monitors sedation of a patient during a surgical procedure and decides when an action to adjust the flow of sedative needs to be taken. There is a concern in the medical community that such reliance on humans being in the loop may compromise patient safety. Caregivers, who are often overworked and operate under severe time pressures, may miss a critical warning sign. Nurses, for example, typically care for multiple patients at a time and can become distracted. Using an automatic controller to provide continuous monitoring of the patient state and handling of routine situations would relieve some of the pressure on the caregiver and might potentially improve patient care and safety. Although the computer will probably never replace the caregiver completely, it can significantly reduce the workload, calling the caregiver's attention only when something out of the ordinary happens.

Scenarios based on physiological closed-loop control have been used in the medical device industry for some time. However, their application has been mostly limited to implantable devices that cover relatively well-understood body organs—for example, the heart, in the case of pacemakers and defibrillators. Implementing closed-loop scenarios in distributed medical device systems is a relatively new idea that has not made its way into mainstream practice as yet.

1.3.1.4 Continuous Monitoring and Care

Due to the high costs associated with in-hospital care, there has been increasing interest in alternatives such as home care, assisted living, telemedicine, and sport-activity monitoring. Mobile monitoring and home monitoring of vital signs and physical activities allow health to be assessed remotely at all times. Also, sophisticated technologies such as body sensor networks to measure training effectiveness and athletic performance based on physiological data such as heart rate, breathing rate, blood sugar level, stress level, and skin temperature are becoming more popular. However, most of the current systems operate in store-and-forward mode, with no real-time diagnostic capability. Physiological closed-loop technology will allow diagnostic evaluation of vital signs in real time and make constant care possible.

1.3.2 Quality Attributes and Challenges of the MCPS Domain

Building MCPS applications requires ensuring the following quality attributes, which in turn pose significant challenges:

- *Safety:* Software is playing an increasingly important role in medical devices. Many functions traditionally implemented in hardware—including safety interlocks—are now being implemented in software. Thus high-confidence software development is critical to ensure the safety and effectiveness of MCPS. We advocate the use of model-based development and analysis as a means of ensuring the safety of MCPS.
- *Interoperability:* Many modern medical devices are equipped with network interfaces, enabling us to build MCPS with new capabilities by combining existing devices. Key to such systems is the concept of interoperability, wherein individual devices can exchange information

facilitated by an application deployment platform. It is essential to ensure that the MCPS built from interoperable medical devices are safe, effective, and secure, and can eventually be certified as such.

- *Context-awareness:* Integration of patient information from multiple sources can provide a better understanding of the state of the patient's health, with the combined data then being used to enable early detection of ailments and generate effective alarms in the event of emergency. However, given the complexity of human physiology and the many variations of physiological parameters over patient populations, developing such computational intelligence is a nontrivial task.
- *Autonomy:* The computational intelligence that MCPS possess can be applied to increase the autonomy of the system by enabling actuation of therapies based on the patient's current health state. Closing the loop in this manner must be done safely and effectively. Safety analysis of autonomous decisions in the resulting closed-loop system is a major challenge, primarily due to the complexity and variability of human physiology.
- *Security and privacy:* Medical data collected and managed by MCPS are very sensitive. Unauthorized access or tampering with this information can have severe consequences to the patient in the form of privacy loss, discrimination, abuse, and physical harm. Network connectivity enables new MCPS functionality by exchanging patient data from multiple sources; however, it also increases the vulnerability of the system to security and privacy violations.
- *Certification:* A report by the U.S. National Academy of Science, titled "Software for Dependable Systems: Sufficient Evidence?," recommends an evidence-based approach to the certification of high-confidence systems such as MCPS using explicit claims, evidence, and expertise [Jackson07]. The complex and safety-critical nature of MCPS requires a cost-effective way to demonstrate medical device software dependability. Certification, therefore, is both an essential requirement for the eventual viability of MCPS and an important challenge to be addressed. An assurance case is a structured argument supported by a documented body of evidence that provides a convincing and consistent argument that a system is adequately safe (or secure) [Menon09]. The notion of assurance cases holds the promise of providing an objective, evidence-based approach to software certification. Assurance cases are increasingly being used as a means

of demonstrating safety in industries such as nuclear power, transportation, and automotive systems, and are mentioned in the recent IEC 62304 development standard for medical software.

1.3.3 High-Confidence Development of MCPS

The extreme market pressures faced by the medical devices industry has forced many companies to reduce their development cycles as much as possible. The challenge is to find a development process that will deliver a high degree of safety assurance under these conditions. Model-based development can be a significant part of such a development process. The case study discussed in this section illustrates the steps of the high-assurance development process using a simple medical device. Each of the steps can be implemented in a variety of ways. The choice of modeling, verification, and code generation technologies depends on factors such as complexity and criticality level of the application. Nevertheless, the process itself is general enough to accommodate a wide variety of rigorous development technologies.

1.3.3.1 Mitigation of Hazards

Most of the new functionality in medical devices is software based, and many functions traditionally implemented in hardware—including safety interlocks—are now being relegated to software. Thus, high-confidence software development is very important for the safety and effectiveness of MCPS.

Figure 1.2 depicts a relatively conventional approach to high-assurance development of safety-critical systems based on the mitigation of hazards. The process starts with the identification of the desired functionality and the hazards associated with the system's operation. The chosen functionality yields the system functional requirements, while hazard mitigation strategies yield the system safety requirements.

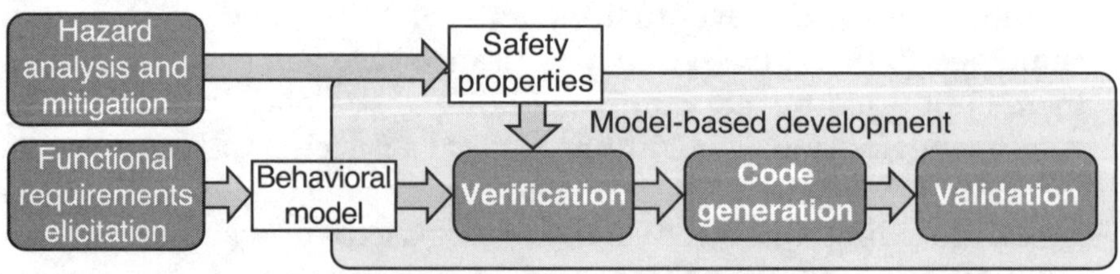

Figure 1.2: *High-assurance development process for embedded software*

The functional requirements are used to build detailed behavioral models of the software modules, while the safety requirements are turned into properties that these models should satisfy. Models and their desired properties are the inputs to the model-based software development, which consists of verification, code generation, and validation phases.

Model-based development has emerged as a means of raising the level of assurance in software systems. In this approach, developers start with declarative models of the system and perform a rigorous model verification with respect to safety and functional requirements; they then use systematic code generation techniques to derive code that preserves the verified properties of the model. Such a development process allows the developers to detect problems with the design and fix them at the model level, early in the design cycle, when changes are easier and cheaper to make. More importantly, it holds the promise of improving the safety of the system through verification. Model-based techniques currently used in the medical device industry rely on semi-formal approaches such as UML and Simulink [Becker09], so they do not allow developers to fully utilize the benefits of model-based design. The use of formal modeling facilitates making mathematically sound conclusions about the models and generating code from them.

1.3.3.2 Challenges of Model-Driven Development of MCPS

Several challenges arise when developing MCPS through the model-driven implementation process. The first challenge is choosing the right level of abstraction for the modeling effort. A highly abstract model makes the verification step relatively easy to perform, but a model that is too abstract is difficult to use in the code generation process, since too many implementation decisions have to be guessed by the code generator. Conversely, a very detailed model makes code generation relatively straightforward, but pushes the limits of the currently available verification tools.

Many modeling approaches rely on the separation of the platform-independent and platform-dependent aspects of development. From the modeling and verification perspective, there are several reasons to separate the platform-independent aspects from the platform-dependent aspects.

First, hiding platform-dependent details reduces the modeling and verification complexity. Consider, for example, the interaction between a device and its sensors. For code generation, one may need to specify the details of how the device retrieves data from sensors. A sampling-based mechanism with a particular sampling interval will yield a very different generated code compared to an interrupt-based mechanism. However, exposing such details in the model adds another level of complexity to the model, which may increase verification time to an unacceptable duration.

In addition, abstracting away from a particular platform allows us to use the model across different target platforms. Different platforms may have different kinds of sensors that supply the same value. For example, consider an empty-reservoir alarm, such as that implemented on many infusion pumps. Some pumps may not have a physical sensor for that purpose and simply estimate the remaining amount of medication based on the infusion rate and elapsed time. Other pumps may have a sensor based on syringe position or pressure in the tube. Abstracting away these details would allow us to implement the same pump control code on different pump hardware. At the same time, such separation leads to integration challenges at the implementation level. The code generated by the platform-independent model needs to be integrated with the code from the various target platforms in such a way that the verified properties of the platform-independent model are preserved.

Second, there is often a semantic gap between the model and the implementation. A system is modeled using the formal semantics provided by the chosen modeling language. However, some of the model semantics may not match well with the implementation. For example, in UPPAAL and Stateflow, the interaction between the PCA pump and the environment (e.g., user or pump hardware) can be modeled by using instantaneous channel synchronization or event broadcasting that has a zero time delay. Such semantics simplifies modeling input and output of the system so that the modeling/verification complexity is reduced. Unfortunately, the correct implementation of such semantics is hardly realizable at the implementation level, because execution of those actions requires interactions among components that have a non-zero time delay.

The following case study concentrates on the development of a PCA infusion pump system and considers several approaches to address these challenges.

1.3.3.3 Case Study: PCA Infusion Pumps

A PCA infusion pump primarily delivers pain relievers, and is equipped with a feature that allows for additional limited delivery of medication, called a bolus, upon patient demand. This type of infusion pump is widely used for pain control of postoperative patients. If the pump overdoses opioid drugs, however, the patient can be at risk of respiratory depression and death. Therefore, these medical devices are subject to stringent safety requirements that aim to prevent overdose.

According to the FDA's Infusion Pump Improvement Initiative [FDA10a], the FDA received more than 56,000 reports of adverse events associated with the use of infusion pumps from 2005 through 2009. In the same period, 87 recalls of infusion pumps were conducted by the FDA, affecting all major pump manufacturers. The prevalence of the problems clearly indicates the need for better development techniques.

The Generic PCA Project

The Generic PCA (GPCA) project, a joint effort between the PRECISE Center at the University of Pennsylvania and researchers at the FDA, aims to develop a series of publicly available artifacts that can be used as guidance for manufacturers of PCA infusion pumps. In the first phase of the project, a collection of documents has been developed, including a hazard analysis report [UPenn-b], a set of safety requirements [UPenn-a], and a reference model of PCA infusion pump systems [UPenn]. Based on these documents, companies can develop PCA infusion pump controller software following a model-driven implementation.

In the case study, software for the PCA pump controller is developed by using the model-driven implementation approach starting from the reference model and the safety requirements. A detailed account of this effort is presented in [Kim11].

The development approach follows the process outlined in Figure 1.2. The detailed steps are shown in Figure 1.3. In addition, the case study included the construction of an assurance case—a structured argument based on the evidence collected during the development process, which aims to convince evaluators that the GPCA-reference implementation complies with its safety requirements. The assurance case development is discussed in more detail in Section 1.3.7.

Modeling

The reference model of the GPCA pump implemented in Simulink/Stateflow is used as the source of functional requirements and converted to UPPAAL [Behrmann04] via a manual but systematic translation process. The model structure follows the overall architecture of the reference model, which is shown in Figure 1.4. The software is organized into two state machines: the state controller and the alarm-detecting component. The user interface has been considered in a follow-up case study [Masci13]. Both state machines interact with sensors and actuators on the pump platform.

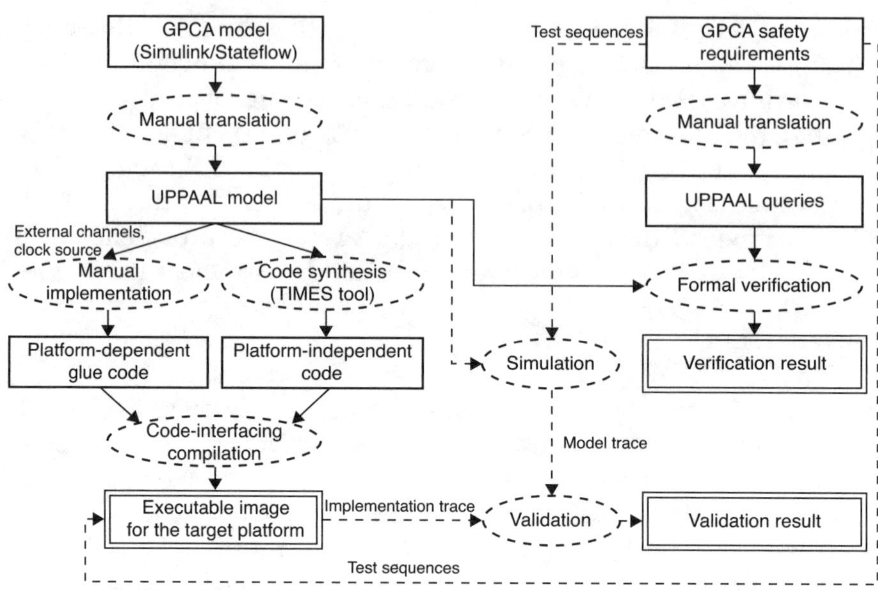

Figure 1.3: *The model-driven development for the GPCA prototype*

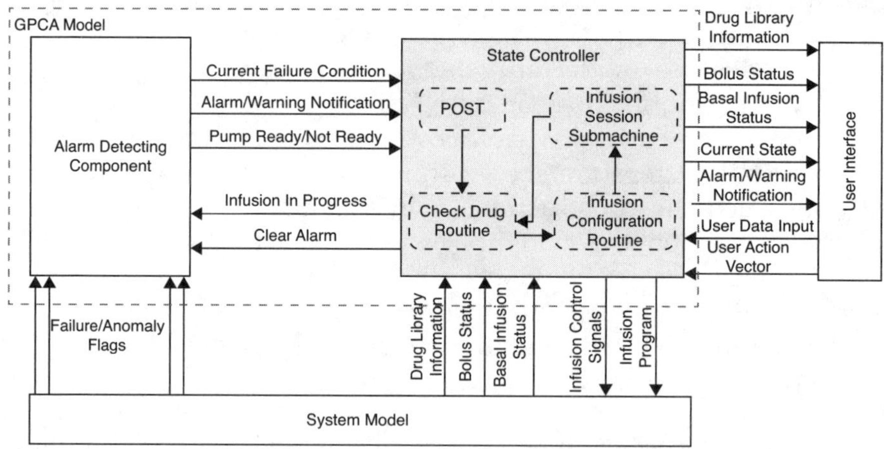

Figure 1.4: *The system architecture of the GPCA model*

The state machines are organized as a set of modes, with each mode captured as a separate submachine. In particular, the state controller contains four modes:

- Power-on self-test (POST) mode is the initial mode that checks system components on start-up.

- The check-drug mode represents a series of checks that the caregiver performs to validate the drug loaded into the pump.
- The infusion configuration mode represents interactions with the caregiver to configure infusion parameters such as infusion rate and volume to be infused (VTBI) and validate them against the limits encoded in the drug library.
- The infusion session is where the pump controls delivery of the drug according to the configuration and the patient's bolus requests.

Model Verification

GPCA safety requirements are expressed in English as "shall" statements. Representative requirements are "No normal bolus doses shall be administered when the pump is alarming" and "The pump shall issue an alert if paused for more than *t* minutes."

Before verification can be performed, requirements need to be formalized as properties to be checked. We can categorize the requirements according to their precision and level of abstraction:

- *Category A:* Requirements that are detailed enough to be formalized and verified against the model
- *Category B:* Requirements that are beyond the scope of the model
- *Category C:* Requirements that are too imprecise to be formalized

Only requirements in Category A can be readily used in verification. Just 20 out of the 97 GPCA requirements fell into this category.

Most of the requirements in Category B concern the functional aspects of the system that are abstracted away at the modeling level. For example, consider the requirement "If the suspend occurs due to a fault condition, the pump shall be stopped immediately without completing the current pump stroke." There is another requirement to complete the current stroke under other kinds of alarms. Thus, the motor needs to be stopped in different ways in different circumstances. These requirements fall into Category B, since the model does not detail the behavior of the pump stroke. Handling of properties in this category can be done in several ways.

One approach is to introduce additional platform-specific details into the model, increasing complexity of the model. However, this would blur the distinction between platform-independent and platform-specific models—a distinction that is useful in the model-based development. An alternative approach is to handle these requirements outside of the model-based process—for example, validating by testing. In this case, however, the benefits of formal modeling are lost.

A better approach is to match the level of detail by further decomposing the requirements. At the platform-independent level, we might check that the system performs two different stop actions in response to different alarm conditions (which would be a Category A requirement). Then, at the platform-specific level, we might check that one stop action corresponds to immediate stopping of the motor, while the other stop action lets the motor complete the current stroke.

An example requirement from Category C is "Flow discontinuity at low flows should be minimal," which does not specify what is a low flow or which discontinuity can be accepted as minimal. This case is a simple example of a deficiency in the requirement specification uncovered during formalization.

Once the categorization of the requirements is complete, requirements in Category A are formalized and verified using a model checker. In the case study, the requirements were converted into UPPAAL queries. Queries in UPPAAL use a subset of timed computation tree logic (CTL) temporal logic and can be verified using the UPPAAL model checker.

Code Generation and System Integration

Once the model is verified, a code generation tool is used to produce the code in a property-reserving manner. An example of such a tool is TIMES [Amnell03] for UPPAAL timed automata. Since the model is platform independent, the resulting code is also platform independent. For example, the model does not specify how the actual infusion pump interacts with sensors and actuators attached to the specific target platform. Input and output actions (e.g., a bolus request by a patient or triggering of the occlusion alarm from the pump hardware) are abstracted as instantaneous transitions subject to input/output synchronization with their environment. On a particular platform, the underlying operating system schedules the interactions, thereby affecting the timing of their execution.

Several approaches may be used to address this issue at the integration stage. In [Henzinger07], higher-level programming abstraction is proposed as a means to model the timing aspects and generate code that is independent from the scheduling algorithms of a particular platform. The platform integration is then performed by verifying time-safety—that is, checking whether the platform-independent code can be scheduled on the particular platform. Another approach is to systematically generate an I/O interface that helps the platform-independent and -dependent code to be integrated in a traceable manner [Kim12]. From a code generation perspective, [Lublinerman09] proposed a way to generate code for a given composite block of the model independently from context and using minimal information about the internals of the block.

Validation of the Implementation

Unless the operation of an actual platform is completely formalized, inevitably some assumptions will be made during the verification and code generation phases that cannot be formally guaranteed. The validation phase is meant to check that these assumptions do not break the behavior of the implementation. In the case study, a test harness systematically exercises the code using test cases derived from the model. A rich literature on model-based test generation exists; see [Dias07] for a survey of the area. The goal of such testing-based validation is to systematically detect deviations of the system behavior from that of the verified model.

1.3.4 On-Demand Medical Devices and Assured Safety

On-demand medical systems represent a new paradigm for safety-critical systems: The final system is assembled by the user instead of the manufacturer. Research into the safety assessment of these systems is actively under way. The projects described in this section represent a first step toward understanding the engineering and regulatory challenges associated with such systems. The success and safety of these systems will depend not only on new engineering techniques, but also on new approaches to regulation and the willingness of industry members to adopt appropriate interoperability standards.

1.3.4.1 Device Coordination

Historically, medical devices have been used as individual tools for patient therapy. To provide complex therapy, caregivers (i.e., physicians and nurses) must coordinate the activities of the various medical devices manually. This is burdensome for the caregiver, and prone to errors and accidents.

One example of manual device coordination in current practice is the X ray and ventilator coordination mentioned in Section 1.2; another example is trachea or larynx surgery performed with a laser scalpel. In this type of surgery, the patient is placed under general anesthesia while the surgeon makes cuts on the throat using a high-intensity laser. Because the patient is under anesthesia, his or her breathing is supported by an anesthesia ventilator that supplies a high concentration of oxygen to the patient. This situation presents a serious hazard: If the surgeon accidentally cuts into the breathing tube using the laser, the increased concentration of oxygen can lead to rapid combustion, burning the patient from the inside out. To mitigate this hazard, the surgeon and the anesthesiologist must be in constant communication: When the surgeon needs to cut, he or she signals the anesthesiologist, who reduces or stops the oxygen being supplied to the patient. If the patient's oxygenation level drops too low, the anesthesiologist signals the surgeon to stop cutting so oxygen can be supplied again.

If medical devices could coordinate their actions, then the surgeon and the anesthesiologist would not have to expend their concentration and effort to ensure that the activities of the medical devices are safely synchronized. Furthermore, the patient would not be exposed to the potential for human error.

Many other clinical scenarios might benefit from this kind of automated medical device coordination. These scenarios involve either *device synchronization*, *data fusion*, or *closed-loop control*. The laser scalpel ventilator safety interlock epitomizes device synchronization: Each device must always be in a correct state relative to the other devices. In data fusion, physiologic readings from multiple separate devices are considered as a collective. Examples of such applications include smart alarms and clinical decision support systems (see Section 1.3.5). Finally, closed-loop control of therapy can be achieved by collecting data from devices that sense the patient's physiological state and then using those data to control actuators such as infusion pumps (see Section 1.3.6).

1.3.4.2 Definition: Virtual Medical Devices

Let us now clarify the concept of virtual medical devices, including why they are considered a different entity. A collection of devices working in unison to implement a given clinical scenario is, in essence, a new medical device. Such collections have been referred to as virtual medical devices (VMDs) because no single manufacturer is producing this device and delivering it fully formed to the clinician. A VMD does not exist until assembled at the patient's bedside. A VMD instance is created each time the clinician assembles a particular set of devices for the VMD and connects them together.

1.3.4.3 Standards and Regulations

Several existing standards are designed to enable medical device interconnectivity and interoperability. These standards include the Health Level 7 standards [Dolin06], IEEE-11073 [Clarke07, ISO/IEEE11073], and the IHE profiles [Carr03]. While these standards enable medical devices to exchange and interpret data, they do not adequately address more complex interactions between medical devices, such as the inter-device coordination and control needed with the laser scalpel and ventilator combination. The notion of a VMD poses one major fundamental question: How does one assure safety in systems that are assembled by their users? Traditionally, most safety-critical cyber-physical systems, such as aircraft, nuclear power plants, and medical devices, are evaluated for safety by regulators before they can be used.

The state of the art in safety assessment is to consider the complete system. This is possible because the complete system is manufactured by a single systems integrator. Virtual medical devices, in contrast, are constructed at bedside, based on the needs of an individual patient and from available devices. This means that a caregiver may instantiate a VMD from a combination of medical devices (i.e., varying in terms of make, model, or feature set) that have never been combined into an integrated system for that particular clinical scenario. Finally, "on-demand" instantiation of the VMD confounds the regulatory pathways for medical devices that are currently available. In particular, there is no consensus on the role of the regulator when it comes to VMDs. Should regulators mandate specific standards? Do regulators need to adopt component-wise certification regimes? What is the role, if any, of third-party certifiers?

1.3.4.4 Case Studies

The subject of safety assessment of on-demand medical systems has been the focus of a number of research projects. These projects have explored different aspects of on-demand medical systems, their safety, and possible mechanisms for regulatory oversight. The Medical Device Plug-and-Play project articulated the need for on-demand medical systems, documented specific clinical scenarios that would benefit, and developed the Integrated Clinical Environment (ICE) architecture, which has been codified as an ASTM standard (ASTM F2761-2009)

[ASTM09]. ICE proposes to approach the engineering and regulatory challenges by building medical systems around a system architecture that supports compositional certification. In such an architecture, each medical system would be composed out of a variety of components (clinical applications, a medical application platform, and medical devices), which would be regulated, certified, and then obtained by the healthcare organization separately [Hatcliff12].

Integrated Clinical Environment
Figure 1.5 shows the primary components of the integrated clinical environment (ICE) architecture. This case study summarizes the intended functionality and goals for each of these components. Note

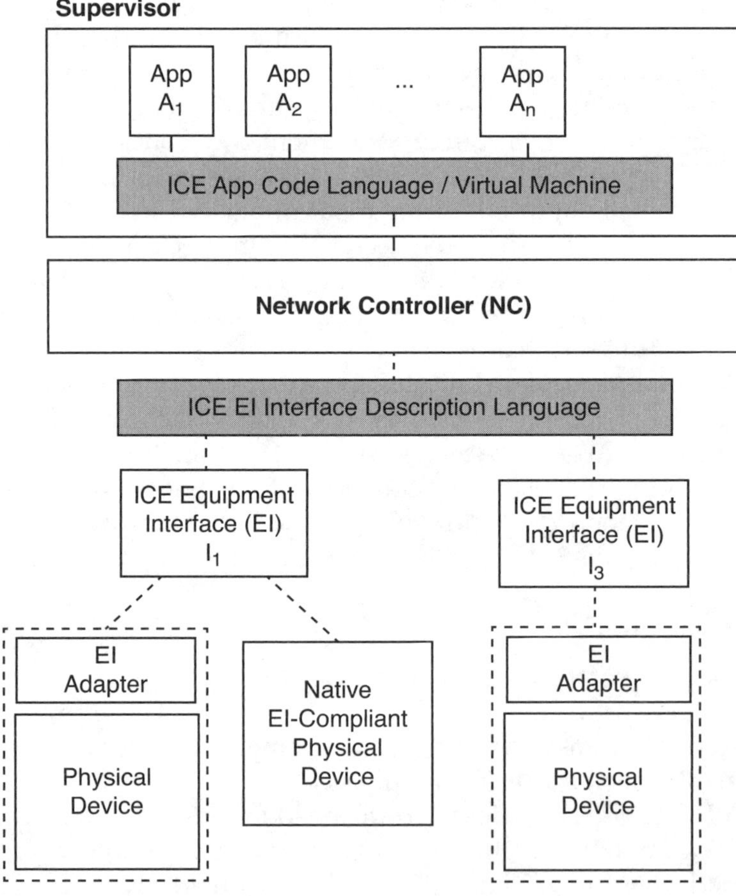

Figure 1.5: *ICE architecture*

that ASTM F2761-2009 does not provide detailed requirements for these components, as it is purely an architectural standard. Nevertheless, the roles of each of the components in the architecture imply certain informal requirements:

- *Apps:* Applications are software programs that provide the coordination algorithm for a specific clinical scenario (i.e., smart alarms, closed-loop control of devices). In addition to executable code, these applications contain device requirements declarations—that is, a description of the medical devices they need to operate correctly. These apps would be validated and verified against their requirements specification before they are marketed.
- *Devices:* Symmetrical to the applications, medical devices used in the ICE architecture would implement an interoperability standard and carry a self-descriptive model, known as a capabilities specification. Each medical device would be certified that it conforms to its specification before it is marketed and sold to end users.
- *Supervisor:* The supervisor provides a secure isolation kernel and virtual machine (VM) execution environment for clinical applications. It would be responsible for ensuring that apps are partitioned in both data and time from each other.
- *Network controller:* The network controller is the primary conduit for physiologic signal data streams and device control messages. The network controller would be responsible for maintaining a list of connected devices and ensuring proper quality of service guarantees in terms of time and data partitioning of data streams, as well as security services for device authentication and data encryption.
- *ICE interface description language:* The description language is the primary mechanism for ICE-compliant devices to export their capabilities to the network controller. These capabilities may include which sensors and actuators are present on the device, and which command set it supports.

Medical Device Coordination Framework

The Medical Device Coordination Framework (MDCF) [King09, MDCF] is an open-source project that aims to provide a software implementation of a medical application platform that conforms to the ICE standard. The modular framework is envisioned as enabling researchers to rapidly prototype systems and explore implementation and engineering issues associated with on-demand medical systems.

The MDCF is implemented as a collection of services that work together to provide some of the capabilities required by ICE as essential for a medical application platform. The functionality of these services also may be decomposed along the architectural boundaries defined in the ICE architecture (see Figure 1.6); that is, the MDCF consists of network controller services, supervisor services, and a global resource management service.

Network controller services are as follows:

- *Message bus:* Abstracts the low-level networking implementation (e.g., TCP/IP) and provides a publish/subscribe messaging service. All communication between medical devices and the MDCF occurs via the message bus, including protocol control messages, exchanges of patient physiologic data, and commands sent from apps to devices. The message bus also provides basic real-time guarantees (e.g., bounded end-to-end message transmission delays) that apps can take as assumptions. Additionally, the message bus supports various fine-grained message and stream access control and isolation

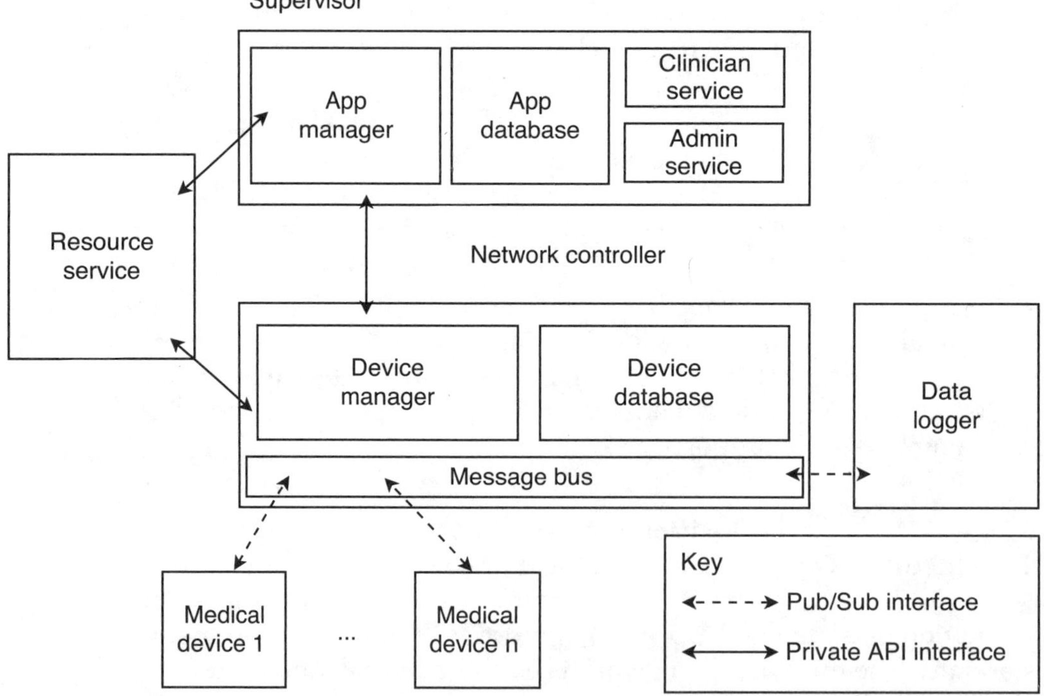

Figure 1.6: *MDCF services decomposed along ICE architectural boundaries*

policies. While the current implementation of the message bus encodes messages using XML, the actual encoding strategy is abstracted away from the apps and devices by the message bus API, which exposes messages as structured objects in memory.

- *Device manager:* Maintains a registry of all medical devices currently connected with the MDCF. The device manager implements the server side of the MDCF device connection protocol (medical devices implement the client side) and tracks the connectivity of those devices, notifying the appropriate apps if a device goes offline unexpectedly. The device manager also serves another important role: It validates the trustworthiness of any connecting device by determining whether the connecting device has a valid certificate.
- *Device database:* Maintains a list of all specific medical devices that the healthcare provider's bioengineering staff has approved for use. In particular, the database lists each allowed device's unique identifier (e.g., an Ethernet MAC address), the manufacturer of the device, and any security keys or certificates that the device manager will use to authenticate connecting devices against.
- *Data logger:* Taps into the flows of messages moving across the message bus and selectively logs them. The logger can be configured with a policy specifying which messages should be recorded. Because the message bus carries every message in the system, the logger can be configured to record any message or event that propagates through the MDCF. Logs must be tamper resistant and tamper evident; access to logs must itself be logged, and be physically and electronically controlled by a security policy.

Supervisor services are as follows:

- *Application manager:* Provides a virtual machine for apps to execute in. In addition to simply executing program code, the application manager checks that the MDCF can guarantee the app's requirements at runtime and provides resource and data isolation, as well as access control and other security services. If the app requires a certain medical device, communications latency, or response time from app tasks, but the MDCF cannot currently make those guarantees (e.g., due to system load or because the appropriate medical device has not been connected), then the app manager will not let the clinician start the app in question. If the resources are available, the application manager will reserve those resources so as to guarantee the required performance to the app. The application manager further detects and flags potential medically meaningful app interactions, since individual apps are isolated and may not be aware which other apps are associated with a given patient.

- *Application database:* Stores the applications installed in the MDCF. Each application contains executable code and requirement metadata used by the application manager to allocate the appropriate resources for app execution.
- *Clinician service:* Provides an interface for the clinician console GUI to check the status of the system, start apps, and display app GUI elements. Since this interface is exposed as a service, the clinician console can be run locally (on the same machine) that is running the supervisor, or it can be run remotely (e.g., at a nurse's station).
- *Administrator service:* Provides an interface for the administrator's console. System administrators can use the administrator's console to install new applications, remove applications, add devices to the device database, and monitor the performance of the system.

1.3.5 Smart Alarms and Clinical Decision Support Systems

Fundamentally, clinical decision support (CDS) systems are a specialized form of MCPS with physical actuation limited to visualization. They take as inputs multiple data streams, such as vital signs, lab test values, and patient history; they then subject those inputs to some form of analysis, and output the results of that analysis to a clinician. A smart alarm is the simplest form of decision support system, in which multiple data streams are analyzed to produce a single alarm for the clinician. More complex systems may use trending, signal analysis, online statistical analysis, or previously constructed patient models, and may produce detailed visualizations.

As more medical devices become capable of recording continuous vital signs, and as medical systems become increasingly interoperable, CDS systems will evolve into essential tools that allow clinicians to process, interpret, and analyze patient data. While widespread adoption of CDS systems in clinical environments faces some challenges, the current efforts to build these systems promise to expose their clinical utility and provide impetus for overcoming those challenges.

1.3.5.1 The Noisy Intensive Care Environment

Hospital intensive care units (ICUs) utilize a wide array of medical devices in patient care. A subset of these medical devices comprises sensors that detect the intensity of various physical and chemical signals in the body. These sensors allow clinicians (doctors, nurses, and other clinical caretakers) to better understand the patient's current state. Examples of such sensors include automatic blood pressure cuffs, thermometers, heart rate monitors, pulse oximeters, electroencephalogram meters, automatic glucometers, electrocardiogram meters, and so on. These sensors range from very simple to very complex in terms of their technology. Additionally, along with the traditional techniques, digital technologies have enabled new sensors to be developed and evaluated for clinical use.

The vast majority of these medical devices act in isolation, reading a particular signal and outputting the result of that signal to some form of visualization technology so it may be accessed by clinicians. Some devices stream data to a centralized visualization system (such as a bedside monitor or nursing station [Phillips10, Harris13]) for ease of use. Each of the signals is still displayed independently, however, so it is up to the clinician to synthesize the presented information to determine the patient's actual condition.

Many of these devices can be configured to alert clinicians to a deterioration in the patient's condition. Most sensors currently in use can be configured with only threshold alarms, which activate when the particular vital sign being measured crosses a predefined threshold. While threshold alarms can certainly be critical in the timely detection of emergency states, they have been shown to be not scientifically derived [Lynn11] and have a high rate of false alarms [Clinical07], often attributable to insignificant random fluctuations in the patient's vital signs or noise caused by external stimuli. For example, patient movement can cause sensors to move, be compressed, or fall off. The large number of erroneous alarms generated by such devices causes alarm fatigue—a desensitization to the presence of these alarms that causes clinicians to ignore them [Commission13]. In an effort to reduce the number of false alarms, clinicians may sometimes improperly readjust settings on the monitor or turn off alarms entirely [Edworthy06]. Both of these actions can lead to missed true alarms and a decrease in quality of care [Clinical07, Donchin02, Imhoff06].

Various efforts have been made to reduce alarm fatigue. These strategies usually focus on improving workflow, establishing appropriate patient-customized thresholds, and identifying situations where alarms are not clinically relevant [Clifford09, EBMWG92, Oberli99, Shortliffe79]. However, isolated threshold alarms cannot capture sufficient nuance in patient state to completely eliminate false alarms. Also, these alarms simply alert clinicians to the fact that some threshold was crossed; they fail to provide any physiologic or diagnostic information about the current state of the patient that might help reveal the underlying cause of the patient's distress.

Clinicians most often use multiple vital signs in concert to understand the patient's state. For example, a low heart rate (bradycardia) can be normal and healthy. However, if a low heart rate occurs in conjunction with an abnormal blood pressure or a low blood oxygen level, this collection of findings can be cause for concern. Thus, it seems pertinent to develop smart alarm systems that would consider multiple vital signs in concert before raising an alarm. This would reduce false alarms, improving the alarm precision and reducing alarm fatigue, thereby leading to improved care.

Such a smart alarm system would be a simple version of a CDS system [Garg05]. Clinical decision support systems combine multiple sources of patient information with preexisting health knowledge to help clinicians make more informed decisions. It has repeatedly been shown that well-designed CDS systems have the potential to dramatically improve patient care, not just by reducing alarm fatigue, but by allowing clinicians to better utilize data to assess patient state.

1.3.5.2 Core Feature Difficulties

As CDS systems are a specialized form of MCPS, the development of CDS systems requires satisfying the core features of cyber-physical system development. In fact, without these features, CDS system development is impossible. The current lack of widespread use of CDS systems in part reflects the difficulty that has been encountered in establishing these features in a hospital setting.

One of the most fundamental of these requirements is the achievement of device interoperability. Even the simplest CDS system (such as a smart alarm system) must obtain access to the real-time vital signs data being collected by a number of different medical devices attached to the patient. To obtain these data, the devices collecting the required vital signs must be able to interoperate—if not with each other, then with a central data repository. In this repository, data could be collected, time synchronized, analyzed, and visualized.

In the past, achieving interoperability of medical devices has been a major hurdle. Due to increased costs, the exponential increase in regulatory difficulty, and the lucrative potential from selling a suite of devices with limited interoperability, individual device manufacturers currently have few incentives to make their devices interoperate. Development of an interoperable platform for device communication would enable MCPS to stream real-time medical information from different devices.

Many other challenges exist. For example, the safety and effectiveness of CDS systems depend on other factors, such as network reliability and real-time guarantees for message delivery. As networks in current hospital systems are often ad hoc, highly complex, and built over many decades, such reliability is rare.

Another challenge is related to data storage. To achieve high accuracy, the parameters of the computational intelligence at the heart of a CDS system must often be tuned using large quantities of retrospective data. Dealing with Big Data, therefore, is a vital component of the development of CDS systems. Addressing this problem will require hospitals to recognize the value of capturing and storing patients' data and to develop a dedicated hospital infrastructure to store and access data as part of routine workflow.

CDS systems require some level of context-aware computational intelligence. Information from multiple medical device data streams must be extracted and filtered, and used in concert with a patient model to create a context-aware clinical picture of the patient. There are three major ways in which context-aware computational intelligence can be achieved: by encoding hospital guidelines, by capturing clinicians' mental models, and by creating models based on machine learning of medical data.

While the majority of hospital guidelines can usually be encoded as a series of simple rules, they are often vague or incomplete. Thus, while they may serve as a useful baseline, such guidelines are often insufficient on their own to realize context-aware computational intelligence. Capturing clinicians' mental models involves interviewing a large number of clinicians about their decision-making processes and then hand-building an algorithm based on the knowledge gleaned from the interviews. This process can be laborious, it can be difficult to quantify in software how a clinician thinks, and the results from different clinicians can be difficult to reconcile. Creating models using machine learning is often the most straightforward approach. However, training such models requires large amounts of retrospective patient data and clear outcome labels, both of which can be difficult to acquire. When such data sets are available, they often prove to be noisy, with many missing values. The choice of learning technique can be a difficult one, too. While algorithm transparency is a good metric (to empower clinicians to understand the underlying process and avoid opaque black-box algorithms), there is no single choice of learning technique that is most appropriate for all scenarios.

1.3.5.3 Case Study: A Smart Alarm System for CABG Patients

Patients who have undergone coronary artery bypass graft (CABG) surgery are at particular risk of physiologic instability, so continuous monitoring of their vital signs is routine practice. The hope is that detection of physiologic changes will allow practitioners to intervene in a timely manner and prevent postsurgery complications. As previously discussed, the continuous vital signs monitors are usually equipped only with simple threshold-based alarms, which, in combination with the rapidly evolving post-surgical state of such patients, can lead to a large number of false alarms. For example, it is common for the finger-clip sensors attached to pulse oximeters to fall off patients as they get situated in their ICU bed, or for changes in the artificial lighting of the care environment to produce erroneous readings.

To reduce these and other erroneous alarms, a smart alarm system was developed that combines four main vital signs routinely collected in the surgical ICU (SICU): blood pressure (BP), heart rate (HR), respiratory rate (RR), and blood oxygen saturation (SpO_2). ICU nurses were interviewed to determine appropriate ranges for binning each vital sign into a number of ordinal sets (e.g., "low," "normal," "high," and "very high," leading to classifying, for example, a blood pressure greater than 107 mm Hg as "high"). Binning vital signs in this way helped overcome the difficulty of establishing a rule set customized to each patient's baseline vital signs. The binning criteria can be modified to address a specific patient with, for example, a very low "normal" resting heart rate, without rewriting the entire rule set.

Afterward, a set of rules was developed in conjunction with nurses to identify combinations of these vital signs statuses that would be cause for concern. The smart alarm monitors a patient's four vital signs, categorizes them according to which ordinal set they belong in, and searches the rule table for the corresponding alarm level to output. To deal with missing data (due to network or sensor faults), rapid drops to zero for a vital sign are conservatively classified as "low" for the duration of the signal drop.

This smart alarm avoided many of the challenges that CDS systems normally face in the clinical environment. The set of vital signs employed was very limited and included only those commonly collected and synchronized by the same medical device. As the "intelligence" of the smart alarm system was a simple rule table based on clinician mental models, it did not require large amounts of retrospective data to calibrate, and it was transparent and easy for clinicians to understand. While network reliability would be a concern for such a system running in the ICU, the classification of missing values as "low" provided a conservative fallback in case of a brief network failure. Additionally, running the system on a real-time middleware product would provide the necessary data delivery guarantees to ensure system safety.

To evaluate the performance of this system, 27 patients were observed while they convalesced in the ICU immediately after their CABG procedure. Of these 27 patients, 9 had the requisite vital signs samples stored in the hospital IT system during the time period of the observation. Each of these patients was observed for between 26 and 127 minutes, totaling 751 minutes of observation. To compare monitor alarm performance with the CABG smart alarm, the minute-by-minute samples of these patients' physiologic state were retroactively retrieved (after the observations) from the UPHS data store. The smart alarm algorithm was applied to the retrieved data streams, resulting in a trace of the smart alarm outputs that would have been produced if the smart alarm were active at the patient's bedside. Because of the relatively slow rate at which a patient can deteriorate and the expected response time of the care staff, an intervention alarm was considered to be covered by a smart alarm if the alarm occurred within 10 minutes of the intervention.

Overall, the smart alarm system produced fewer alarms. During the study, the smart alarm was active 55% of the time that the standard monitor alarms were active, and of the 10 interventions during the observation time period, 9 were covered by the smart alarm. The significant alarm was likely deemed "significant" not due to the absolute values of the vital signs being observed, but rather by their trend. An improved version of this smart alarm system would include rules concerning the trend of each of the vital signs.

1.3.6 Closed-Loop System

Given that medical devices are aimed at controlling a specific physiological process in a human, they can be viewed as a closed loop between the device and the patient. In this section, we discuss clinical scenarios from this point of view.

1.3.6.1 A Higher Level of Intelligence

A clinical scenario can be viewed as a control loop: The patient is the plant, and the controller collects information from sensors (e.g., bedside monitors) and sends configuration commands to actuators (e.g., infusion pumps) [Lee12]. Traditionally, caregivers act as the controller in most scenarios. This role imposes a significant decision-making burden on them, as one caregiver is usually caring for several patients and can check on each patient only sporadically. Continuous monitoring, whereby the patient's condition is under constant surveillance, is an active area of research [Maddox08]. However, to improve patient safety further, the system should be able to continuously react to changes in patient condition as well.

The smart alarm systems and decision support systems, discussed in the previous section, facilitate the integration and interpretation of clinical information, helping caregivers make decisions more efficiently. Closed-loop systems aim to achieve a higher level of intelligence: In such systems, a software-based controller automatically collects and interprets physiological data, and controls the therapeutic delivery devices. Many safety-critical systems utilize automatic controllers—for example, autopilots in airplanes and adaptive cruise control in vehicles. In patient care, the controller can continuously monitor the patient's state and automatically reconfigure the actuators when the patient's condition stays within a predefined operation region. It will alert and hand control back to caregivers if the patient's state starts veering out of the safe range. Such physiological closed-loop systems can assume part of the caregivers' workload, enabling them to better focus on handling critical events, which would ultimately improve patient safety. In addition, software controllers can run advanced decision-making algorithms (e.g., model-predictive control in blood glucose regulation [Hovorka04]) that are too computationally complicated for human caregivers to apply, which may improve both the safety and the effectiveness of patient care.

The concept of closed-loop control has already been introduced in medical applications—for example, in implantable devices such as cardioverter defibrillators and other special-purpose stand-alone devices. A physiological closed-loop system can also be built by networking multiple existing devices, such as infusion pumps and vital sign monitors. The networked physiological closed-loop system can be modeled as a VMD.

1.3.6.2 Hazards of Closed-Loop Systems

The networked closed-loop setting introduces new hazards that could compromise patient safety. These hazards need to be identified and mitigated in a systematic way. Closed-loop MCPS, in particular, raise several unique challenges for safety engineering.

First, the plant (i.e., the patient) is an extremely complex system that usually exhibits significant variability and uncertainty. Physiological modeling has been a decade-long challenge for biomedical engineers and medical experts, and the area remains at the frontier of science. Unlike in many other engineering disciplines, such as mechanical engineering or electronic circuit design, where high-fidelity first-principle models are usually directly applicable to theoretical controller design, the physiological models are usually nonlinear and contain parameters that are highly individual dependent, time varying, and not easily identifiable given the technologies available. This imposes a major burden on control design as well as system-level safety reasoning.

Second, in the closed-loop medical device system, complex interactions occur between the continuous physiology of the patient and the discrete behavior of the control software and network. Since most closed-loop systems require supervision from users (either caregivers or patients themselves), the human behavior must be considered in the safety arguments.

Third, the control loop is subject to uncertainties caused by sensors, actuators, and communication networks. For example, some body sensors are very sensitive to patient movements—vital signs monitors may alert faulty readings due to a dropped finger-clip—and due to technological constraints, some biosensors have non-negligible error even when they are used correctly (e.g., the continuous glucose monitor) [Ginsberg09]. The network behavior also has a critical impact on patient safety: Patients can be harmed by the actuators if packets that carry critical control commands are dropped as they travel across the network.

1.3.6.3 Case Study: Closed-Loop PCA Infusion Pump

One way to systematically address the challenges faced by closed-loop systems is to employ a model-based approach similar to the one outlined in Section 1.3.3. This effort involves extending the high-confidence approach based on hazard identification and mitigation from individual devices to a system composed of a collection of devices and a patient.

This section briefly describes a case study of the use of physiological closed loop in pain control using a PCA infusion pump, introduced in Section 1.3.3.3. The biggest safety concern that arises with the use of PCA pumps for pain control is the risk of overdose of an opioid analgesic, which can cause respiratory failure. Existing safety mechanisms built into PCA pumps include limits on bolus amounts, which are programmed by a caregiver before the infusion starts, and minimum time intervals between consecutive bolus doses. In addition, nursing manuals prescribe periodic checks of the patient condition by a nurse, although these mechanisms are considered insufficient to cover all possible scenarios [Nuckols08].

The case study [Pajic12] presents a safety interlock design for PCA infusion, implemented as an on-demand MCPS as described in Section 1.3.4. The pulse oximeter continuously monitors heart rate and blood oxygen saturation. The controller receives measurements from the pulse oximeter, and it may stop the PCA infusion if the HR/SpO_2 readings indicate a dangerous decrease in respiratory activity, thereby preventing overdosing.

Safety requirements for this system are based on two regions in the space of possible patient states as reported by the two sensors, as illustrated in Figure 1.7. The critical region represents imminent danger to the patient and must be avoided at all times; the alarming region is not immediately dangerous but raises clinical concerns.

The control policy for the safety interlock may be to stop the infusion as soon as the patient state enters the alarming region. The immediate challenge is to define the alarming region to be large enough so that the pump can always be stopped before the patient enters the critical region. At the same time, the region should not be too large, so as to avoid false alarms that would decrease the effectiveness of pain control unnecessarily. Finding the right balance and defining exact boundaries of the two regions was beyond the scope of the case study.

The goal of the case study was to verify that the closed-loop system satisfies its patient requirements. To achieve this goal, one needs models of the infusion pump, the pulse oximeter, the control algorithm, and the physiology of the patient.

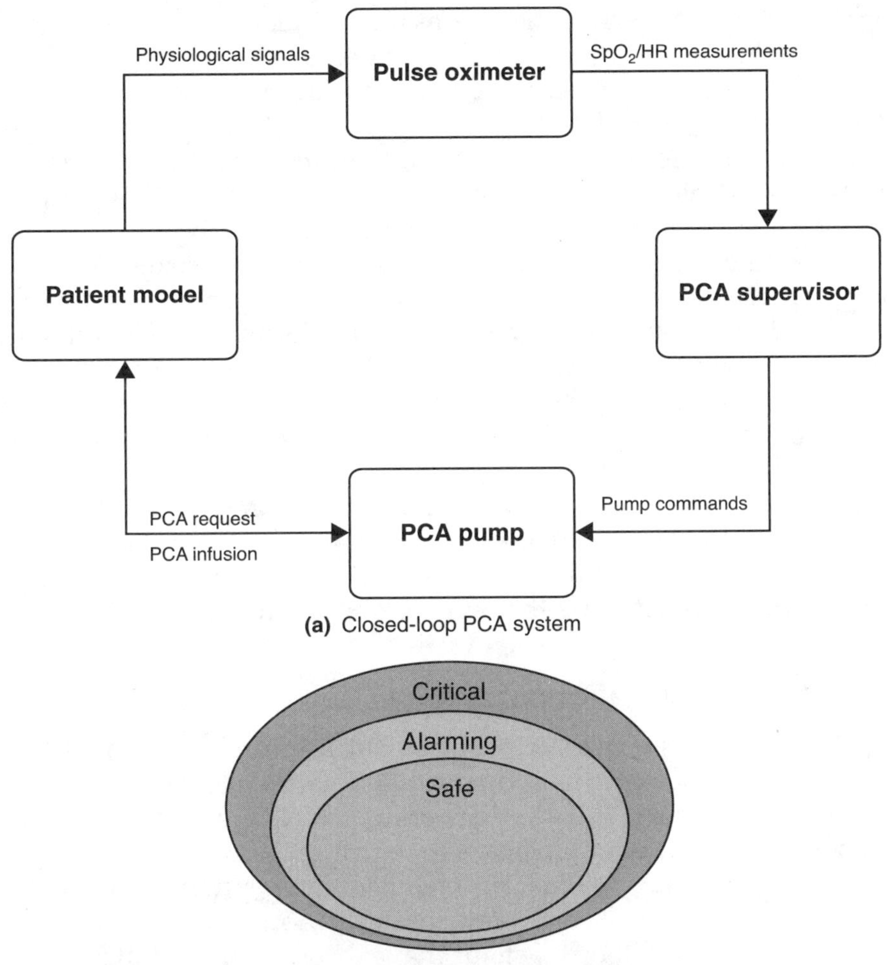

(a) Closed-loop PCA system

(b) Regions of patient's possible conditions

Figure 1.7: *PCA safety interlock design*

Patient modeling is the critical aspect in this case. Both pharmacokinetic and pharmacodynamics aspects of physiology should be considered [Mazoit07]. Pharmacokinetics specifies how the internal state of the patient, represented by the drug concentration in the blood, is affected by the rate of infusion. Pharmacodynamics specifies how the patient's internal state affects observable outputs of the model—that is, the relationship between the drug concentration and oxygen saturation levels measured by the pulse oximeter. The proof-of-concept approach taken in the case study relies on the simplified pharmacokinetic model of [Bequette03]. To make the model applicable to a diverse patient population, parameters of the model were taken to be ranges, rather than fixed values. To avoid the complexity of pharmacodynamics, a linear relationship between the drug concentration and the patient's vital signs was assumed.

Verification efforts concentrated on the timing of the control loop. After a patient enters the alarming region, it takes time for the controller to detect the danger and act on it. There are delays involved in obtaining sensor readings, delivering the readings from the pulse oximeter to the controller, calculating the control signal, delivering the signal to the pump, and finally stopping the pump motor. To strengthen confidence in the verification results, the continuous dynamics of the patient model were used to derive t_{crit}, the minimum time over all combinations of parameter values in the patient model that can pass from the moment the patient state enters the alarming region to the moment it enters the critical region. With this approach, the verification can abstract away from the continuous dynamics, significantly simplifying the problem. Using a timing model of the components in the system, one can verify that the time it takes to stop the pump is always smaller than t_{crit}.

1.3.6.4 Additional Challenging Factors

The PCA system is a relatively simple but useful use case of closed-loop medical devices. Other types of closed-loop systems, by comparison, may introduce new engineering challenges due to their functionalities and requirements. For example, blood glucose control for patients with diabetes has garnered a lot of attention from both the engineering and clinical communities, and various concepts of closed-loop or semi-closed-loop systems have been proposed [Cobelli09, Hovorka04, Kovatchev09]. Compared to the PCA system, the closed-loop glucose control system is substantially more complex and opens up many opportunities for new research.

The fail-safe mode in the PCA system is closely related to the clinical objective: Overdosing is the major concern. While the patient may suffer from more pain when PCA is stopped, stopping the infusion is considered a safe action, at least for a reasonable time duration. This kind of fail-safe mode may not exist in other clinical scenarios. For example, in the glucose control system, the goal is to keep the glucose level within a target range. In this case, stopping the insulin pump is not a default safe action, because high glucose level is also harmful.

The safety criteria in the PCA system are defined by delineating a region in the state space of the patient model (such as the critical region in the previous case study). Safety violations are then detected as threshold crossings in the stream of patient vital signs. Such crisp, threshold-based rules are often crude simplifications. Physiological systems have a certain level of resilience, and the true relationship between health risks and physiological variables is still not completely understood. Time of exposure is also important: A short spike in the drug concentration may be less harmful than a lower-level concentration that persists over a longer interval.

The pulse oximeter—the sensor used in the PCA system—is relatively accurate with respect to the ranges that clinicians would consider in their decision making. In some other scenarios, however, sensor accuracy is a non-negligible factor. For example, a glucose sensor can have a relative error of as much as 15% [Ginsberg09]; given that the target range is relatively narrow, such an error may significantly impact system operation and must be explicitly considered in the safety arguments.

Even if the sensor is perfectly accurate, it may not be predictive enough. While oxygen saturation can be used to detect respiratory failure, for example, this value may not decline until a relatively late point, after harm to the patient is already done. Capnography data, which measure levels of carbon dioxide exhaled by the patient, can be used to detect the problem much sooner, but this technique is more expensive and involves invasive technology compared to pulse oximetry. This example highlights the need to include more accurate pharmacodynamics data into the patient model, which can be used to account for the detection delay.

Another important factor in the closed-loop medical system is the human user's behavior. In the PCA system, the user behavior is relatively simple: The clinicians are alerted when certain conditions arise, and most of the times they do not need to intervene in the operation of the control loop. In other applications with more complicated requirements, however, the user may demand a more hands-on role in the control. For example, in the glucose control application, a user will need to take back the control authority when the glucose level is significantly out of range; even when the automatic controller is running, the user may choose to reject certain control actions for various reasons (e.g., the patient is not comfortable with a large insulin dose). This kind of more complicated user interaction pattern introduces new challenges to the model-based validation and verification efforts.

1.3.7 Assurance Cases

Recently, safety cases have become popular and acceptable ways for communicating ideas and information about the safety-critical systems among the system stakeholders. In the medical devices domain, the FDA issued draft guidance for medical infusion pump manufacturers indicating that they should provide a safety case with their premarket submissions [FDA10]. In this section, we briefly introduce the concept of safety cases and the notations used to describe them. Three aspects of safety cases that can be manipulated to make them practically useful are discussed—namely, facilitating safety case construction, justifying the existence of sufficient trust in safety arguments and cited evidence, and providing a framework for safety case assessment for regulation and certification.

Safety case patterns can help both device manufacturers and regulators to construct and review the safety cases more efficiently while improving confidence and shortening the period in which a device's application is in FDA-approval limbo. Qualitative reasoning for having confidence in a device is believed to be more consistent with the inherited subjectivity in safety cases than the quantitative reasoning. The separation between safety and confidence arguments reduces the size of the core safety argument. Consequently, this structure is believed to facilitate the development and reviewing processes for safety cases. The constructed confidence arguments should be used in the appraisal process for assurance arguments as illustrated in [Ayoub13, Cyra08, Kelly07].

Given the subjective nature of safety cases, the review methods cannot hope to replace the human reviewer. Instead, they form frameworks that lead safety case reviewers through the evaluation process. Consequently, the result of the safety case review process is always subjective.

1.3.7.1 Safety Assurance Cases

The safety of medical systems is of great public concern—a concern that is reflected in the fact that many such systems must adhere to government regulations or be certified by licensing bodies [Isaksen97]. For example, medical devices sold in the United States are regulated by the FDA. Some of these medical devices, such as infusion pumps, cannot be commercially distributed before receiving an approval from the FDA. There is a need to communicate, review, and debate the trustworthiness of systems with a range of stakeholders (e.g., medical device manufacturers, regulatory authorities).

Assurance cases can be used to justify the adequacy of medical device systems. The assurance case is a method for arguing that a body of evidence justifies a claim. An assurance case addressing safety is called a *safety case*. A safety assurance case presents an argument, supported by a body of evidence, that a system is acceptably safe when used in a given context [Menon09]. The notion of safety cases is currently embraced by several European industry sectors (e.g., aircraft, trains, nuclear power). More recently in the United States, the FDA issued draft guidance indicating that medical infusion pump manufacturers should provide a safety case with their premarket submissions [FDA10]. Thus, an infusion pump manufacturer is expected not only to achieve safety, but also to convince regulators that it has been achieved [Ye05] through the submitted safety case. The manufacturer's role is to develop and submit a safety case to regulators showing that its product is acceptably safe to operate in the intended context [Kelly98]. The regulator's role, in turn, is to assess the submitted safety case and make sure that the system is really safe.

Many different approaches are possible for the organization and presentation of safety cases. Goal Structuring Notation (GSN) is one description technique that has proved useful for constructing safety cases [Kelly04]. GSN is a graphical argumentation notation developed at the University of York. A GSN diagram includes elements that represent goals, argument strategies, contexts, assumptions, justifications, and evidence. The principal purpose of any goal structure in GSN is to show how goals—that is, claims about the system specified with text within rectangular elements—are supported by valid and convincing arguments. To this end, goals are successively decomposed into subgoals through implicit or explicit strategies. Strategies, specified with text within parallelograms, explicitly define how goals are decomposed into subgoals. The decomposition continues until a point is reached where claims are supported by direct reference to available evidence, and the solution specified with text within circles. Assumptions/justifications, which define the rationale of the decomposition approach, are represented with ellipses. The context in which goals are stated is given in rectangles with rounded sides.

Another popular description technique is called Claims–Arguments–Evidence (CAE) notation [Adelard13]. While this notation is less standardized than GSN, it shares the same element types as GSN.

The primary difference is that strategy elements are replaced with argument elements. In this work, we use GSN notation in presenting safety cases.

1.3.7.2 Justification and Confidence

The objective of a safety case development process is to provide a justifiable rationale for the design and engineering decisions and to instill confidence in those design decisions (in the context of system behavior) in stakeholders (e.g., manufacturers and regulatory authorities). Adopting assurance cases necessarily requires the existence of proper reviewing mechanisms. These mechanisms address the main aspects of assurance cases—that is, building, trusting, and reviewing assurance cases.

All three aspects of assurance cases bring own challenges. These challenges need to be addressed to make safety cases practically useful:

- *Building assurance cases:* The *Six-Step method* [Kelly98a] is a widely used method for systematically constructing safety cases. Following the Six-Step method or any other method does not prevent safety case developers from making some common mistakes, such as leaping from claims to evidence. Even so, capturing successful (i.e., convincing, sound) arguments used in safety cases and reusing them in constructing new safety cases can minimize the mistakes that may be made during the safety case development. The need for argument reusability motivates the use of the pattern concept (where *pattern* means a model or original used as an archetype) in the safety case constructions. Predefined patterns can often provide an inspiration or a starting point for new safety case developments. Using patterns may also help improve the maturity and completeness of safety cases. Consequently, patterns can help medical device manufacturers to construct safety cases in a more efficient way in terms of completeness, thereby shortening the development period. The concept of safety case patterns is defined in [Kelly97] as a way to capture and reuse "best practices" in safety cases. Best practices incorporate company expertise, successfully certified approaches, and other recognized means of assuring quality. For example, patterns extracted from a safety case built for a specific product can be reused in constructing safety cases for other products that are developed via similar processes. Many safety case patterns were introduced in [Alexander07, Ayoub12, Hawkins09, Kelly98, Wagner10, Weaver03] to capture best practices.

- *Trusting assurance cases:* Although a structured safety case explicitly explains how the available evidence supports the overall claim of acceptable safety, it cannot ensure that the argument itself is good (i.e., sufficient for its purpose) or that the evidence is sufficient. Safety arguments typically have some weaknesses, so they cannot be fully trusted on their own. In other words, there is always a question about the level of trust for the safety arguments and cited evidence, which makes a justification for the sufficiency of confidence in safety cases essential. Several attempts have been to quantitatively measure confidence in safety cases, such as in [Bloomfield07, Denney11].

  A new approach for creating clear safety cases was introduced in [Hawkins11] to facilitate their development and increase confidence in the constructed cases. This approach basically separates the major components of safety cases into a safety argument and a confidence argument. A safety argument is limited to arguments and evidence that directly target the system safety—for example, explaining why a specific hazard is sufficiently unlikely to occur and arguing this claim by testing results as evidence. A confidence argument is given separately; it seeks to justify the sufficiency of confidence in the safety argument. For example, questions about the level of confidence in the given testing result evidence (e.g., whether that testing was exhaustive) should be addressed in the confidence argument. These two components, while presented explicitly and separately, are interlinked so that the justification for having sufficient confidence in individual aspects of the safety component is clear and readily available but not confused with the safety component itself.

  Any gap that prohibits perfect confidence in safety arguments is referred to as an assurance deficit [Hawkins11]. Argument patterns for confidence arguments are given in [Hawkins11]. Those patterns are defined based on identifying and managing the assurance deficits so as to show sufficient confidence in the safety argument. To this end, it is necessary to identify the assurance deficits as completely as practicable. Following a systematic approach (such as the one proposed by [Ayoub12a]) would help in effectively identifying assurance deficits. In [Menon09, Weaver03], lists of major factors that should be considered in determining the confidence in arguments are defined. Questions to be considered when determining the sufficiency of each factor are given as well.

  To show sufficient confidence in a safety argument, the developer of a confidence argument first explores all concerns about the level of confidence in this argument, and then makes claims that these concerns are addressed. If a claim cannot be supported by convincing evidence, then a deficit is identified. The list of the recognized assurance deficits can be then used when instantiating the confidence pattern given in [Hawkins11] to show that the residual deficits are acceptable.

- *Reviewing assurance cases:* Safety case arguments are rarely provable deductive arguments, but rather are more commonly inductive. In turn, safety cases are, by their nature, often subjective [Kelly07]. The objective of safety case evaluation, therefore, is to assess whether there is a mutual acceptance of the subjective position. The human mind does not deal well with complex inferences based on uncertain sources of knowledge [Cyra08], which are common in safety arguments. Therefore, reviewers should be required to express their opinions about only the basic elements in the safety case. A mechanism should then provide a way to aggregate the reviewers' opinions about the basic elements in the safety case so as to communicate a message about its overall sufficiency.

Several approaches to assessing assurance cases have been proposed. The work in [Kelly07] presents a structured approach to assurance case review by focusing primarily on assessment of the level of assurance offered by the assurance case argument. The work in [Goodenough12] outlines a framework for justifying confidence in the truth of assurance case claims. This framework is based on the notion of eliminative induction—the principle that confidence in the truth of a claim increases as reasons for doubting its truth are identified and eliminated. Defeaters, in contrast, offer possible reasons for doubting. The notion of Baconian probability is then used to provide a measure of confidence in assurance cases based on how many defeaters have been identified and eliminated.

A structured method for assessing the level of sufficiency and insufficiency of safety arguments was outlined in [Ayoub13]. The reviewer assessments and the results of their aggregation are represented in the Dempster-Shafer model [Sentz02]. The assessing mechanism given in [Ayoub13] can be used in conjunction with the step-by-step review approach proposed in [Kelly07] to answer the question given in the last step of this reviewing approach, which deals with the overall sufficiency of the safety argument. In other words, the approach in [Kelly07] provides a skeleton for a systematic review process; by comparison, the mechanism in [Ayoub13] provides a systematic procedure to measure the sufficiency and insufficiency of the safety arguments. An appraisal mechanism is proposed in [Cyra08] to assess the trust cases using the Dempster-Shafer model.

Finally, linguistic scales are introduced in [Cyra08] as a means to express the expert opinions of reviewers and the aggregation results. Linguistic scales are appealing in this context, as they are closer to human nature than are numbers. They are based on qualitative values such as "high," "low," and "very low" and are mapped into the interval for evaluation.

1.3.7.3 Case Study: GPCA Safety

This section builds on the case study of the GPCA infusion pump, which was presented in Section 1.3.3.3. Assurance cases for medical devices have been discussed in [Weinstock09]. The work in [Weinstock09] can be used as starting point for the GPCA safety case construction. A safety case given in [Jee10] is constructed for a pacemaker that is developed following a model-based approach similar to the one used in the GPCA case study.

Safety Case Patterns

Similarities in development approach are likely to lead to similarities in safety arguments. In keeping with this understanding, safety case patterns [Kelly97] have been proposed as means of capturing similarities between arguments. Patterns allow the common argument structure to be elaborated with device-specific details. To capture the common argument structure for systems developed in a model-based fashion, a safety case pattern, called the *from_to* pattern, has been proposed in [Ayoub12]. In this section, the *from_to* pattern is illustrated and instantiated for the GPCA reference implementation.

A safety case for the GPCA reference implementation would claim that the PCA implementation software does not contribute to the system hazards when used in the intended environment. To address this claim, one needs to show that the PCA implementation software satisfies the GPCA safety requirements in the intended environment. This is the starting point for the pattern. The context for this claim is that GPCA safety requirements are defined to mitigate the GPCA hazards, which would be argued separately in another part of the safety case.

Figure 1.8 shows the GSN structure of the proposed *from_to* pattern. Here, {to} refers to the system implementation and {from} refers to a model of this system. The claim (G1) about the implementation correctness (i.e., satisfaction of some property [referenced in C1.3]) is justified

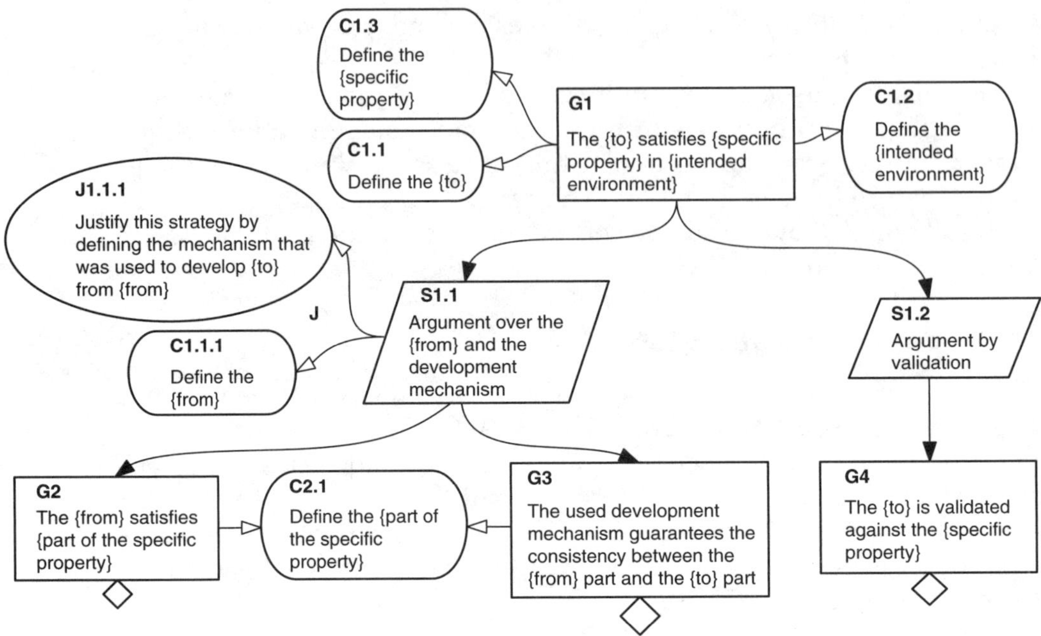

Figure 1.8: *The proposed from_to pattern*

A. Ayoub, B. Kim, I. Lee, O. Sokolsky. *Proceedings of NASA Formal Methods: 45th International Symposium*, pp. 141–146. With permission from Springer.

not only by validation (G4 through S1.2), but also by arguing over the model correctness (G2 through S1.1), and the consistency between the model and the implementation created based on it (G3 through S1.1). The model correctness (i.e., further development for G2) is guaranteed through the model verification (i.e., the second step of the model-based approach). The consistency between the model and the implementation (i.e., further development for G3) is supported by the code generation from the verified model (i.e., the third step of the model-based approach). Only part of the property of concern (referenced in C2.1) can be verified at the model level due to the differing abstraction levels between the model and the implementation. However, the validation argument (S1.2) covers the entire property of concern (referenced in C1.3). The additional justification (given in S1.1) increases the assurance in the top-level claim (G1).

Figure 1.9 shows an instantiation of this pattern that is part of the PCA safety case. Based on [Kim11], for this pattern instance, the {to} part is the PCA implementation software (referenced in C1.1), the {from} part is the GPCA timed automata model (referenced in C1.1.1), and the GPCA safety requirements (referenced in C1.3) represent the

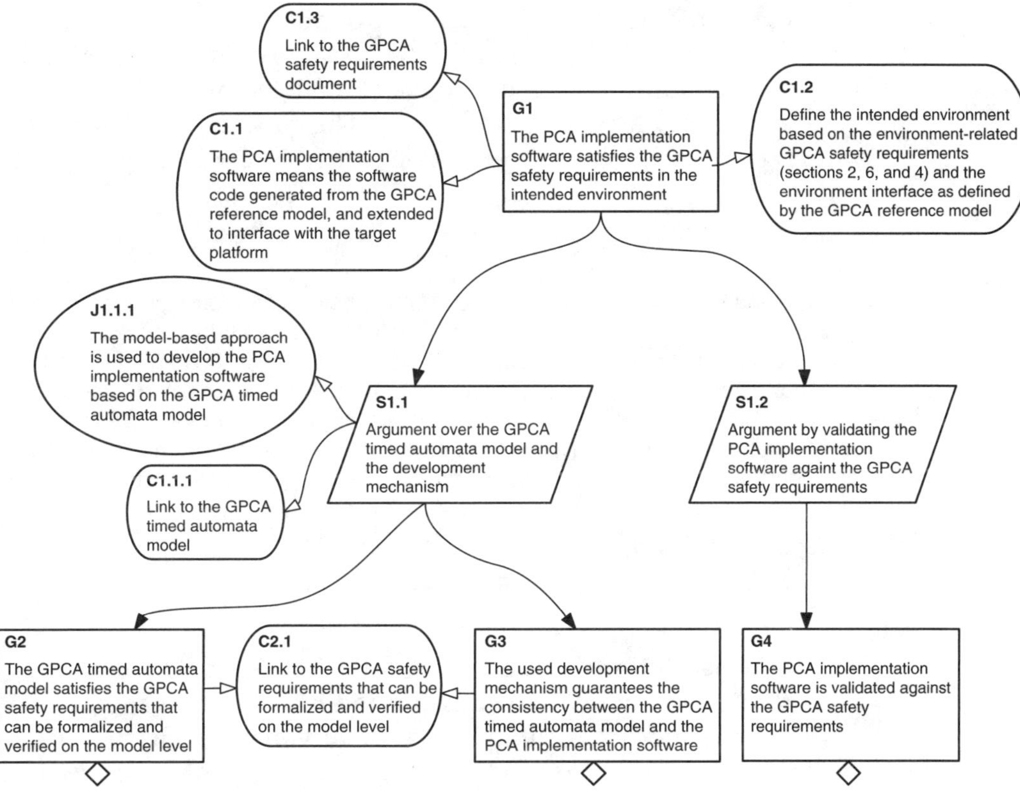

Figure 1.9: *An instance of the from_to pattern.*

A. Ayoub, B. Kim, I. Lee, O. Sokolsky. *Proceedings of NASA Formal Methods: 45th International Symposium*, pp. 141–146. With permission from Springer.

concerned property. In this case, correct PCA implementation means it satisfies the GPCA safety requirements that were defined to guarantee the PCA safety. The satisfaction of the GPCA safety requirements in the implementation level (G1) is decomposed by two strategies (S1.1 and S1.2). The argument in S1.1 is supported by the correctness of the GPCA timed automata model (G2) as well as by the consistency between the model and the implementation (G3). The correctness of the GPCA timed automata model (i.e., further development for G2) is proved by applying the UPPAAL model-checker against the GPCA safety requirements, which can be formalized (referenced in C2.1). The consistency between the model and the implementation (i.e., further development for G3) is supported by the code synthesis from the verified GPCA timed automata model.

Note that not all of the GPCA safety requirements (referenced in C1.3) can be verified against the GPCA timed automata model [Kim11].

Only the part referenced in C2.1 can be formalized and verified in the model level (e.g., "No bolus dose shall be possible during the power-on self-test"). Other requirements cannot be formalized or verified against the model given its level of detail (e.g., "The flow rate for the bolus dose shall be programmable" cannot be formalized meaningfully and then verified on the model level).

Note

Generally, using safety case patterns does not necessarily guarantee that the constructed safety case will be sufficiently compelling. Thus, when instantiating the *from_to* pattern, it is necessary to justify each instantiation decision to guarantee that the constructed safety case is sufficiently compelling. Assurance deficits should be identified throughout the construction of a safety argument. Where an assurance deficit is identified, it is necessary to demonstrate that the deficit is either acceptable or addressed such that it becomes acceptable. An explicit justification should be provided as to why the residual assurance deficit is considered acceptable. This can be done by adopting appropriate approaches such as ACARP (As Confident As Reasonably Practical) [Hawkins09a].

Assurance Deficit Example

As discussed in Section 1.3.3.3 and shown in Figure 1.3, the GPCA Simulink/Stateflow model was transformed into an equivalent GPCA timed automata model. Although it is relatively straightforward to translate the original GPCA model written in Simulink/Stateflow into a UPPAAL timed automata model, there is no explicit evidence to show the equivalence between the two models at the semantic level. A potential assurance deficit associated with the GPCA timed automata model (context C1.1.1 in Figure 1.9) can be stated as "There are semantic differences between the Simulink/Stateflow and the UPPAAL timed automata model." To mitigate this residual assurance deficit, exhaustive conformance testing between the GPCA Simulink/Stateflow model and the GPCA timed automata model may suffice.

1.4 Practitioners' Implications

One can distinguish the following groups of stakeholders in MCPS:

- MCPS developers, including manufacturers of medical devices and integrators of medical information technologies
- MCPS administrators—typically clinical engineers in hospitals, who are tasked with deploying and maintaining MCPS

- MCPS users—clinicians who perform treatment using MCPS
- MCPS subjects—patients
- MCPS regulators, who certify the safety of MCPS or approve their use for clinical purposes

In the United States, the FDA is the regulatory agency charged with assessing the safety and effectiveness of medical devices and approving them for specific uses.

All of the stakeholder groups have a vested interest in MCPS safety. However, each group has additional drivers that need to be taken into account when designing or deploying MCPS in a clinical setting. In this section we consider each group of stakeholders and identify specific concerns that apply to them as well as unique challenges that they pose.

1.4.1 MCPS Developer Perspective

Dependence of MCPS on software, as well as complexity of software used in medical devices, has been steadily increasing over the past three decades. In recent years, the medical device industry has been plagued with software-related recalls, with 19% of all recalls of medical devices in the United States being related to software problems [Simone13].

Many other safety-regulated industries, such as avionics and nuclear power, operate on relatively long design cycles. By contrast, medical device companies are under intense market pressure to quickly introduce additional features into their products. At the same time, medical devices are often developed by relatively small companies that lack the resources for extensive validation and verification of each new feature they introduce. Model-based development techniques, such as the ones described in Section 1.3.3, hold the promise of more efficient verification and validation, leading to shorter development cycles.

At the same time, many medical device companies complain about the heavy regulatory burden imposed by the FDA and similar regulatory agencies in other countries. Formal models and verification results, introduced by the model-based development approaches, provide evidence that MCPS is safe. Combined with the assurance cases that organize this evidence into a safety argument, these rigorous development methods may help reduce the regulatory burden for MCPS developers.

1.4.2 MCPS Administrator Perspective

Clinical engineers in hospitals are charged with maintaining the wide variety of medical devices that constitute the MCPS used in patient treatment. Most clinical scenarios today involve multiple medical devices. A clinical engineer needs to ensure that the devices used in treating a patient can all work together. If an incompatibility is discovered after treatment commences, the patient may be harmed. Interoperability techniques, described in Section 1.3.4, may help to ensure that more devices are compatible with one another, making the job of maintaining the inventory and the assembly of clinical scenarios easier. This, in turn, reduces treatment errors and improves patient outcomes and, at the same time, saves the hospital money.

1.4.3 MCPS User Perspective

Clinicians use MCPS as part of delivering patient treatments. A specific treatment can, in most cases, be performed with different MCPS implementations using similar devices from different vendors. A primary concern, then, is ensuring that clinicians are equally familiar with the different implementations. The concepts of clinical scenarios and virtual medical devices, introduced in Section 1.3.4, can help establish a common user interface for the MCPS, regardless of the specific devices used to implement it. Such an interface would help to reduce clinical errors when using these devices. Furthermore, the user interface can be verified as part of the analysis of the MCPS model, as suggested by [Masci13].

MCPS development must take existing standards of care into consideration. Clinical personnel need to be involved in the analysis of the scenario models to ensure that they are consistent with extant clinical guidelines for the respective treatment and are intuitive for caregivers to use.

A particular challenge in modern health care is the high workload faced by caregivers. Each healthcare provider is likely to be caring for multiple patients and must keep track of multiple sources of information about each patient. On-demand MCPS have the potential to control cognitive overload in caregivers by offering virtual devices that deliver intelligent presentation of clinical information or smart alarm functionality. Smart alarms, which can correlate or prioritize alarms from individual devices, can be of great help to caregivers, by giving a more accurate picture of the patient's condition and reducing the rate of false alarms [Imhoff09].

1.4.4 Patient Perspective

Of all the various stakeholder groups, patients stand to gain the most from the introduction of MCPS. In addition to the expected improvements in the safety of treatments achieved through higher reliability of individual devices and their bedside assemblies, patients would get the benefit of improvements in treatments themselves. These improvements may come from several sources.

On the one hand, MCPS can offer continuous monitoring that caregivers, who normally must attend to multiple patients as part of their workload, cannot provide by themselves. Clinical guidelines often require caregivers to obtain patient data at fixed intervals—for example, every 15 minutes. An MCPS may collect patient data as frequently as allowed by each sensor and alert caregivers to changes in the patient's condition earlier, thereby enabling them to intervene before the change leads to a serious problem. Furthermore, continuous monitoring, combined with support for predictive decision making, similar to the system discussed in Section 1.3.5, will allow treatment to be proactive rather than reactive.

Probably the biggest improvement in the quality of care for patients will come with the transition from general guidelines meant to apply to all patients within a certain population to personalized approaches, in which treatment is customized to the individual needs of the patient and takes into account his or her specific characteristics. Personalized treatments, however, cannot be effected without detailed patient models. Such models can be stored in patient records and interpreted by the MCPS during treatment.

1.4.5 MCPS Regulatory Perspective

Regulators of the medical devices industry are tasked with assessing the safety and effectiveness of MCPS. The two main concerns that these regulators face are improving the quality of the assessment and making the best use of the limited resources that agencies have available for performing the assessment. These two concerns are not independent, because more efficient ways of performing assessments would allow regulators more time to conduct deeper evaluations. The safety case technologies discussed in Section 1.3.7 may help address both. The move toward evidence-based assessment may allow regulators to perform more accurate and reliable assessments. At the same time, organizing evidence into a coherent argument will help them perform these assessments more efficiently.

1.5 Summary and Open Challenges

This chapter presented a broad overview of trends in MCPS and the design challenges that these trends present. It also discussed possible approaches to address these challenges, based on recent results in MCPS research.

The first challenge is related to the prevalence of software-enabled functionality in modern MCPS, which makes assurance of patient safety a much harder task. Model-based development techniques provide one way to ensure the safety of a system. Increasingly, model-based development is embraced by the medical devices industry. Even so, the numerous recalls of medical devices that have occurred in recent years demonstrate that the problem of device safety is far from being solved.

The next-level challenge arises from the need to organize individual devices into a system of interconnected devices that collectively treat the patient in a complex clinical scenario. Such multi-device MCPS can provide new modes of treatment, give enhanced feedback to the clinician, and improve patient safety. At the same time, additional hazards can arise from communication failures and lack of interoperability between devices. Reasoning about safety of such on-demand MCPS, which are assembled at the bedside from available devices, creates new regulatory challenges and requires medical application platforms—that is, trusted middleware that can ensure correct interactions between the devices. Research prototypes of such middleware are currently being developed, but their effectiveness needs to be further evaluated. Furthermore, interoperability standards for on-demand MCPS need to be further improved and gain wider acceptance.

To fully utilize the promise inherent in multi-device MCPS, new algorithms need to be developed to process and integrate patient data from multiple sensors, provide better decision support for clinicians, produce more accurate and informative alarms, and so on. This need gives rise to two kinds of open challenges. On the one hand, additional clinical research as well as data analysis needs to be performed to determine the best ways of using the new information made available through combining multiple rich data sources. On the other hand, new software tools are needed to facilitate fast prototyping and deployment of new decision support and visualization algorithms.

MCPS promises to enable a wide array of physiological closed-loop systems, in which information about the patient's condition, collected from multiple sensors, can be used to adjust the treatment process or its parameters. Research on such closed-loop control algorithms is gaining prominence, especially as means to improve glycemic control in patients with diabetes. However, much research needs to be performed to better understand patient physiology and develop adaptive control algorithms that can deliver personalized treatment to each patient.

In all of these applications, patient safety and effectiveness of treatment are the two paramount concerns. MCPS manufacturers need to convince regulators that systems they build are safe and effective. The growing complexity of MCPS, the high connectivity, and the prevalence of software-enabled functionality make evaluation of such systems' safety quite difficult. Construction of effective assurance cases for MCPS, as well as for CPS in general, remains a challenge in need of further research.

References

[Adelard13]. Adelard. "Claims, Arguments and Evidence (CAE)." http://www.adelard.com/asce/choosing-asce/cae.html, 2013.

[Alexander07]. R. Alexander, T. Kelly, Z. Kurd, and J. Mcdermid. "Safety Cases for Advanced Control Software: Safety Case Patterns." Technical Report, University of York, 2007.

[Amnell03]. T. Amnell, E. Fersman, L. Mokrushin, P. Pettersson, and W. Yi. "TIMES: A Tool for Schedulability Analysis and Code Generation of Real-Time Systems." In *Formal Modeling and Analysis of Timed Systems*. Springer, 2003.

[Arney09]. D. Arney, J. M. Goldman, S. F. Whitehead, and I. Lee. "Synchronizing an X-Ray and Anesthesia Machine Ventilator: A Medical Device Interoperability Case Study." *Biodevices*, pages 52–60, January 2009.

[ASTM09]. ASTM F2761-2009. "Medical Devices and Medical Systems—Essential Safety Requirements for Equipment Comprising the Patient-Centric Integrated Clinical Environment (ICE), Part 1: General Requirements and Conceptual Model." ASTM International, 2009.

[Ayoub13]. A. Ayoub, J. Chang, O. Sokolsky, and I. Lee. "Assessing the Overall Sufficiency of Safety Arguments." Safety Critical System Symposium (SSS), 2013.

[Ayoub12]. A. Ayoub, B. Kim, I. Lee, and O. Sokolsky. "A Safety Case Pattern for Model-Based Development Approach." In *NASA Formal Methods*, pages 223–243. Springer, 2012.

[Ayoub12a]. A. Ayoub, B. Kim, I. Lee, and O. Sokolsky. "A Systematic Approach to Justifying Sufficient Confidence in Software Safety Arguments." International Conference on Computer Safety, Reliability and Security (SAFECOMP), Magdeburg, Germany, 2012.

[Becker09]. U. Becker. "Model-Based Development of Medical Devices." *Proceedings of the Workshop on Computer Safety, Reliability, and Security (SAFECERT)*, Lecture Notes in Computer Science, vol. 5775, pages 4–17, 2009.

[Behrmann04]. G. Behrmann, A. David, and K. Larsen. "A Tutorial on UPPAAL." In *Formal Methods for the Design of Real-Time Systems*, Lecture Notes in Computer Science, pages 200–237. Springer, 2004.

[Bequette03]. B. Bequette. *Process Control: Modeling, Design, and Simulation*. Prentice Hall, 2003.

[Bloomfield07]. R. Bloomfield, B. Littlewood, and D. Wright. "Confidence: Its Role in Dependability Cases for Risk Assessment." 37th Annual IEEE/IFIP International Conference on Dependable Systems and Networks, pages 338–346, 2007.

[Carr03]. C. D. Carr and S. M. Moore. "IHE: A Model for Driving Adoption of Standards." *Computerized Medical Imaging and Graphics*, vol. 27, no. 2–3, pages 137–146, 2003.

[Clarke07]. M. Clarke, D. Bogia, K. Hassing, L. Steubesand, T. Chan, and D. Ayyagari. "Developing a Standard for Personal Health Devices Based on 11073." 29th Annual International Conference of the IEEE Engineering in Medicine and Biology Society, pages 6174–6176, 2007.

[Clifford09]. G. Clifford, W. Long, G. Moody, and P. Szolovits. "Robust Parameter Extraction for Decision Support Using Multimodal Intensive Care Data." *Philosophical Transactions of the Royal Society A: Mathematical, Physical and Engineering Sciences*, vol. 367, pages 411–429, 2009.

[Clinical07]. Clinical Alarms Task Force. "Impact of Clinical Alarms on Patient Safety." *Journal of Clinical Engineering*, vol. 32, no. 1, pages 22–33, 2007.

[Cobelli09]. C. Cobelli, C. D. Man, G. Sparacino, L. Magni, G. D. Nicolao, and B. P. Kovatchev. "Diabetes: Models, Signals, and Control." *IEEE Reviews in Biomedical Engineering*, vol. 2, 2009.

[Commission13]. The Joint Commission. "Medical Device Alarm Safety in Hospitals." *Sentinel Event Alert*, no. 50, April 2013.

[Cyra08]. L. Cyra and J. Górski. "Expert Assessment of Arguments: A Method and Its Experimental Evaluation." International Conference on Computer Safety, Reliability and Security (SAFECOMP), 2008.

[Denney11]. E. Denney, G. Pai, and I. Habli. "Towards Measurement of Confidence in Safety Cases." International Symposium on Empirical Software Engineering and Measurement (ESEM), Washington, DC, 2011.

[Dias07]. A. C. Dias Neto, R. Subramanyan, M. Vieira, and G. H. Travassos. "A Survey on Model-Based Testing Approaches: A Systematic Review." *Proceedings of the ACM International Workshop on Empirical Assessment of Software Engineering Languages and Technologies*, pages 31–36, 2007.

[Dolin06]. R. H. Dolin, L. Alschuler, S. Boyer, C. Beebe, F. M. Behlen, P. V. Biron, and A. Shvo. "HL7 Clinical Document Architecture, Release 2." *Journal of the American Medical Informatics Association*, vol. 13, no. 1, pages 30–39, 2006.

[Donchin02]. Y. Donchin and F. J. Seagull. "The Hostile Environment of the Intensive Care Unit." *Current Opinion in Critical Care*, vol. 8, pages 316–320, 2002.

[Edworthy06]. J. Edworthy and E. Hellier. "Alarms and Human Behaviour: Implications for Medical Alarms." *British Journal of Anaesthesia*, vol. 97, pages 12–17, 2006.

[EBMWG92]. Evidence-Based Medicine Working Group. "Evidence-Based Medicine: A New Approach to Teaching the Practice of Medicine." *Journal of the American Medical Association*, vol. 268, pages 2420–2425, 1992.

[FDA10]. U.S. Food and Drug Administration, Center for Devices and Radiological Health. "Infusion Pumps Total Product Life Cycle: Guidance for Industry and FDA Staff." Premarket Notification [510(k)] Submissions, April 2010.

[FDA10a]. U.S. Food and Drug Administration, Center for Devices and Radiological Health. "Infusion Pump Improvement Initiative." White Paper, April 2010.

[Garg05]. A. X. Garg, N. K. J. Adhikari, H. McDonald, M. P. Rosas-Arellano, P. J. Devereaux, J. Beyene, J. Sam, and R. B. Haynes. "Effects of Computerized Clinical Decision Support Systems on Practitioner Performance and Patient Outcomes: A Systematic

Review." *Journal of the American Medical Association*, vol. 293, pages 1223–1238, 2005.

[Ginsberg09]. B. H. Ginsberg. "Factors Affecting Blood Glucose Monitoring: Sources of Errors in Measurement." *Journal of Diabetes Science and Technology*, vol. 3, no. 4, pages 903–913, 2009.

[Goldman05]. J. Goldman, R. Schrenker, J. Jackson, and S. Whitehead. "Plug-and-Play in the Operating Room of the Future." *Biomedical Instrumentation and Technology*, vol. 39, no. 3, pages 194–199, 2005.

[Goodenough12]. J. Goodenough, C. Weinstock, and A. Klein. "Toward a Theory of Assurance Case Confidence." Technical Report CMU/SEI-2012-TR-002, Software Engineering Institute, Carnegie Mellon University, Pittsburgh, PA, 2012.

[Harris13]. Harris Healthcare (formerly careFX). www.harris.com.

[Hatcliff12]. J. Hatcliff, A. King, I. Lee, A. Macdonald, A. Fernando, M. Robkin, E. Vasserman, S. Weininger, and J. M. Goldman. "Rationale and Architecture Principles for Medical Application Platforms." *Proceedings of the IEEE/ACM 3rd International Conference on Cyber-Physical Systems (ICCPS)*, pages 3–12, Washington, DC, 2012.

[Hawkins09]. R. Hawkins and T. Kelly. "A Systematic Approach for Developing Software Safety Arguments." *Journal of System Safety*, vol. 46, pages 25–33, 2009.

[Hawkins09a]. R. Hawkins and T. Kelly. "Software Safety Assurance: What Is Sufficient?" 4th IET International Conference of System Safety, 2009.

[Hawkins11]. R. Hawkins, T. Kelly, J. Knight, and P. Graydon. "A New Approach to Creating Clear Safety Arguments." In *Advances in Systems Safety*, pages 3–23. Springer, 2011.

[Henzinger07]. T. A. Henzinger and C. M. Kirsch. "The Embedded Machine: Predictable, Portable Real-Time Code." *ACM Transactions on Programming Languages and Systems (TOPLAS)*, vol. 29, no. 6, page 33, 2007.

[Hovorka04]. R. Hovorka, V. Canonico, L. J. Chassin, U. Haueter, M. Massi-Benedetti, M. O. Federici, T. R. Pieber, H. C. Schaller, L. Schaupp, T. Vering, and M. E. Wilinska. "Nonlinear Model Predictive Control of Glucose Concentration in Subjects with Type 1 Diabetes." *Physiological Measurement*, vol. 25, no. 4, page 905, 2004.

[Imhoff06]. M. Imhoff and S. Kuhls. "Alarm Algorithms in Critical Care Monitoring." *Anesthesia and Analgesia*, vol. 102, no. 5, pages 1525–1536, 2006.

[Imhoff09]. M. Imhoff, S. Kuhls, U. Gather, and R. Fried. "Smart Alarms from Medical Devices in the OR and ICU." *Best Practice and Research in Clinical Anaesthesiology*, vol. 23, no. 1, pages 39–50, 2009.

[Isaksen97]. U. Isaksen, J. P. Bowen, and N. Nissanke. "System and Software Safety in Critical Systems." Technical Report RUCS/97/TR/062/A, University of Reading, UK, 1997.

[ISO/IEEE11073]. ISO/IEEE 11073 Committee. "Health Informatics—Point-of-Care Medical Device Communication Part 10103: Nomenclature—Implantable Device, Cardiac." http://standards.ieee.org/findstds/standard/11073-10103-2012.html.

[Jackson07]. D. Jackson, M. Thomas, and L. I. Millett, editors. *Software for Dependable Systems: Sufficient Evidence?* Committee on Certifiably Dependable Software Systems, National Research Council. National Academies Press, May 2007.

[Jee10]. E. Jee, I. Lee, and O. Sokolsky. "Assurance Cases in Model-Driven Development of the Pacemaker Software." 4th International Conference on Leveraging Applications of Formal Methods, Verification, and Validation, Volume 6416, Part II, ISoLA'10, pages 343–356. Springer-Verlag, 2010.

[Jeroeno4]. J. Levert and J. C. H. Hoorntje. "Runaway Pacemaker Due to Software-Based Programming Error." *Pacing and Clinical Electrophysiology*, vol. 27, no. 12, pages 1689–1690, December 2004.

[Kelly98]. T. Kelly. "Arguing Safety: A Systematic Approach to Managing Safety Cases." PhD thesis, Department of Computer Science, University of York, 1998.

[Kelly98a]. T. Kelly. "A Six-Step Method for Developing Arguments in the Goal Structuring Notation (GSN)." Technical Report, York Software Engineering, UK, 1998.

[Kelly07]. T. Kelly. "Reviewing Assurance Arguments: A Step-by-Step Approach." Workshop on Assurance Cases for Security: The Metrics Challenge, Dependable Systems and Networks (DSN), 2007.

[Kelly97]. T. Kelly and J. McDermid. "Safety Case Construction and Reuse Using Patterns." International Conference on Computer Safety, Reliability and Security (SAFECOMP), pages 55–96. Springer-Verlag, 1997.

[Kelly04]. T. Kelly and R. Weaver. "The Goal Structuring Notation: A Safety Argument Notation." DSN 2004 Workshop on Assurance Cases, 2004.

[Kim11]. B. Kim, A. Ayoub, O. Sokolsky, P. Jones, Y. Zhang, R. Jetley, and I. Lee. "Safety-Assured Development of the GPCA Infusion

Pump Software." *Embedded Software (EMSOFT)*, pages 155–164, Taipei, Taiwan, 2011.

[Kim12]. B. G. Kim, L. T. Phan, I. Lee, and O. Sokolsky. "A Model-Based I/O Interface Synthesis Framework for the Cross-Platform Software Modeling." 23rd IEEE International Symposium on Rapid System Prototyping (RSP), pages 16–22, 2012.

[King09]. A. King, S. Procter, D. Andresen, J. Hatcliff, S. Warren, W. Spees, R. Jetley, P. Jones, and S. Weininger. "An Open Test Bed for Medical Device Integration and Coordination." *Proceedings of the 31st International Conference on Software Engineering*, 2009.

[Kovatchev09]. B. P. Kovatchev, M. Breton, C. D. Man, and C. Cobelli. "In Silico Preclinical Trials: A Proof of Concept in Closed-Loop Control of Type 1 Diabetes." *Diabetes Technology Society*, vol. 3, no. 1, pages 44–55, 2009.

[Lee06]. I. Lee, G. J. Pappas, R. Cleaveland, J. Hatcliff, B. H. Krogh, P. Lee, H. Rubin, and L. Sha. "High-Confidence Medical Device Software and Systems." *Computer*, vol. 39, no. 4, pages 33–38, April 2006.

[Lee12]. I. Lee, O. Sokolsky, S. Chen, J. Hatcliff, E. Jee, B. Kim, A. King, M. Mullen-Fortino, S. Park, A. Roederer, and K. Venkatasubramanian. "Challenges and Research Directions in Medical Cyber-Physical Systems." *Proceedings of the IEEE*, vol. 100, no. 1, pages 75–90, January 2012.

[Lofsky04]. A. S. Lofsky. "Turn Your Alarms On." *APSF Newsletter*, vol. 19, no. 4, page 43, 2004.

[Lublinerman09]. R. Lublinerman, C. Szegedy, and S. Tripakis. "Modular Code Generation from Synchronous Block Diagrams: Modularity vs. Code Size." *Proceedings of the 36th Annual ACM SIGPLAN-SIGACT Symposium on Principles of Programming Languages (POPL 2009)*, pages 78–89, New York, NY, 2009.

[Lynn11]. L. A. Lynn and J. P. Curry. "Patterns of Unexpected In-Hospital Deaths: A Root Cause Analysis." *Patient Safety in Surgery*, vol. 5, 2011.

[Maddox08]. R. Maddox, H. Oglesby, C. Williams, M. Fields, and S. Danello. "Continuous Respiratory Monitoring and a 'Smart' Infusion System Improve Safety of Patient-Controlled Analgesia in the Postoperative Period." In K. Henriksen, J. Battles, M. Keyes, and M. Grady, editors, *Advances in Patient Safety: New Directions and Alternative Approaches*, Volume 4 of *Advances in Patient Safety*, Agency for Healthcare Research and Quality, August 2008.

[Masci13]. P. Masci, A. Ayoub, P. Curzon, I. Lee, O. Sokolsky, and H. Thimbleby. "Model-Based Development of the Generic PCA Infusion Pump User Interface Prototype in PVS." *Proceedings of the 32nd International Conference on Computer Safety, Reliability and Security (SAFECOMP)*, 2013.

[Mazoit07]. J. X. Mazoit, K. Butscher, and K. Samii. "Morphine in Postoperative Patients: Pharmacokinetics and Pharmacodynamics of Metabolites." *Anesthesia and Analgesia*, vol. 105, no. 1, pages 70–78, 2007.

[McMaster13]. Software Quality Research Laboratory, McMaster University. Pacemaker Formal Methods Challenge. http://sqrl.mcmaster.ca/pacemaker.htm.

[MDCF]. Medical Device Coordination Framework (MDCF). http://mdcf.santos.cis.ksu.edu.

[MDPNP]. MD PnP: Medical Device "Plug-and-Play" Interoperability Program. http://www.mdpnp.org.

[Menon09]. C. Menon, R. Hawkins, and J. McDermid. Defence "Standard 00-56 Issue 4: Towards Evidence-Based Safety Standards." In *Safety-Critical Systems: Problems, Process and Practice*, pages 223–243. Springer, 2009.

[Nuckols08]. T. K. Nuckols, A. G. Bower, S. M. Paddock, L. H. Hilborne, P. Wallace, J. M. Rothschild, A. Griffin, R. J. Fairbanks, B. Carlson, R. J. Panzer, and R. H. Brook. "Programmable Infusion Pumps in ICUs: An Analysis of Corresponding Adverse Drug Events." *Journal of General Internal Medicine*, vol. 23 (Supplement 1), pages 41–45, January 2008.

[Oberli99]. C. Oberli, C. Saez, A. Cipriano, G. Lema, and C. Sacco. "An Expert System for Monitor Alarm Integration." *Journal of Clinical Monitoring and Computing*, vol. 15, pages 29–35, 1999.

[Pajic12]. M. Pajic, R. Mangharam, O. Sokolsky, D. Arney, J. Goldman, and I. Lee. "Model-Driven Safety Analysis of Closed-Loop Medical Systems." *IEEE Transactions on Industrial Informatics*, PP(99):1–1, 2012.

[Phillips10]. Phillips eICU Program. http://www.usa.philips.com/healthcare/solutions/patient-monitoring.

[Rae03]. A. Rae, P. Ramanan, D. Jackson, J. Flanz, and D. Leyman. "Critical Feature Analysis of a Radiotherapy Machine." International Conference of Computer Safety, Reliability and Security (SAFECOMP), September 2003.

[Sapirstein09]. A. Sapirstein, N. Lone, A. Latif, J. Fackler, and P. J. Pronovost. "Tele ICU: Paradox or Panacea?" *Best Practice and*

Research Clinical Anaesthesiology, vol. 23, no. 1, pages 115–126, March 2009.

[Sentz02]. K. Sentz and S. Ferson. "Combination of Evidence in Dempster-Shafer Theory." Technical report, Sandia National Laboratories, SAND 2002-0835, 2002.

[Shortliffe79]. E. H. Shortliffe, B. G. Buchanan, and E. A. Feigenbaum. "Knowledge Engineering for Medical Decision Making: A Review of Computer-Based Clinical Decision Aids." *Proceedings of the IEEE*, vol. 67, pages 1207–1224, 1979.

[Simone13]. L. K. Simone. "Software Related Recalls: An Analysis of Records." *Biomedical Instrumentation and Technology*, 2013.

[UPenn]. The Generic Patient Controlled Analgesia Pump Model. http://rtg.cis.upenn.edu/gip.php3.

[UPenn-a]. Safety Requirements for the Generic Patient Controlled Analgesia Pump. http://rtg.cis.upenn.edu/gip.php3.

[UPenn-b]. The Generic Patient Controlled Analgesia Pump Hazard Analysis. http://rtg.cis.upenn.edu/gip.php3.

[Wagner10]. S. Wagner, B. Schatz, S. Puchner, and P. Kock. "A Case Study on Safety Cases in the Automotive Domain: Modules, Patterns, and Models." *International Symposium on Software Reliability Engineering*, pages 269–278, 2010.

[Weaver03]. R. Weaver. "The Safety of Software: Constructing and Assuring Arguments." PhD thesis, Department of Computer Science, University of York, 2003.

[Weinstock09]. C. Weinstock and J. Goodenough. "Towards an Assurance Case Practice for Medical Devices." Technical Report, CMU/SEI-2009-TN-018, 2009.

[Ye05]. F. Ye and T. Kelly. "Contract-Based Justification for COTS Component within Safety-Critical Applications." PhD thesis, Department of Computer Science, University of York, 2005.

Chapter 2

Energy Cyber-Physical Systems

Marija Ilic[1]

This chapter concerns new modeling, analysis, and design challenges and opportunities in energy cyber-physical systems (CPS), with the main emphasis being on the role of smart electric grids in enabling performance of broader energy systems. The chapter briefly summarizes today's operations and contrasts them with what might become possible given rapidly changing technology and needs for energy. It then describes one possible new paradigm, referred to as Dynamic Monitoring and Decision System (DyMonDS), as the basis for sustainable end-to-end CPS for electric energy systems. These systems are fundamentally different because operators and planners make decisions based on proactive and binding information exchange with system users; this approach makes it possible

1. This chapter draws on the work of many researchers in the Electric Energy Systems Group (EESG) at Carnegie Mellon University. Due to space limitations, only the major ideas underlying the work by Kevin Bachovchin, Stefanos Baros, Milos Cvetkovic, Sanja Cvijic, Andrew Hsu, Jhi-Young Joo, Soummya Kar, Siripha Kulkarni, Qixing Liu, Rohit Negi, Sergio Pequito, and Yang Wang are summarized.

to account for choice of value from the system, while at the same time meeting difficult societal goals. The information exchange is repeated periodically, as decisions are being made either for the distant future or for closer to real time.

With DyMonDS, the process becomes a win–win situation for all, and future uncertainties are distributed over many stakeholders according to the risks they are willing to take and according to the expected value from the selected technologies. They ultimately lead to qualitatively new reliable, efficient, and clean services, and technologies which bring value that survives beyond subsidy stages. Implementing the proposed CPS requires qualitatively different modeling, estimation, and decision-making methods, which lend themselves to multilayered interactive information processing in support of sustainable energy processing. Basic principles of such a new approach are discussed.

2.1 Introduction and Motivation

The electric power industry is a major critical national infrastructure. It represents a huge portion of the U.S. economy (a more than $200 billion business) and is, at the same time, currently a major source of the country's carbon footprint. This industry has the potential to become one of the largest users of cyber-technologies; more generally, the industrialized economy depends on low-cost electricity service. While the common thinking is that the power grids work well, and that not much fundamental innovation is needed, several key hidden problems currently need fixing:

- The increased frequency and duration of service interruption (loss measured in billions)
- Major hidden inefficiencies in today's system (an estimated 25% economic inefficiency according to the Federal Energy Regulatory Commission [FERC] and supported by case studies)
- Unsustainable operation and planning of high-penetration renewable resources if the system is operated and planned as in the past
- Lack of seamless electrification of the transportation systems

More generally, the long-term resource mix must serve the country's long-term demand needs well, and it is not clear at this point that this feat will be achieved. Notably, all of the previously mentioned problems are systems-type problems, which are difficult to fix by simply adding more generation and/or transmission and distribution (T&D) capacity; instead, their solutions require fundamental rethinking of the power industry's approaches. Innovation needed is primarily in cyber-physical systems, so as to better utilize the existing hardware and to support effective end-to-end integration of new resources into the legacy electric power grids. The boundaries between hardware and software improvements have become very blurry in this industry. This, then, raises questions about the major pitfalls in today's electric power industry in which the use of cyber-technologies is implemented in a piecemeal fashion and never designed with system-level objectives.

2.2 System Description and Operational Scenarios

At present, there are no agreed-on performance metrics necessary to integrate and operate various industry stakeholders at their maximum value. Defining the rights, rules, and responsibilities between traditional utilities, on the one hand, and the electricity users, generation providers, and delivery entities, on the other hand, is a work in progress, at best. This situation creates major impediments to deploying new technologies so as to realize their full value.

Another way of thinking about this problem is to note that there are no seamless information technology (IT)–enabled protocols for interfacing the objectives of generation providers, users, and T&D operators within a utility (the company that sells electricity to the final users)—nor, for that matter, between electrically interconnected utilities across larger geographical regions. This lack of coordination has major implications for how large-scale wind- and solar-power systems are built and utilized, as well as for the infrastructure necessary to deploy and utilize these new resources. Moreover, the lack of integrated protocols between different industry stakeholders and layers makes it difficult to begin to give incentives to customers to respond to the system needs based on value. In particular, inclusion of electric vehicles, smart buildings, and groups of residential customers willing to participate in the broader system-level supply–demand balancing is currently on hold, by and large.

While government has recently invested in pilot experiments to demonstrate proof-of-concept small-scale readiness of these new untapped resources, there are no major government-supported research and development projects focusing on the design of the missing protocols for large-scale integration of new technologies according to well-defined and quantifiable objectives. Some efforts are under way related to extra-high-voltage (EHV) transmission planning for supporting long-term, large-scale integration of these technologies. Unfortunately, the recent planning studies are primarily scenario analysis-type studies that do not deal with explicit IT protocols for incentivizing long-term sustainable solutions; likewise, they do not explicitly consider the role and value of cyber-technologies in enabling more efficient, more reliable, and cleaner electric power grid operations. In particular, they do not take advantage of major opportunities to sense and process information in support of dynamic decision making given the huge uncertainties within the system.

For the purposes of this chapter, it is important to recognize that a systematic use of cyber-technologies is needed to enhance the performance of the existing system as well as to integrate new resources; the industry objectives in the old physical systems and the changing physical systems may be different. Notably, the new system-level objectives could be shaped significantly by the sub-objectives of system users.

The operations of the electric power grids that rely on IT have evolved over time. They are a result of a careful mix of engineering insights about the power grids and specific-purpose computer algorithms used in an advisory manner by the human operators. At present, most modern utility control centers routinely use computer applications, developed using model-based feed-forward techniques, to evaluate and schedule the cheapest predictable power generation for a system where both the demand and the supply vary over time. The only utility-level closed-loop feedback coordinated scheme is the automatic generation control (AGC), which is dedicated to balancing quasi-static, hard-to-predict deviations from real power forecasts by adjusting the setpoints of governors on fast-responding power plants. The control logic of these governors and other primary controllers (e.g., automatic voltage regulators [AVRs]) and some T&D-controllable equipment (e.g., transformers and capacitor banks) is used to ensure local stabilization of frequency and voltage to their setpoints. This combination of utility-level power scheduling for predictable power imbalance, quasi-static AGC for frequency regulation, and local primary control of generators and some T&D equipment forms the basis of today's hierarchical control of power grids [Illic00].

More saliently, power grids are complex, large-scale, dynamic networks. Their effective operations and planning require one to view the power grid as a complex system in which energy and information processing are intricately intertwined. No ready-to-use tools are available for general CPS that can ensure their large-scale dynamic optimization under uncertainties, nor are there means of ensuring that the dynamics will meet, in a guaranteed way, the performance specifications required for reliability. Many general CPS concepts are applicable, but creating formal definitions of utility problems for purposes of applying more general CPS concepts has proved difficult. This chapter attempts to provide some examples of such formulations.

Perhaps the biggest challenge is ensuring provable performance in systems of such high complexity as are seen in electric power grids; this poses major challenges to the overall state of the art in systems engineering. Notably, the system dynamics in the electric power industry are highly nonlinear. The time–space scales over which acceptable performance is required are vast and range from milliseconds to decades, and from an appliance or building to the eastern or western U.S. power grid interconnections, respectively. Fast storage remains scarce and expensive, and it is still not readily available—a factor that limits the controllability of the grid. In addition, the current systems are not dynamically observable. Large-scale optimization under the uncertainties inherent in these grids remains a major challenge, though there are targeted efforts under way to introduce more computationally effective algorithms using powerful computers.

2.3 Key Design Drivers and Quality Attributes

We are currently witnessing the emergence of both massive opportunities and enormous challenges concerning the national-scale integration of coordinated complex systems. In the electric power industry, these trends involve a mix of innovations in physical energy systems (power grid, power electronics, energy resources), communication systems (hardware and protocols), control systems (algorithms and massive computation), and economic and policy systems (regulations and stimuli). It has been estimated that just the online IT implementations of these innovations would enable the following benefits:

- Approximately 20% increase in economic efficiency (FERC estimate)
- Cost-effective integration of renewable energy sources and reduction of emissions
- Differentiated quality of service (QoS) without sacrificing of basic reliability of service
- Seamless prevention of blackouts
- Infrastructure expansion (e.g., generation, T&D, demand side) for maximum benefit and minimum intrusion

To date, there has not been widespread recognition of this major potential for enhancing the performance of the power grid by means of online IT. This chapter attempts to partially fill this void.

Achieving the system-level objectives by implementing online IT in complex power grids represents a huge intellectual challenge. Thus it must be a university-led transformation, providing a once-in-a-lifetime opportunity for academics. Efforts to define and model the problem of future electric energy systems as a heterogeneous technical, social, and institutional complex system are long overdue. To fill this gap, it is necessary to represent in sufficient detail the multilayered and multidirectional dynamic interactions involving these systems, which are driven by often-unaligned societal needs and the needs of various stakeholder groups, and which are constrained by the ability of the energy systems to integrate them according to choice and at value. The existing physical energy system, including its communications, control, and computer methods, does not readily lend itself to enabling choice and multidirectional interactions for aligning often conflicting goals. Instead, it is essential to design intelligence for T&D operations that would be capable of aligning these goals and, consequently, getting the most out of the available resources, while simultaneously providing acceptable QoS, including resilient response to major uncertainties.

The CPS technologies will play a key role in the transitioning of the existing system to the new system as new sensors and actuators are deployed into legacy power grids. In this chapter, we consider how the recent trend of deploying smart sensors and actuators into the legacy electric power grids creates both challenges and opportunities. These myriad challenges will require systems thinking if cyber-technologies are to be ultimately designed, implemented, and proven useful in major ways.

2.3.1 Key Systems Principles

One way of defining the underlying principles for supporting the design of the physical electric power grid and its cyber-physical infrastructure is to view this system in terms of its role as the enabler of sustainable energy services [Ilic11, Ilic11a]. Today's primarily grid-centric design is rapidly becoming outdated and inadequate. It is no longer possible to assume that utilities can forecast and manage demand to the level of coarseness necessary to respond effectively to ever-changing conditions and customers' preferences, and to account for the effects of intermittent resources as they shape utility-level supply–demand imbalances. Similarly, it is no longer possible to assume full knowledge and control of a rapidly growing number of small, diverse distributed energy resources (DERs) connected to the lower distribution grid voltage levels.

Instead, given different technology developers and owners, it will be necessary to introduce both planning and operating protocols for exchanging information about the functionalities of the DERs equipped with cyber "smarts." In this environment, participants in the electric power industry must be able to differentiate the effects of ordinary buildings and smart buildings, fast-charging and smart-charging electric vehicles (EVs), smart and passive wires, and much more. Viewed from the utility level, it will no longer be possible to know all the details internal to the generation, demand, or delivery components of electric power. Based on these observations, a possible unifying principle of designing future smart grids (both their physical and cyber infrastructure) is described next.

2.3.1.1 Sustainable Socio-ecological Energy Systems

We can consider socio-ecological energy systems (SEES) to be similar to any other socio-ecological systems, comprising resources, users, and governance [Ostrom09]. A smart grid has the key overall function of matching the characteristics of the SEES resources, users, and governance system, which are defined in terms of their temporal, spatial, and contextual properties. The more closely these characteristics are aligned,

Table 2.1: *Architectures for Socio-ecological Energy Systems*

Architecture	Description	Operating Environment
1	Bulk SEES	Regulated
2	Bulk SEES	Restructured
3	Hybrid SEES	Restructured
4	Fully distributed SEES	Developed countries
5	Fully distributed SEES	Developing countries

the more sustainable the system will be [Ostrom09, Ilic00].[2] Notably, the resources, users, and governance system can align these properties either internally in a distributed way or by managing interactions between them via a human-made physical power grid and its cyber-smarts.

Posing the problem of CPS design for a sustainable SEES this way sets the stage for introducing meaningful performance specifications for its design. The performance objectives of different SEES architectures are fundamentally different; therefore, they require qualitatively different cyber-designs. Table 2.1 shows several representative SEES architectures.

In Table 2.1, Architectures 1 and 2 represent large-scale systems whose major energy resources generate constant power at peak capacity, such as nuclear plants. The governance in place considers these energy services to be a public good and social right; the design of such systems evolves around unconditional service to relatively passive energy users. Figure 2.1 depicts Architecture 1. The larger circles at the top of this figure denote conventional bulk power resources, the smaller circles denote DERs (renewable generation, aggregate responsive loads), and the interconnecting lines represent the transmission grid. Architectures 1 and 2 differ mainly in terms of their governance systems. In Architecture 2, energy resources are generally privately owned and controlled; energy is provided through electricity market arrangements.

Architecture 3, illustrated in Figure 2.2, represents a combination of hybrid resources (large-scale, fully controllable generation and a

2. The notion of sustainability is not standardized. Here we use the same notion as in [Ostrom09], meaning a combination of acceptable QoS, and the ability to sustain the economic, environmental, and business goals desired by the members of a given socio-ecological system (SES). This is not an absolute metric; rather, sustainability is defined by the SES members within the governance rules.

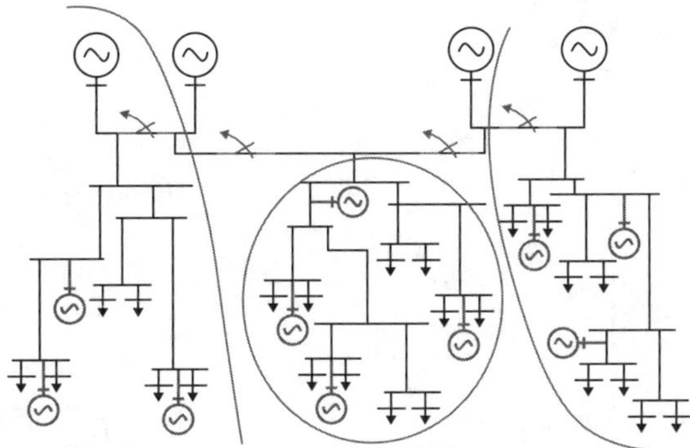

Figure 2.1: *SEES Architectures 1 and 2. Circles with wiggles on top represent generators, arrows represent end users, and interconnections represent controllable wires.*

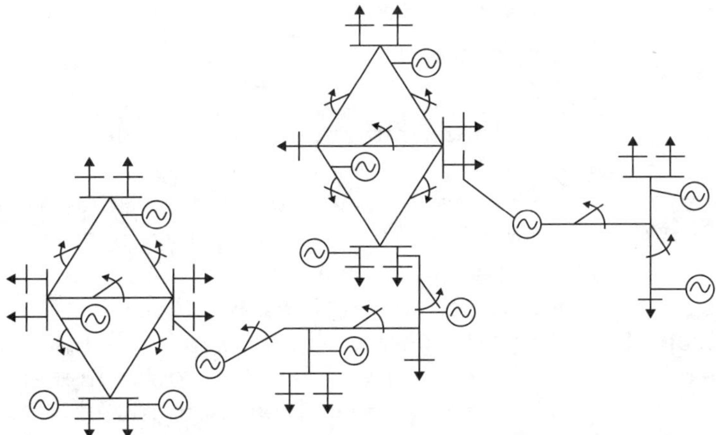

Figure 2.2: *SEES Architecture 3. Circles represent already controlled power plants, arrows represent end users, and the interconnections are controllable wires.*

certain percentage of intermittent resources), a mix of passive and responsive demand, governance requiring high-quality energy service, and compliance with high emission rules.

Architectures 4 and 5, shown in Figure 2.3, comprise a stand-alone distribution-level micro-grid—for example, an island; a large computer data center; military, navy, or air base; or a shopping mall completely disconnected from the local utility. This kind of SEES is

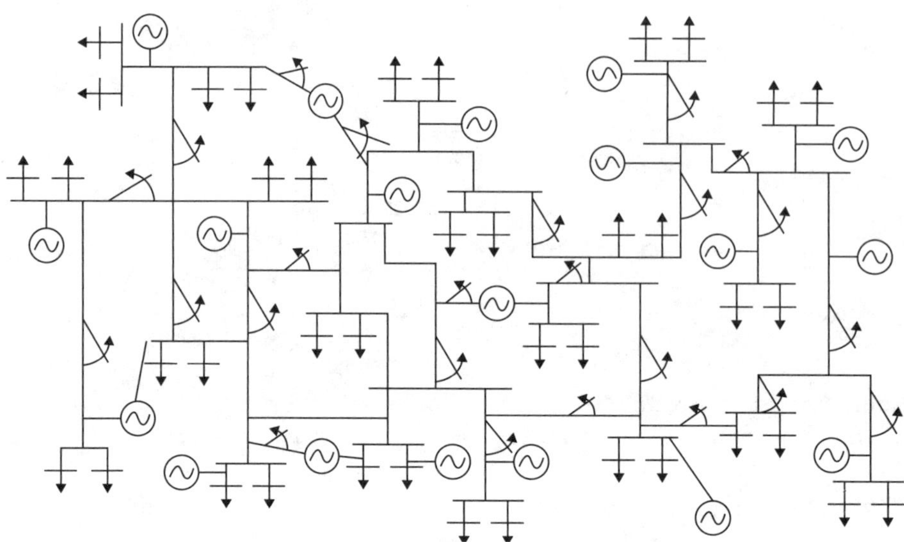

Figure 2.3: *SEES Architectures 4 and 5*

supplied by many small intermittent, highly variable DERs, such as rooftop photovoltaics (PVs), bioresources, small-scale wind power generation, small pumped hydropower sources, and the like. In addition, such systems often count on storage of power in highly responsive smart EV vehicles. The governance here is completely private and unregulated for all practical purposes. Architectures 4 and 5 have qualitatively different performance objectives for reliability and efficiency. Architecture 4 is typically characterized by higher requirements for QoS and lower emphasis on cost minimization than Architectures 1 through 3. In contrast, Architecture 5 is typical of the low-cost, low-QoS systems found in developing countries, in which the objective is to provide electricity services within the financial constraints of electricity users.

Table 2.2 highlights the major differences between the objectives of cyber-physical systems in regulated bulk electric energy systems on one side, and the evolving hybrid and distributed energy systems on the other. In Architectures 3 through 5, the performance objectives of the system as a whole reflect the system users' needs and preferences. In contrast, in Architecture 1, and to a lesser extent in Architecture 2, the performance objectives for the power grid are defined in a top-down way by the utility and the governance system.

Table 2.2: *Qualitatively Different Performance Objectives in SEES Architectures*

Performance Objectives in Architectures 1 and 2	Performance Objectives in Architectures 3–5
Deliver supply to meet given demand	Both supply and demand have preferences (bids)
Deliver power assuming to predefined tariff	QoS given willingness to pay for it
Deliver power subject to predefined emission constraints	QoS given willingness to pay for emissions (preferences)
Deliver supply to meet given demand subject to given grid constraints	QoS including delivery at value
Use storage to balance fast-varying supply and demand	Build storage given grid users' preferences for stable service
Build new grid components for forecast demand	Build new grid components according to longer-term ex-ante contract for service

2.3.1.2 Critical System-Level Characteristics

Given the specific characteristics of the system resources, users, and governance system, the human-made smart grid needs to be designed so that it enables energy services that are as sustainable as possible. Any given SEES is more sustainable when the characteristics of the users, resources, and governance system are more closely aligned. Once this relationship is understood, it becomes possible to arrive at basic principles of cyber-physical systems for smart grid design. A smart CPS electric power grid is a physical electrical network in which cyber-technologies are used to meet the specific performance objectives shown in Table 2.2. In Table 2.2, the sustainability of a bulk power system (i.e., Architecture 1) critically depends on system-level optimization of the supply–demand balance subject to the many constraints listed in the left column. In contrast, in Architectures 3 through 5, CPS becomes necessary to enable users and resources to manage themselves in a more distributed way and to characterize their participation in system-level functions. In Architecture 2, social objectives are defined by the governance system (regulators) and implemented by the system operators, planners, and electricity markets based on the specifications identified in system users' bids.

The emerging Architectures 4 and 5 shown in Figure 2.3 are defined by system users and resources themselves, as these are neither regulated by the governance system nor operated by the traditional utilities. A small-system illustration of the evolving architectures is shown in Figure 2.4. The black lines represent the existing system, and the light gray lines represent new technologies being connected to the existing system. An end-to-end CPS design for such evolving systems needs to be developed so that the temporal, spatial, and contextual characteristics of both newly added components and the existing system work well together.

Aligning the temporal characteristics basically means that supply and demand must be balanced at all times; if resources and users do not have such properties, storage becomes critical. In Architecture 1, for example, base-load power plants cannot follow time-varying system load variations. This problem can be remedied by adding faster power plants and/or fast storage. Similarly, aligning spatial characteristics requires that supply can be delivered to the location where users are. Since electric energy systems have minimal mobility, aligning on this principle fundamentally requires having a power grid to interconnect users and resources. Finally, aligning contextual characteristics implies developing requirements to meet different performance objectives established by different participants in a SEES. The utility, for example, has its own objectives and is supported by customized hardware and software to meet the objectives identified in Table 2.2. Similarly, users have their own sub-objectives, which comprise a combination of cost, environmental, and QoS preferences. More examples of such contextual preferences can be seen in Table 2.2; it is important to note that the type of architecture largely determines the contextual preferences since the governance characteristics are qualitatively different.

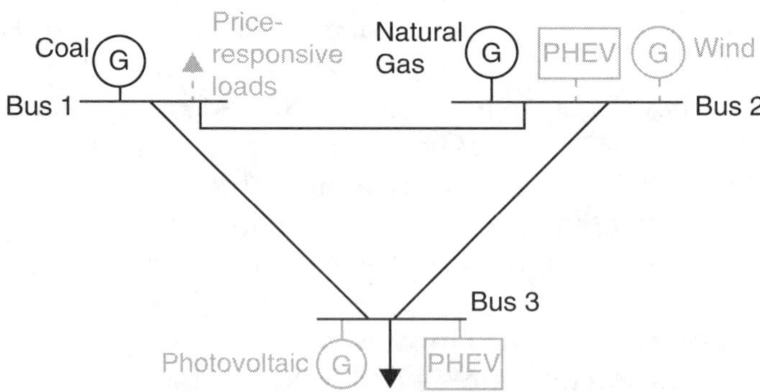

Figure 2.4: *A small-system illustration of the evolving SEES*

Understanding these somewhat subtle distinctions between the objectives of CPS design in different SEES architectures is essential for understanding which technologies may best fit the needs of the various human-made smart grids. We next consider the state-of-the-art cyber aspects of different architectures and identify which improvements are needed for the specific architectures to meet the objectives shown in Table 2.2.

2.3.2 Architecture 1 Performance Objectives

Given that it operates under the aegis of fully regulated utility governance, the main performance objective of Architecture 1 is network-centric. That is, each utility is required to supply energy to its users all the time and at the lowest cost possible.

This outcome is achieved by scheduling the available generation P_G produced by the controllable plants within their given capacity limits and delivering this power to supply the system-level utility load. This load must be forecast, as it varies with weather conditions and temporal cycles (i.e., daily, weekly, seasonally). Over the long term, the utility must plan to increase its generation capacity K_G to provide sufficient energy to meet the long-term load growth and/or plans for decommissioning/retiring the existing, polluting power plants as they become outdated. Both for short-term scheduling and for long-term power grid investments, load is assumed to be a completely exogenous, nonresponsive system input. Moreover, the utility has the responsibility to enhance physical transmission capacity K_T necessary to ensure reliable power delivery continues as the new generation is connected to the grid. The mathematical algorithms underlying both the scheduling and capacity investments are large-scale optimization methods subject to the power grid's constraints and to the physical and control limits of all equipment. For a mathematical formulation of the optimization objectives and underlying constraints, see [Ilic98], Chapter 2, and [Yu99].

While the theoretical formulations allow for K_G, K_T, and P_G as equally valid decision variables, common practice has been to first plan for the generation capacity expansion K_G given the long-term system load forecast, and then to build sufficient transmission capacity K_T so that delivery is feasible; the transmission investment decisions are often based on analyses for the worst-case scenario and are not optimized [Ilic98, Chapter 8]. Notably, utilities do not proactively plan for capital investments in demand efficiency improvements K_D, and the effects of such investments are not compared to the effects of building more generation capacity K_G. Instead, system demand P_D is assumed to be known, and planning of K_G and K_T is done to have sufficient capacity during the worst-case equipment failures in which one or two large generators and/or transmission components are disconnected from the grid. This case is known as the $(N - 1)$ or $(N - 2)$ reliability criterion.

Notably, so-called performance-based regulation (PBR) represents an effort to provide incentives to utilities to consider innovation that might be beneficial in reducing the overall cost of service, while meeting the customers' QoS expectations [Barmack03, Hogan00]. The basic idea of PBR is for a utility regulator to assess periodically—say, every five years—the total cost incurred when providing the service and to allow for tariff increases in accordance with the overall cost reduction. Unfortunately, a PBR design for complex energy networks remains an open problem at present.

2.3.2.1 Systems Issues in Architecture 1

Several fundamental assumptions underlying today's system operations and planning can prevent one from enhancing system-level performance by means of cyber-technologies. They can be classified as follows:

- *Nonlinear dynamics-related problems:* As yet, there are no nonlinear models that would support the sensor and actuator logic needed to ensure the desired performance both during normal operations and during abnormal conditions [Ilic05].
- *Use of models that do not control instability adaptively:* All controllers are constant gain and decentralized (local). The system is not fully controllable or observable.
- *Time–space network complexity-related problems:* Problems occur when attempting to avoid fast dynamic instabilities following major equipment outages by means of conservative resource scheduling. The system is generally operated inefficiently by requiring standby fast reserves just in case large equipment fails.

When tuning controllers, it is assumed that the interconnections between the dynamic components are weak and, therefore, the fast response is localized. Coordinated economic scheduling is done under the assumptions that the network has linear network flow and the effects of nonlinear constraints are approximated by defining proxy thermal limits to transferring power through major "flow gates." As a result, reserves are always planned so that the load can be served in an uninterrupted way even during the worst-case equipment failures; there is very little (close to none) reliance on data-driven online resource scheduling during such equipment failures. This effectively amounts to planning sufficient resource capacity but operating the available resources inefficiently and not counting on dynamic demand response, scheduling of setpoints on other available controllable generation and T&D equipment, grid reconfiguration, and the like. In the early stages of a typical blackout scenario, no systematic adjustments are made to ensure that the cascading events do not take place—even though doing this could prevent blackouts from occurring [Ilic05a].

Today's models underlying hierarchical control systems do not support interactions between higher and lower levels within the power grid. For example, scheduling is done by modeling the forecasted demand as real demand; therefore, such a system does not lend itself to integrating responsive demand or the effects of new distributed energy resources (DERs). AGC is designed to manage quasi-static deviations from this forecast under the assumptions that scheduling is done well and the fast system dynamics are stable. Primary controllers are tuned once for the supposed worst-case condition, but are known to lead to instabilities when operating conditions and/or topology vary within the regions not studied.

This state of computer industry applications requires rethinking, as the power grids of the future are not expected to operate in the conditions for which the grid and its hierarchical control were designed. Integrating new technologies that bring more resiliency and efficiency through flexibility is essential but most assuredly will not be easy. New best-effort solutions are generally not acceptable by the industry and are viewed as threats to reliable operations.

2.3.2.2 Enhanced Cyber Capabilities in Architecture 1 Systems

As a general rule, utilities need to learn many more details about their customers' load profiles. They should begin to invest in smart generation and/or smart delivery and consumption so that they offset the high capital cost investments that will be necessary to transition the existing systems to the CPS-based systems of the future, or at least delay these investments by becoming able to better manage service under uncertainties. Advanced metering infrastructures (AMIs) are a recent step in this direction. The information gathered using AMIs can be applied to implement adaptive load management (ALM) that benefits both customers and utilities, enabling the latter to better manage the tradeoff between large capital investments and the cost of necessary cyber-technologies [Joo10]. As system demand becomes less predictable, the value of CPS in enabling just-in-time (JIT) and just-in-place (JIP) supply–demand balancing will only increase [Ilic00, Ilic11].

It is particularly important to recognize the need for online resource management using cyber-technologies. For example, many "critical" contingencies are not time critical. Having robust state estimator (SE) and scheduling software for adjusting other controllable equipment as the loading conditions vary could go a very long way toward avoiding—or at least offsetting—the high capital costs of the large-scale equipment that is currently viewed as necessary for providing reliable service. Ultimately, utilities should be able to implement very fast power-electronically switched automation for stabilizing system dynamics even during time-critical equipment failures [Cvetkovic11, Ilic12]. Estimates of the potential benefits from such automation are based on the cumulative savings of using less expensive and cleaner fuel during normal conditions and the avoidance of maintaining "must-run" generation in case a very rare event happens.

2.3.2.3 CPS Design Challenges for Architecture 1

While it is somewhat obvious that CPS-based enhancements for Architecture 1 are possible, innovations in this area and the adoption of new cyber-based solutions have been rather slow. Data-driven operations require enhanced utility-level state estimators; computationally robust optimization tools for computing the most effective adjustments of controllable generation and T&D equipment, as new System Control and Data Acquisition System (SCADA)–supported state estimates are made available online; and remote control for adjusting this equipment. One of the major computational challenges comes from the highly intertwined temporal and spatial complexities within the large-scale electric power grid. The controllable equipment has its own dynamics, and the power grid is spatially complex and nonlinear. At present, there are no general-purpose, large-scale software applications that are ready to deploy for the power grid complexity usually seen in typical Architecture 1 and that offer proven performance in such applications.

2.3.2.4 CPS Design Challenges for Architecture 2

The system-level objectives for Architecture 2 are very similar to the system-level objectives for Architecture 1. In this type of system, however, the generation costs result from non-utility-owned bids and may not be the same as the actual generation costs. Bidders naturally seek to optimize their own profits without exposing themselves to excessive risks. Also, at least in principle, load-serving utilities are being formed as competition to today's utilities and are offering to provide service to groups of smaller users by participating in the wholesale market. In this scenario, the objective of these load-serving entities (LSEs) is profit maximization by managing load served.

The main challenge in Architecture 2 is to develop a cyber design that ensures electricity markets are consistent with power grid operations.

2.3.2.5 CPS Design Challenges for Architectures 3–5

The system-level objectives of the emerging Architectures 3 through 5 are not well defined at present. Architectures 4 and 5 are basically at the green-field design stage; that is, they are evolving as new technologies are researched. In particular, there has been a major trend toward large electricity users separating from the utility grid and becoming effectively stand-alone micro-grids; this trend is taking place in smart data centers, military bases, and navy and airspace centers. In addition, some municipalities and other groups comprising relatively small users managed by energy service providers (ESPs) are beginning to consider the benefits of stand-alone resources and lower dependence on the local utility providers.

In contrast with Architectures 1 and 2, Architectures 3 through 5 are stand-alone, self-managed relatively small power grids, as shown in Figures 2.2 and 2.3. They can be completely disconnected from the backbone power grids, or they can be connected to them for ensuring reliable service at times when their own resources are insufficient or too costly. Collectively, these architectures are referred to as micro-grids [Lopes06, Xie09]. The process whereby users previously served by utilities are connecting their own small resources to the backbone grid is currently evolving. For their part, the utilities, which are still required to remain the providers of last resort, have found it difficult to plan for and interoperate with an increasing number of micro-grids.

Very little formal mathematical modeling of the emerging architectures, as complex large-scale dynamical systems, has been performed to date [Doffler14, Smart14]. Instead, much effort has been devoted to modeling stand-alone (groups of) components, such as wind power plants, smart buildings, solar power plants, and their local controllers and sensors. The lack of models in turn leads to a lack of methods for the systematic estimation, communications, and control design needed to account for the presence of new resources and their value to the system. For example, the lack of methods for smart charging of EVs may lead to a larger power peak seen by the utility at the planning stage; this increased demand then has at least two undesired effects—namely, the need to build new, often polluting plants, and a lack of economic incentives to the EVs to reduce the utility load.

These problems are fundamentally caused by the poor end-to-end multidirectional information exchanges among system users, and can be seen across various technologies and types of system users (wind power plants, PVs, smart buildings, and users with other responsive demands). The same problem becomes evident when the T&D grid owners and the system operators responsible for ultimate services to the end users fail to coordinate their actions: T&D owners of controllable voltage equipment do not adjust online the setpoints of these controllers to support the most efficient and clean power generation delivery and utilization. More generally, without multidirectional information exchange, it is impossible to provide incentives for smart grids to help meet either the objectives of different stakeholders or the objectives of the system as a whole.

2.3.3 A Possible Way Forward

No single method will be sufficient to support truly resilient and efficient operations of future power grids. Similarly, no single cyber magic architecture will be best. Instead, non-unique combinations of the existing cyber-technologies, new embedded sensors and actuators, and minimal multilayered interactive online information exchange protocols will be necessary to enable smart DERs to work in synchrony with the existing physical grid and resources. As suggested earlier in this chapter, how much sensing, how many actuations, and which type of information exchange protocols will be needed depend fundamentally on how well the existing resources and users are aligned within the existing legacy power system, as well as on the changing governance and societal objectives. Notably, the rights, rules, and responsibilities of different stakeholders are changing, and these changes are likely to bring about major changes in the necessary cyber-physical systems to support the operation of the evolving system [Ilic09]. More generally, proactive participation of electricity users in power grid supply–demand balancing for environmental and economic reasons is likely to make it almost impossible for utilities to rely on the same system demand forecasting techniques and schedule generation as they used in the past.

New IT companies are emerging and entering the business of data gathering and processing on behalf of system users [helio14, nest14]. When more DERs are owned and operated by the non-utility stakeholders, this raises major questions concerning ultimate responsibility for electricity services, particularly during equipment failures. Many of the DERs and large loads that are self-sufficient during normal operations might still wish to call upon utility support when their own resources are not available. This pattern puts major pressure on utilities as the last-resort providers of electricity service. Ironically, the efficiency of utilities has been based on economies of scale supported by their large power generation and relatively smooth system load. A paradigm in which smaller DERs provide local, often dynamically volatile loads brings up issues regarding the overall efficiency of the newly emerging patterns of electricity provision. To induce efficiency and robustness, cyber-enabled aggregation of resources will be essential.

In the remainder of this chapter, we describe our approach to new modeling, estimation, and decision-making challenges in the changing electric energy systems. The examples and references are based primarily on the work done by the Electric Energy Systems Group (EESG) at Carnegie Mellon University and, as such, are not a full representation of the state of the art.

2.4 Cyber Paradigm for Sustainable SEES

In this section we introduce one possible modeling, estimation, and decision-making paradigm—DyMonDS—and consider it as a basis for new cyber design and implementation to overcome the previously described problems in the electric power industry. We first point out that the underlying principles of this framework are the same for all of the different physical architectures described previously. To see this, consider the typical electric energy system and its embedded intelligence shown in Figure 2.5 [Ilic00, Ilic11]. Two distinct CPS

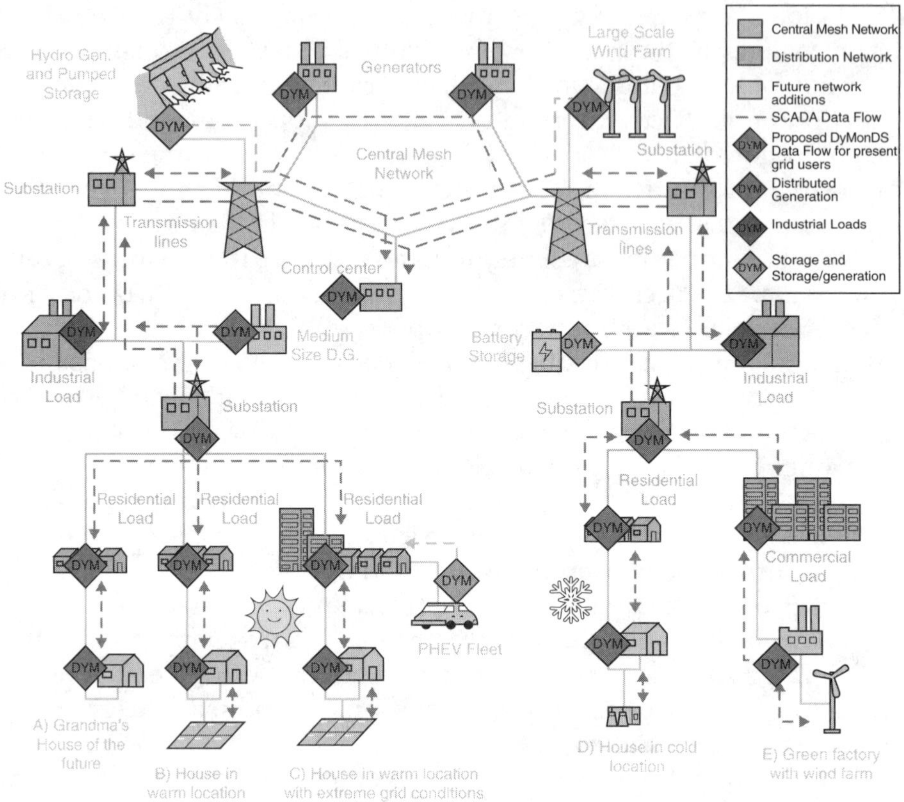

Figure 2.5: *CPS electric energy system with its embedded DyMonDS [Ilic11, Ilic11a]*

processes can be observed: (1) the physical (groups of) components with their own local DyMonDS, and (2) the information and energy flow between the (groups of) components, shown in dotted multidirectional dotted lines.

DyMonDS effectively represents (groups of) physical components with their embedded sensing, learning of system signals, and local decision making and control. The components share common objectives (cooperate) and jointly compete within the complex electric energy system for achieving their objectives. They also respond to the pre-agreed binding information signals and send pre-agreed information back to the system according to well-defined protocols. Notably, our design of both embedded cyber and information exchange is based on sound physics-based models and well-defined performance objectives. A smart and strong grid is strongly dependent on the type of SEES architecture it is supposed to serve. Figure 2.6 is a sketch of a general socio-ecological system (SES) as introduced in [Ostrom09]. Figure 2.7 is a sketch of our proposed DyMonDS-based end-to-end CPS grid enabling sustainability of a socio-ecological energy system (SEES) [Ilic00, Ilic11]. Depending on the basic SEES architecture, the need for the end-to-end CPS design smart grid varies. Nevertheless, the underlying principle is the same: As closely as possible, align the temporal, spatial, and contextual characteristics with the objectives of the resources, users, and governance system. Once this is understood, it becomes possible to design the CPS grid with well-defined objectives, and the value of many diverse grid technologies can be quantified.

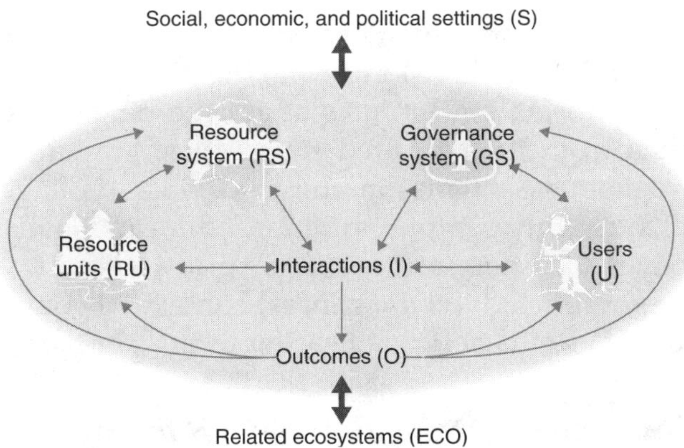

Figure 2.6: *The core subsystems in a framework for analyzing socio-ecological systems (SES) [Ostrom09]*

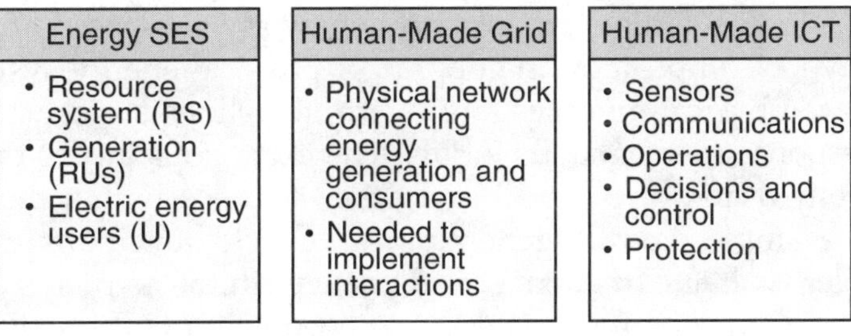

Energy SES	Human-Made Grid	Human-Made ICT
• Resource system (RS) • Generation (RUs) • Electric energy users (U)	• Physical network connecting energy generation and consumers • Needed to implement interactions	• Sensors • Communications • Operations • Decisions and control • Protection

Figure 2.7: *Smart grid: end-to-end CPS for enabling sustainable SEES [Ilic00, Ilic11]*

Recall from Section 2.3.1 that any given SEES will be more sustainable (by all metrics used) when there is better alignment of the temporal, spatial, and contextual properties of the (groups of) physical grid users. This outcome can be realized in two qualitatively different ways: by embedding local intelligence into grid users themselves and/or by coordinating unaligned grid users. In simple words, if users of electric power are very volatile and not responsive, it is necessary either to build faster generation to supply them or to ensure that the users can respond and adjust locally in response to the external information from the rest of the system. If this is not done, the interconnected grid may fail to meet users' needs—a situation characterized as unreliable QoS; alternatively, the overall utilization of resources may be quite inefficient.

When the problem of sustainable SEES is envisioned as the problem of cyber design in complex dynamic systems, one begins to understand that this system is driven by various exogenous inputs and disturbances; its state dynamics evolve at vastly different rates determined by the natural response of its resources, users, and T&D grid, as well as by their embedded DyMonDS and their interactions. Controllers respond to the information updates about the deviations of relevant system outputs from their desired trajectories. Similarly, the spatial effects of exogenous systems on the grid state (voltages, currents, power flows) are highly complex and non-uniform. Finally, the performance objectives of different (groups of) grid users are very diverse and sometimes conflict given the general contextual complexity of the evolving energy systems described earlier in this chapter.

2.4.1 Physics-Based Composition of CPS for an SEES

One of the long-standing theoretical problems underlying complex network systems concerns the systematic composition of sub-networks. This problem is common to both physical designs and cyber designs [Lee00, Zecevic09]. In this section, we propose a new physics-based approach to cyber-network composition in support of physical energy systems. The dynamics of physical modules and their interactions are driven by their natural response to exogenous inputs. The exogenous inputs themselves are generally a mix of physical and cyber changes within the complex SEES. For example, a power generator responds to deviations in its physical variables (voltage, frequency) and changes in the setpoints of its controllers (e.g., governor and exciter). A wind power plant is driven by the exogenous changes in wind speed, as well as by changes in the setpoints of its controllers (e.g., blade position and voltage).

To generalize, each CPS module within an SEES can be thought of as having the general structure shown in Figure 2.8 [Ilic14]. The dynamics of each physical module is characterized by the module's own state variables $x_i(t)$, with the number of states being n_i; the local primary controllers injecting input signals $u_i(t)$; exogenous disturbances $M_i(t)$; and interaction variables $z_i(t)$. The composition challenge is the one of

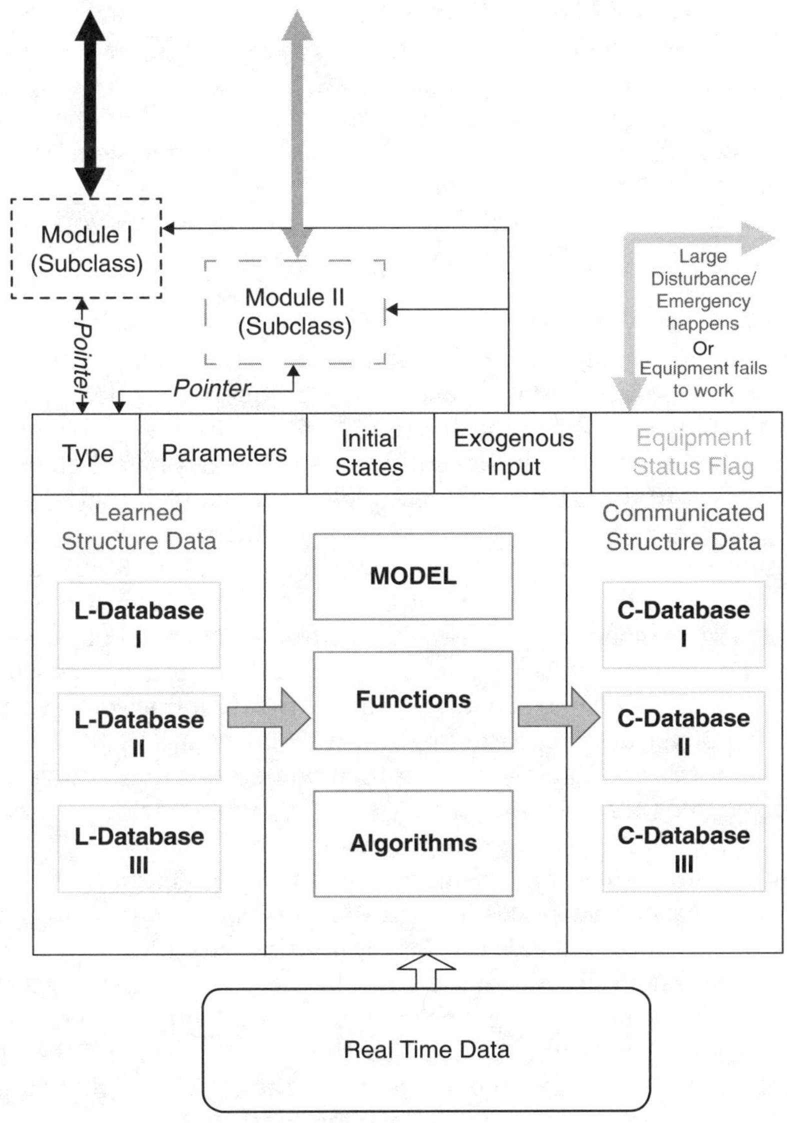

Figure 2.8: *General representation of a CPS module within a SEES*

combining all modules into a complex interconnected system; it arises because the interaction variables of each specific module depend on the dynamics of other modules. The open-loop dynamics of an interconnected SEES are determined by the exogenous inputs $M_i(t)$ to all modules, and by the initial state of all modules $x_i(0)$. Superposed on this process are discrete events reflecting the change of module status (the lightest gray arrow in Figure 2.8). Closed-loop dynamics are further driven by changes in set-points of local controllers $u_i^{ref}(t)$ and the local control $u_i(t)$ reacting to a combination of deviations in local outputs $y_i(t) = C_i\ x_i(t)$ and interaction variables $z_i(t)$ from the desired values. System regulation and local control designs are very different in the five architectures described earlier. Together with the System Control and Data Acquisition System (SCADA), they form the cyber network for an SEES.

To formalize the systematic composition of a cyber network, we first observe that the actual continuous dynamics of states $x_i(t)$ has physics-based structure. In particular, the dynamics of any given module depends explicitly on the dynamics of the neighboring states only, and not on the states of farther away modules [Prasad80]. The idea of observation-decoupled state space (ODSS) frequency modeling, estimation, and control was introduced some time ago based on this inherent structure [Whang81]. More recently, a new physics-based state space model was introduced for fully coupled voltage–frequency dynamics [Xie09]. The transformed state space for each module comprises the remaining $(n_i - 2)$ states and the interaction variable whose physical interpretation is the net energy stored in the module and the rate of stored energy exchanged between the module and the rest of the system. In this newly transformed state space, a multilayered composition of modular dynamics can be characterized by the dynamics of the internal remaining physical states forming the lowest layer, and by the dynamics of the interaction variables forming the zoomed-out higher layer. This, in turn, defines the relevant cyber variables, which are the minimum necessary for characterizing the dynamics of the complex grid.

This multilayered representation is shown in Figure 2.9. In this model, the learned variables in Figure 2.8 for each zoomed-in module are the interaction variables of the neighboring modules, and the communicated variables between the module and the rest of the system are the module's own interaction variables. This modeling step encourages one to view electric energy systems as genuine cyber-physical systems; their physics defines the cyber-structure (the variables that must be known and communicated), and the cyber performance is designed to support the physical performance. The open-loop dynamics can be

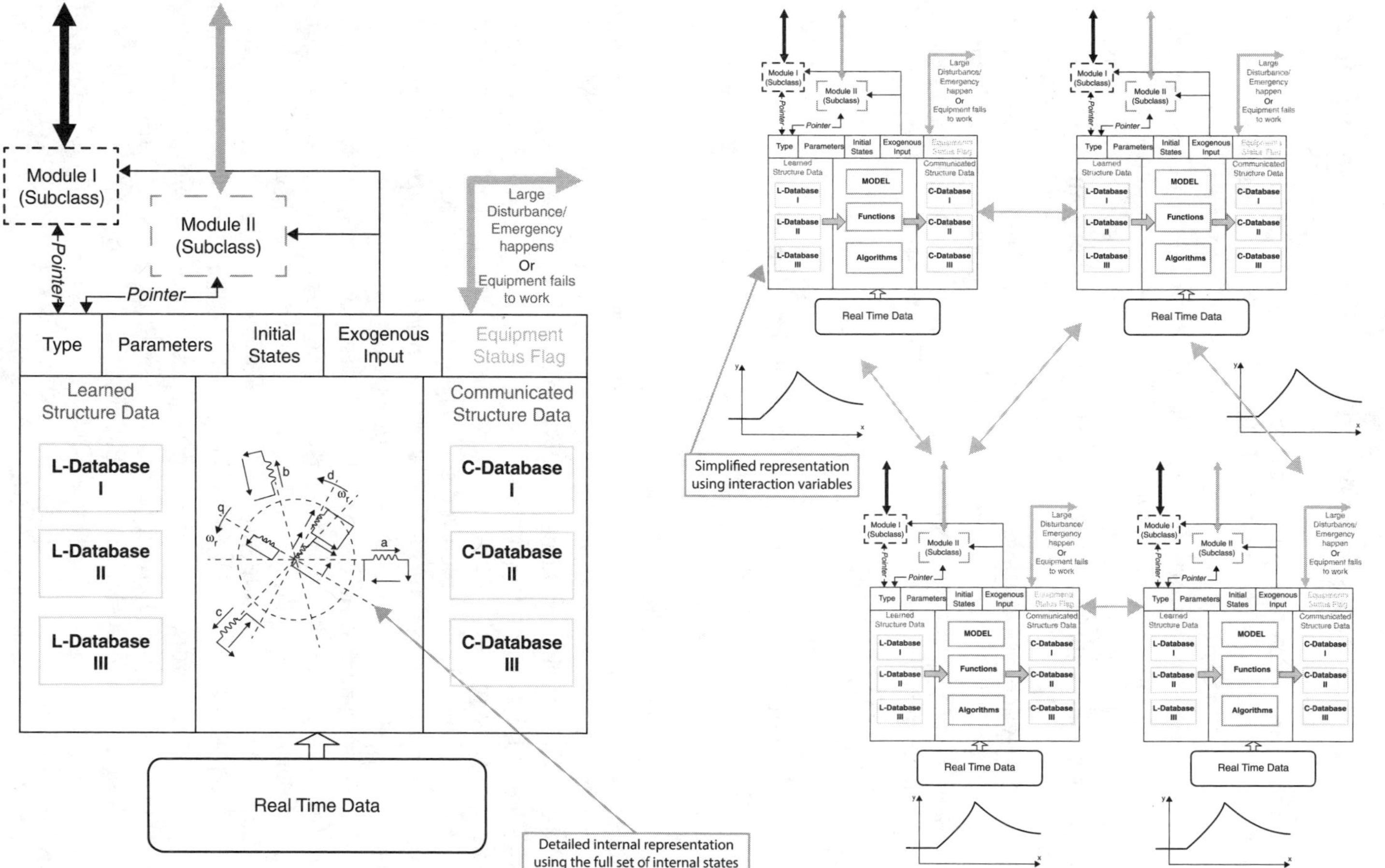

Figure 2.9: *Interaction variables–based CPS composition within an SEES*

composed by exchanging information about the interaction variables. This interaction supports the efficient numerical methods needed to simulate these otherwise very complex grids [Xie09].

The design of closed-loop CPS is based on the same dynamic structure. Given sufficient control of each module to keep the internal states stable, the most straightforward distributed control responds to a combination of dynamics of the internal variables and of the interaction variables. This amounts to fully decentralized competitive control when using the transformed state space. The cyber aspects will amount to fast sampling in support of often high-gain local control and no communications; full system synchronization is achievable this way [Ilic12a]. This cyber-structure would be very effective in an ideal world, such as in Architecture 4, in which all dynamic components, including wires, have fast typically power-electronically switched controllers [Doffler14, Lopes06]. None of the other physical architectures described earlier have so much distributed control, and, even if they did, a very real problem with control saturation might occur. Technical standards/protocols are, therefore, generally needed for more realistic cyber design [Ilic13].

2.4.2 DyMonDS-Based Standards for CPS of an SEES

Our Dynamic Monitoring and Decision Systems framework can be best understood by using the new modeling approach in which modules are characterized by their internal and interaction variables. In DyMonDS embedded in physical modules (shown in Figure 2.5), the design enables modules to be inserted into an existing grid in a plug-and-play manner as long as they meet the standards—that is, as long as they do not create dynamic problems over the time scales of interest in electric energy systems. The multilayered modeling and control design approach sets the basis for the standards that must be met by different modules.

The simplest approach is to require each module to meet the dynamic standards such that the interactions with the rest of the system are cancelled in closed-loop dynamics. This approach is probably acceptable in Architectures 4 and 5, which comprise many small distributed components, each of which has a local cyber-structure. For example, a PV panel might have local storage and fast power electronics capable of cancelling the effects of the stored energy exchange (interaction variables) coming from the rest of the system to the PV. Such a PV would have sufficient control and storage that it could behave as an ideal AC voltage source, at least as seen by the rest of the system; the micro-grid designs envisioned in [Doffler14, Lasseter00] all make this very strong assumption. In contrast, Architectures 1 through 3 require more complicated standards for plug-and-play dynamics. Notably, not every dynamic component in these architectures has a controller.

We have proposed a standard for dynamics in such systems, which must be met by a group of modules, rather than by each individual module [Baros14, Ilic13]. This standard requires cooperative and coordinated control within each group of modules, and minimal or no coordination among the groups. A real-world example of a quasi-static standard for a group of users is the notion of control areas. The interactions between the two control areas are measured in terms of inadvertent energy exchange (IEE), which is determined by the area control error (ACE) power imbalance; this is the underpinning of today's AGC [Ilic00]. A dynamic interaction variable associated with a group of components responsible for ensuring there are no instability problems is a further generalization of this concept. Detailed derivations have shown that the dynamics of these interaction variables is determined by the interaction variables themselves and by some states of the neighboring modules [Cvetkovic11, Ilic00, Ilic12, Ilic12a]. Therefore, it becomes possible to sense and control the interaction variables in a distributed way as well, because their dynamics are fully determined by the locally measurable variables.

2.4.2.1 The Role of Data-Driven Dynamic Aggregation

The preceding discussion could lead one to conclude that design of cyber elements for desired performance of each layer can be done in an entirely distributed manner. Unfortunately, this claim is not always true. Generally, if the effect of interaction variables dominates the effect of local modular dynamics, it becomes necessary to design complex high-gain controllers locally to cancel the interactions and make the modules behave as if their closed-loop dynamics have become weakly coupled. This approach generally calls for lots of local sensing and fast controllable storage, and effectively no communications. The difference between this design and the conventional hierarchical control in power systems is that now it becomes possible to have provable performance that is measured in terms of how well the modules meet their objectives and how closely they are aligned with the objectives of the interconnected system.

The overwhelming complexity is managed by closing the loop and making the modules and layers self-sufficient. Although some savings are achieved through aggregation when groups of users form themselves into modules with common objectives, the modules become competitive in this scenario. This brings up a very difficult system question concerning the system-level efficiency of such CPS. Here again, the conventional hierarchical control thinking would suggest that this approach would lead to a less efficient solution than when there is coordination among the modules. In reality, this conclusion is no longer automatically the case, because the boundaries of modules are not predefined based on administrative divisions within the complex system. Instead, they are formed in a bottom-up aggregated manner by the components themselves until there is very little return from aggregating into a module. If such aggregation becomes data driven and boundaries of cooperating modules are dynamically formed in response to the changing system conditions, such modules can be operated in a decoupled way in the actual operations; they can still achieve equally good performance as if the coordination of interaction variables is done for modules with prefixed boundaries, as is common in today's hierarchical systems.

The optimal size of the modules depends on the natural dynamics of the components belonging to the module itself as well as on the strength of the interconnections between the modules. The physics at play ultimately determine the near-optimal balance between the complexity of the local cyber-structure (DyMonDS) and the communication requirements for coordinating interaction variables. Recent work has just begun to scratch the surface of our knowledge about these multilayered electric energy systems, and the interdependencies between the physical and cyber designs needed to serve the SEES best.

2.4.2.2 Coordination in Systems with Predefined Subsystems

Today, it is very difficult to dynamically aggregate groups of users in a bottom-up fashion so that their closed-loop dynamics is very loosely coupled with the other groups of users. When an electric power system loses a large generator or a transmission line, the relevant model is inherently nonlinear. As a consequence, all bets are off as to whether aggregating the system can be accomplished so that the effects of such topological change are not seen far away from the fault. One possible solution is to use adaptive high-gain control of the remaining components, such that the system appears linear and is differentially flat in the closed loop [Miao14, Murray95]. Power-electronically controlled switching of controllable storage at the kilohertz (kHz) rate to keep the electromechanical system synchronized following a large component failure is one promising path to implement this solution; such technological solutions have become plausible only recently with the major breakthroughs in high-voltage power electronics and with the development of accurate, fast GPS-synchronization [Cvetkovic11, Ilic12].

The mathematics of interconnected power systems with massive fast power-electronically switching storage is worth exploring: The technology naturally makes the closed-loop dynamics of internal states of controlled equipment much faster and decoupled from the slower-evolving interactions of modules comprising electromechanical dynamics of rotating machinery; therefore, fully distributed stabilization that arrests the propagation of the effects of faults through the system's backbone becomes possible. The major challenges are how to dynamically aggregate many components into a single intelligent balancing authority (iBA) for minimizing the cost of storage [Baros14, Ilic00]. Determining when conventional fast controllers shaping very little storage in reactive components of Flexible AC Transmission Systems (FACTS) would have to be supplemented by the real power energy storage devices, such as various batteries and flywheels, is a challenging control design problem at present [Bachovchin15]. Nevertheless, the first steps toward implementation of such an approach are being made.

With more progress, it becomes possible to define standards for dynamics of such iBAs, and the system can be operated in a plug-and-play manner by the weakly coupled iBAs [Baros14, Ilic00]. Figure 2.10 depicts the IEEE 24-node RTS system [Baros14] and the dynamic response of power flows in the area close to the loss of generator at node 1 with and without iBA-based high-gain control (constant gain AVRs and PSSs). Figure 2.11 shows the system response with carefully aggregated iBAs; the dynamic response to the same fault with the closed-loop nonlinear controller is also shown. While the system's response to losing generator oscillates with the existing excitation controllers, such as automatic voltage regulators (AVRs) and constant-gain power system stabilizers (PSSs), the high-gain nonlinear feedback-linearizing excitation controllers would stabilize the closed dynamics of the power plants in the iBA, almost completely eliminating the inter-area oscillations with the rest of the system.

Much work must be done on designing high-gain power-electronically switched controllers to ensure provable performance when the system is in danger of losing its synchronism. It was recently shown that the choice of energy function for keeping the system

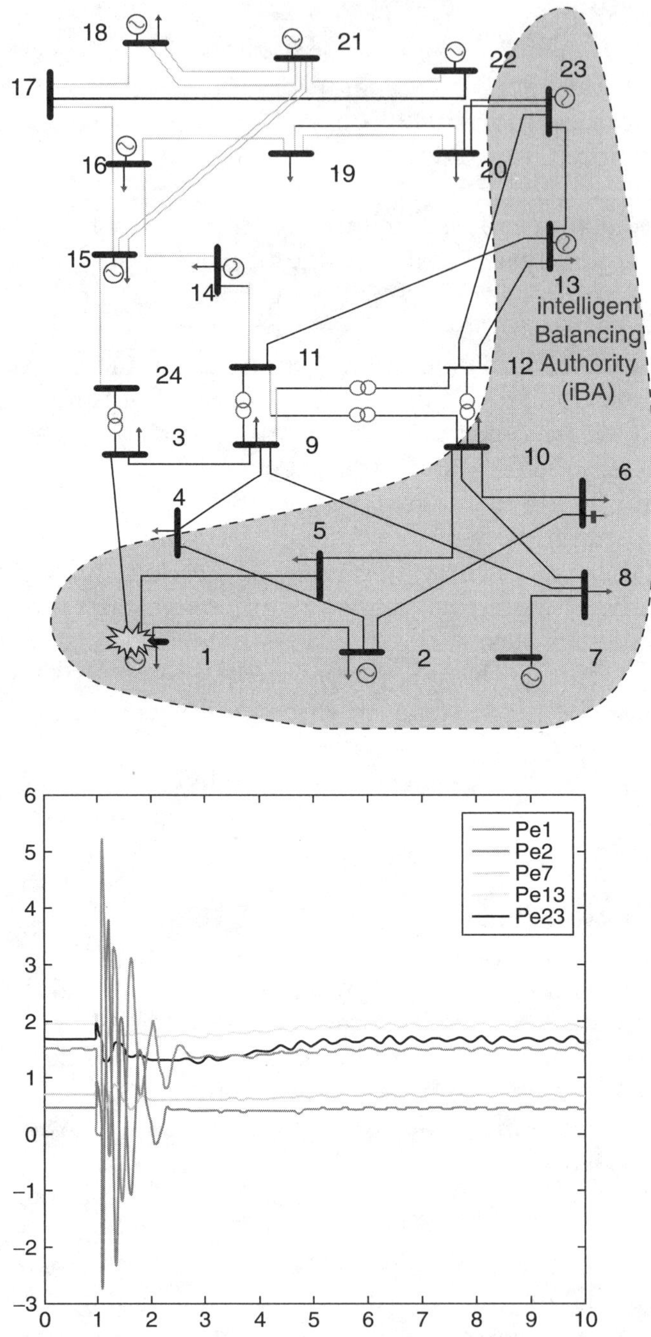

Figure 2.10: *Dynamic response with high-gain iBA distributed control [Baros14]*

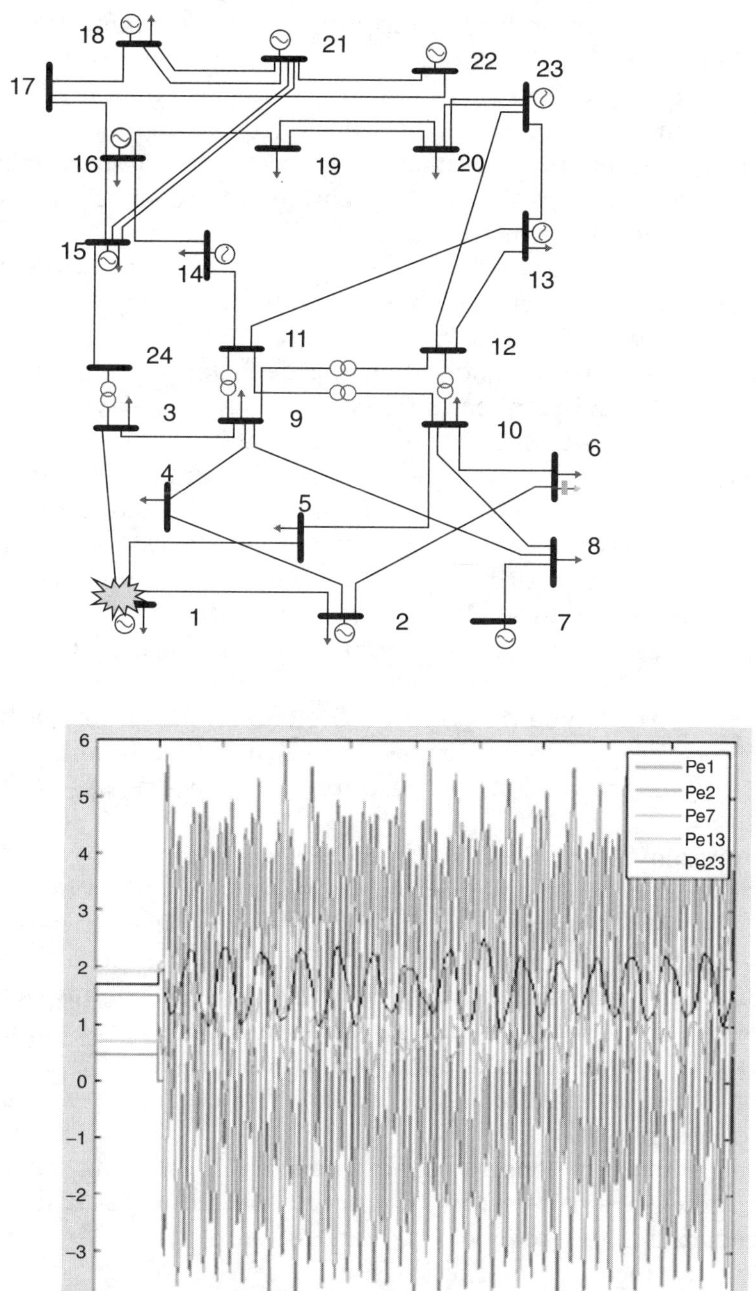

Figure 2.11: *Inter-area oscillations in the system with conventional controllers*

synchronized is critical under these conditions. Here again, the selection of the energy function is fundamentally based on the physics: The best energy function found was the ectropy of the remaining system when the module controller is being designed [Cvetkovic11, Ilic12]. Thus, if the action of the high-gain controller of the module creates minimal disorder in the rest of the system, the interconnected system will not see the effects of the disturbance. Figures 2.12 and 2.13 illustrate a small example in which the system experiences a short-circuit fault, along with its response with and without the nonlinear controller proposed in [Cvetkovic11, Ilic12].

A general multilayered modeling approach for representing the dynamics of any SEES architecture modeled using low-layer detailed models to represent the diverse and unique characteristics of (groups of) users, as well as the higher-layer models representing their interactions only, can be used for coordinated control of interaction variables when necessary to reduce the high-gain control and fast storage requirements. When system disturbances occur on a continuous basis, the well-established hierarchical electric power system modeling of today cannot cope effectively; such a system is typically based on strong assumptions that the higher layers are quasi-stationary and the lower layers only are truly dynamic. When those conditions no longer hold, viewing the entire complex system as being driven by dynamic temporal and spatial interactions of groups of components becomes essential.

The upper layer in the cyber-structure then senses and communicates the dynamics of interaction variables between different system modules; the information exchange required is defined in terms of these interaction variables. The local variables and their local controllers are fundamentally distributed as long as the effects of interactions do not overpower the local dynamics of distributed modules. Figure 2.14 provides a zoomed-in sketch of a local distributed DyMonDS for given interaction variables, and a zoomed-out sketch of the higher-level coordination of the modules.

Several communications and control designs for ensuring stable frequency response to small perturbations can be found in [Ilic12a]. These designs ensure that the frequency deviations return to zero in response to step disturbance or, alternatively, remain within the acceptable threshold in response to persistent small disturbances caused by wind power fluctuations. The designs use the structure-preserving modeling framework based on the modularization of the system into two

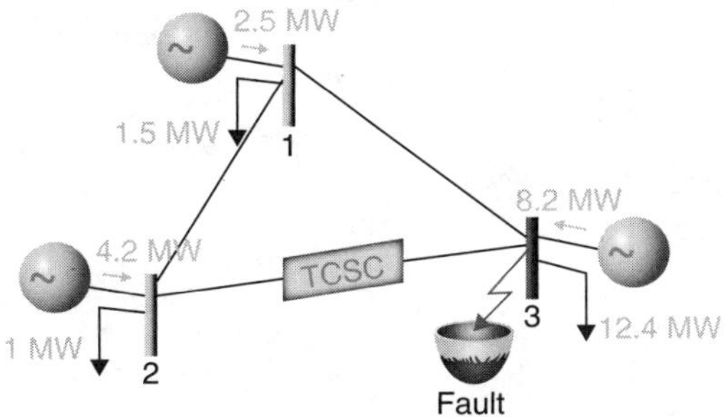

Figure 2.12: *A short-circuit fault in a small power system [Cvetkovic11, Ilic12]*

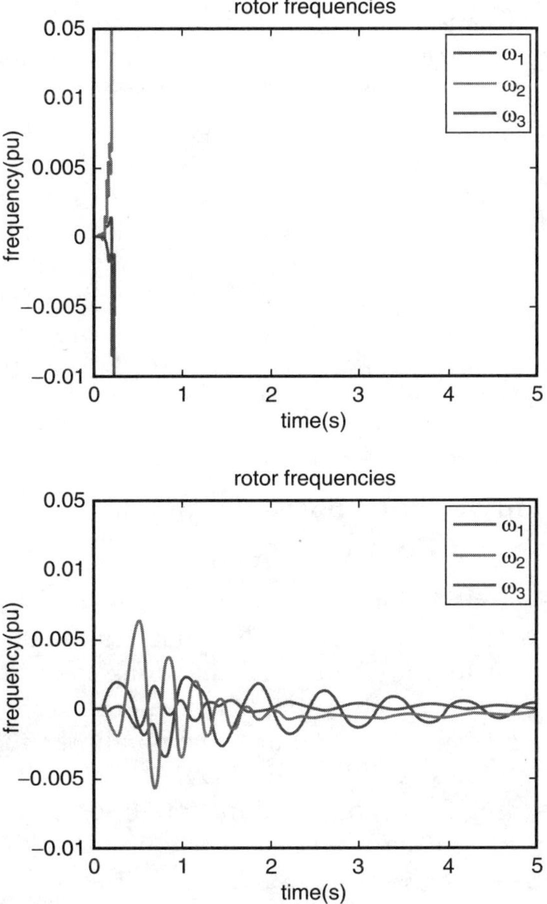

Figure 2.13: *Frequency response without and with ectropy-based controller [Cvetkovic11, Ilic12]*

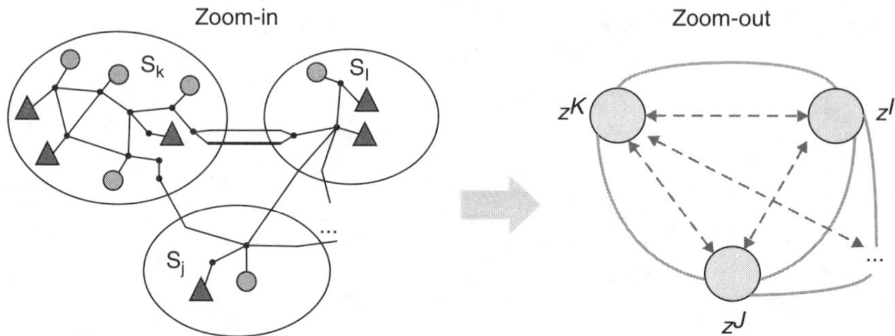

Figure 2.14: *Multilayered approach to coordinating interactions [Ilic12a]*

layers—the coordinating layer and the component-level layer. Such a systematic framework for determining the control responsibilities for each layer is essential for ensuring system-level stability by means of enhanced distributed and coordinated controllers, even when the system is subjected to persistent disturbances. For intriguing assessments of the tradeoffs between the communications complexity needed for coordination, the complexity of local controllers, and the QoS, see [Ilic12a]. Researchers are currently investigating generalizations of this multilayered approach to ensuring transient stability of complex power grids in response to large faults and/or sudden large-amplitude wind disturbances; this approach may become critical for systematic integration of power-electronically switched controllers for stabilization of system frequency and voltage [Baros14].

2.4.3 Interaction Variable–Based Automated Modeling and Control

Anyone working with large-scale electric power systems is well aware of the fact that deriving dynamic models that capture the physical phenomena of interest in sufficient detail has long been the major challenge. Adopting the perspective that future energy systems will consist of multilayered interactive modular systems allows for representing highly diverse modules and their local DyMonDS in as much detail as desired and defining input/output functionalities in terms of interaction variables between the modules for shaping the dynamics of the interconnected system. Coordination of interaction variables generally requires much coarser models.

To meet the need for good modeling, an automated approach for symbolically deriving the standard state space model of an electric power system using the Lagrangian formulation was recently introduced in [Bachovchin14]. The value of this automated approach is that deriving the state space model that is then given to the system control designer has often been a major roadblock in power systems, particularly for large systems, since the governing equations are complex and nonlinear. Since the governing equations for power systems are nonlinear, symbolically solving for the state space model of a large power system using the Lagrangian formulation, as described in [Ilic12a], can be very computationally intensive.

For this reason, an automated modular approach for deriving the state space model was implemented, as described in [Bachovchin14]. Using this approach, the system is divided into modules: The state space model for each module is determined separately, and then the state space models from each module are combined in an automated procedure. This modular approach is particularly useful for power systems because large systems contain many of the same types of components.

Additionally, this approach can be used to obtain dynamic models suitable for control design. In particular, it can be used to derive a model that includes Hamiltonian dynamics. The resulting model significantly simplifies controller design by explicitly capturing the dynamics of the outputs of interest, such as the Hamiltonian, which is the total accumulated energy.

The dynamic equations of several common power system components are given in [Bachovchin14]. These dynamic equations can be found using the Lagrangian approach, or they can be calculated using force laws combined with conservation of energy to determine the coupling between the electrical and mechanical subsystems. Most recently, this automated modeling approach has been used to design power-electronically controlled switches of flywheels to ensure synchronism in power grids with induction-type wind power plants, which are otherwise vulnerable to prolonged fast power imbalances.

As a general note, the overall interaction variables–based modeling and cyber design for sustainable electric energy systems described here remain a work in progress. For the first time we are beginning to relate modeling for sustainability in terms of coarse interaction variables to the modeling in support of making physical grids stronger and smarter. While the initial idea was appealing but provided only the first step in this direction, it is becoming clear that effective design of cyber-physical systems for sustainability must be based on deep understanding of the underlying physics.

2.5 Practitioners' Implications

The new SCADA described in this chapter is fundamentally different from today's SCADA. With the new framework, operators and planners make decisions based on proactive and binding information exchange with system users; this makes it possible to account for choices based on value, while simultaneously meeting difficult societal goals. The information exchange is repeated periodically, as decisions are being made for very long future, or for closer to real time. This process becomes a win–win situation for all, and future uncertainties are distributed over many stakeholders according to the risks they are willing to bear and according to the expected value from the selected technologies. Ultimately, they lead to qualitatively new, reliable, efficient, and clean services, and those technologies that bring value survive beyond the subsidy stages.

2.5.1 IT-Enabled Evolution of Performance Objectives

Multilayered modeling that requires coordination of interaction variables, as described earlier, assumes that boundaries of cooperative (groups of) components with common objectives are given; in addition, the information exchange patterns within the complex system are constrained by various governance rules, rights, and responsibilities. Over the longer term, one might consider bottom-up interactive dynamic aggregation for optimizing common goals, as well as for providing quantifiable feedback to the governance system regarding the effects of its rules on system performance. With this approach, performance objectives are shaped through interactions, and the system evolves into a more sustainable SEES. Much research remains to be done to extend interaction variables–based modeling and design so that they influence the governance system for the electric power industry, the preferences of system users, and the responsiveness of the T&D grid at value.

2.5.2 Distributed Optimization

This chapter has stressed the fact that optimizing resources in highly uncertain environments and ensuring that intertemporal and spatial dependencies (constraints) are taken into account are computationally extremely challenging. To overcome this complexity, our DyMonDS framework can be used to internalize time-scale and uncertainty-related complexities that might prevent controllable generation and transmission from responding as needed. In particular, this model proposes to embed sequential decision making into local DyMonDS that face uncertainties, with the aim of enabling the functionalities of controllable equipment needed for online resource management by the utility's control center without imposing temporal constraints.

The simplest example might be a power plant that commits to a time-varying range of power capacity limits that it can produce without requesting the system operator to observe its ramp rates; the ramp rates are accounted for and internalized when the commitment is made. The electricity users who participate in demand response can do the same. For the theoretical foundations of such distributed management by internalizing intertemporal complexities and uncertainty, see [Ilic13a, Joo10]. The interactions between system users and the system operator can be either pointwise or functional. Most of the distributed optimization approaches are fundamentally pointwise in nature: Each user performs the necessary optimization for the assumed system conditions, and the optimal request for power quantity is submitted. The system operator then collects the requests optimized by the users, and computes the resulting power mismatch. The electricity price is updated accordingly. The larger the supply–demand shortage, the higher the system price is in the next iterative step. Conversely, the larger the supply–demand excess, the lower the price is in the next iterative step.

Under strong convexity assumptions, this process is known to result in system-wide optimum [Joo10]. However, the implementation of pointwise distributed balancing of supply–demand would require a great deal of communication between the system users and the operator. As an alternative, the DyMonDS framework is based on exchanging functions between layers (demand and supply functions for generation and even transmission) [Ilic11, Ilic11a]. This approach eliminates the need for multiple iterations between the modules and makes it feasible to arrive at the consensus relatively rapidly. Figure 2.5 illustrated the newly proposed SCADA with embedded DyMonDS.

2.6 Summary and Open Challenges

Due to space limitations, we are not able to even briefly discuss many different aspects of the CPS for the emerging electric energy systems. We hope that the reader is beginning to realize that the systems approach to designing end-to-end CPS is critical to the integration of many candidate technologies and realization of their maximum value. Loosely speaking, any given technology has a value to the system, which can be measured in terms of cumulative improvements in the desired system performance with and without the candidate technology of interest over the specified decision time. Most of the cyber-technologies offer flexibility, which is essential in conditions of uncertain operation, and contribute to the savings in capacity investments. Figure 2.15 shows what a sample CPS on a small system might look like, briefly summarizing our vision of the future. Various sensors (e.g., PMUs, DLRs) are embedded at the appropriate locations to make the system more observable than it is today, and to use the online estimation about the state to adjust many physical resources so as to make the most out of what one has.

Much remains to be done before systematic methods will become available that decide how many sensors are needed and where they should be placed to realize the largest information gain in the uncertain environment. State estimation based on learning from historic and incoming data holds the promise of providing much better wide-area situational awareness about the system [Weng14]. Nevertheless, there are many unanswered questions about the integration of smart wires, controllers, and fast storage with well-understood end-to-end functionalities. Typical disturbances to this kind of system are shown in Figure 2.16. Decomposing these signals into their predictable components, which can then be controlled in a model-predictive feed-forward way with slower, less expensive

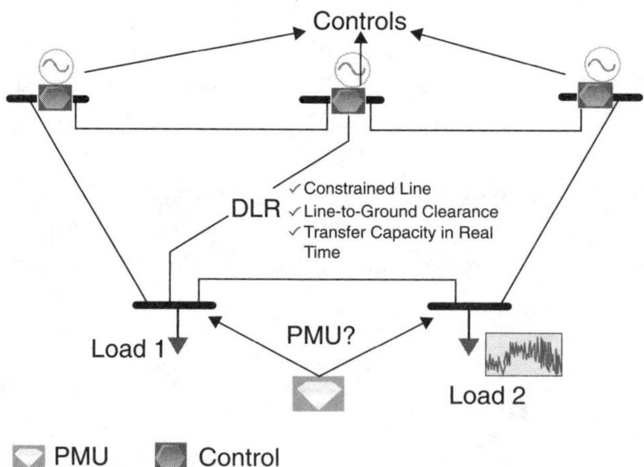

Figure 2.15: *A future CPS for sustainable SEES*

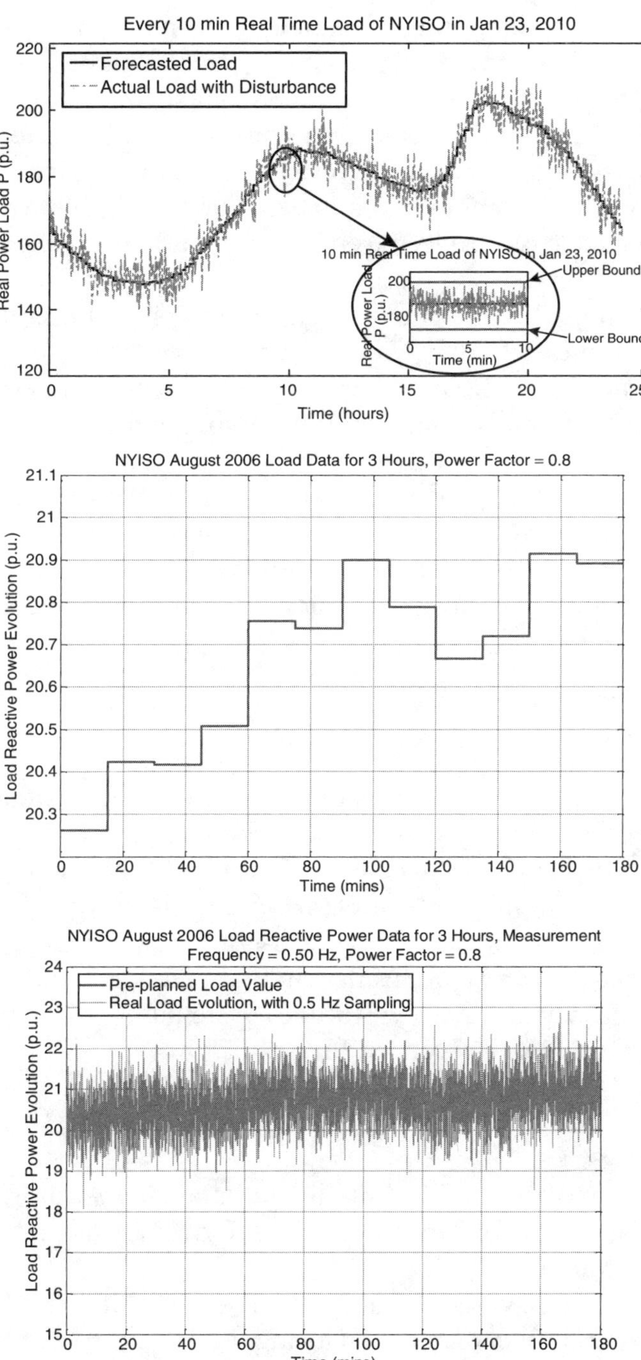

Figure 2.16: *Actual signal: how much can we learn and how much is unpredictable?*

resources, and handling the remaining hard-to-predict disturbances, which require fast automation and storage, will be highly dependent on how good the predictions are. Data-driven, online, multilayered learning and dynamic aggregation is likely to become a major cyber tool supporting sustainable energy services. Nevertheless, the biggest remaining challenge is likely how to combine physics-based knowledge with data-driven learning and enable future electric energy systems to make the most out of what is available in the long run.

Once unthinkable concepts, such as assessing feasibility of the power flow in a distributed way and adjusting smart wires in a complex grid to ensure feasible power flow, actually may make sense and may be possible to bring to the market. Also, the discussion regarding the inherent structure underlying physics-based multilayered interactive power grids could lead to breakthroughs in the form of less complex cyber requirements. Electric energy systems, because of their overwhelming complexity and diversity, will continue to represent a challenge to the state of art in CPS for quite some time to come.

References

[Bachovchin14] K. D. Bachovchin and M. D. Ilic. "Automated Modeling of Power System Dynamics Using the Lagrangian Formulation." EESG, Working Paper No. R-WP-1-2014, February 2014. *International Transactions on Electrical Energy Systems*, 2014.

[Bachovchin15] K. Bachovchin. "Design, Modeling, and Power Electronic Control for Transient Stabilization of Power Grids Using Flywheel Energy Storage Systems." PhD Thesis, ECE Carnegie Mellon University, June 2015.

[Barmack03] M. Barmack, P. Griffes, E. Kahn, and S. Oren. "Performance Incentives for Transmission." *Electricity Journal*, vol. 16, no. 3, April 2003.

[Baros14] S. Baros and M. Ilic "Intelligent Balancing Authorities (iBAs) for Transient Stabilization of Large Power Systems." IEEE PES General Meeting, 2014.

[Cvetkovic11] M. Cvetkovic and M. Ilic. "Nonlinear Control for Stabilizing Power Systems During Major Disturbances." IFAC World Congress, Milan, Italy, August 2011.

[Doffler14] F. Doffler. "LCCC Presentation." Lund, Sweden, October 2014.

[helio14] HelioPower. http://heliopower.com/energy-analytics/.

[Hogan00] W. W. Hogan. "Flowgate Rights and Wrongs." Kennedy School of Government, Harvard University, 2000.

[Ilic05] M. Ilic. "Automating Operation of Large Electric Power Systems Over Broad Ranges of Supply/Demand and Equipment Status." In *Applied Mathematics for Restructured Electric Power Systems*. Kluwer Academic Publishers, pages 105–137, 2005.

[Ilic09] M. Ilic. "3Rs for Power and Demand." *Public Utilities Fortnightly Magazine*, December 2009.

[Ilic11] M. Ilic. "Dynamic Monitoring and Decision Systems for Enabling Sustainable Energy Services." *Proceedings of the IEEE*, January 2011.

[Ilic13] M. Ilic. "Toward Standards for Dynamics in Electric Energy Systems." PSERC Project S-55, 2013–2015.

[Ilic14] M. Ilic. "DyMonDS Computer Platform for Smart Grids." *Proceedings of Power Systems Computation Conference*, 2014.

[Ilic05a] M. Ilic, H. Allen, W. Chapman, C. King, J. Lang, and E. Litvinov. "Preventing Future Blackouts by Means of Enhanced Electric Power Systems Control: From Complexity to Order." *Proceedings of the IEEE*, vol. 93, no. 11, pages 1920–1941, November 2005.

[Ilic12a] M. D. Ilic and A. Chakrabortty, Editors. *Control and Optimization Methods for Electric Smart Grids*, Chapter 1. Springer, 2012.

[Ilic12] M. Ilic, M. Cvetkovic, K. Bachovchin, and A. Hsu. "Toward a Systems Approach to Power-Electronically Switched T&D Equipment at Value." IEEE Power and Energy Society General Meeting, San Diego, CA, July 2012.

[Ilic98] M. Ilic, F. Galiana, and L. Fink. *Power Systems Restructuring: Engineering and Economics*. Kluwer Academic Publishers, 1998.

[Ilic11a] M. Ilic, J. Jho, L. Xie, M. Prica, and N. Rotering. "A Decision-Making Framework and Simulator for Sustainable Energy Services." *IEEE Transactions on Sustainable Energy*, vol. 2, no. 1, pages 37–49, January 2011.

[Ilic13a] M. D. Ilic, X. Le, and Q. Liu, Editors. *Engineering IT-Enabled Sustainable Electricity Services: The Tale of Two Low-Cost Green Azores Islands*. Springer, 2013.

[Ilic00] M. Ilic and J. Zaborszky. *Dynamics and Control of Large Electric Power Systems*. Wiley Interscience, 2000.

[Joo10] J. Joo and M. Ilic. "Adaptive Load Management (ALM) in Electric Power Systems." 2010 IEEE International Conference on Networking, Sensing, and Control, Chicago, IL, April 2010.

[Lasseter00] R. Lasseter and P. Piagi. "Providing Premium Power Through Distributed Resources." *Proceedings of the 33rd HICSS*, January 2000.

[Lee00] E. A. Lee and Y. Xiong. "System-Level Types for Component-Based Design." Technical Memorandum UCB/ERL M00/8, University of California, Berkeley, 2000.

[Lopes06] J. A. P. Lopes, C. L. Moreira, and A. G. Madureira. "Defining Control Strategies for MicroGrids Islanded Operation." *IEEE Transactions on Power Systems*, May 2006.

[Murray95] R. Murray, M. Rathinam, and W. Sluis. "Differential Flatness of Mechanical Control Systems: A Catalog of Prototype Systems." ASME, 1995.

[nest14] Nest Labs. https://nest.com/.

[Ostrom09] E. Ostrom, et al. "A General Framework for Analyzing Sustainability of Socio-ecological Systems." *Science*, vol. 325, page 419, July 2009.

[Prasad80] K. Prasad, J. Zaborszky, and K. Whang. "Operation of Large Interconnected Power System by Decision and Control." IEEE, 1980.

[Smart14] Smart Grids and Future Electric Energy Systems. Course, 18-618, Carnegie Mellon University, Fall 2014.

[Weng14] W. Yang. "Statistical and Inter-temporal Methods Using Embeddings for Nonlinear AC Power System State Estimation." PhD Thesis, Carnegie Mellon University, August 2014.

[Whang81] K. Whang, J. Zaborszky, and K. Prasad. "Stabilizing Control in Emergencies." Part 1 and Part 2, IEEE, 1981.

[Xie09] L. Xie and M. D. Ilic. "Module-Based Interactive Protocol for Integrating Wind Energy Resources with Guaranteed Stability." In *Intelligent Infrastructures*. Springer, 2009.

[Yu99] C. Yu, J. Leotard, M. Ilic. "Dynamics of Transmission Provision in a Competitive Electric Power Industry." In *Discrete Event Dynamic Systems: Theory and Applications*. Kluwer Academic Publishers, 1999.

[Zecevic09] A. Zecevic and D. Siljak. "Decomposition of Large-Scale Systems." In *Communications and Control Engineering*. Springer, 2009.

Chapter 3

Cyber-Physical Systems Built on Wireless Sensor Networks

John A. Stankovic

In this chapter, wireless sensor network (WSN) technologies are discussed, with emphasis on several of the key themes found in cyber-physical systems (CPS) research. These themes include dependencies on the physical world, the multidisciplinary nature of CPS, open systems, systems of systems, and humans in the loop. To understand the motivation behind these technologies, several operational scenarios are given, relating each of them to the corresponding technologies they have produced. Examples include the basic results that exist in WSN for medium access control, routing, localization, clock synchronization, and power management. These topics are considered core services found in a WSN. The results presented in this chapter are not comprehensive, but rather have been chosen to relate WSN to CPS issues.

The key drivers and quality attributes are then presented so that the reader can understand the criteria that researchers in this field use to guide their work. These factors include aspects such as the dependence of the cyber world on the physical world, as well as more traditional aspects of systems such as real-time operation, runtime validation, and security. Finally, this chapter describes the implications for practitioners with respect to programming abstractions.

3.1 Introduction and Motivation

The confluence of the right technologies at the right time can give rise to disruptive and transforming innovations. This is the case with cyber-physical systems [Stankovic05]. While many current systems can be considered CPS, it is the exciting prospect of combining low-cost sensing and actuation, wireless communications, and ubiquitous computing and networking that has the potential to lead to truly revolutionary CPS. This combination of technologies has been under development in the wireless sensor networking community for a number of years. Consequently, one important CPS domain is building upon wireless sensor networks as a CPS infrastructure.

A central issue in CPS research is the question of how the cyber world depends on the physical world, and vice versa. The physical world imposes many difficult and uncertain situations that must be handled in the cyber world in a safe and real-time manner, and often with an adaptive and evolutionary approach. As CPS becomes more open, there is an increasing need for better descriptions of the environmental effects and the means by which software must deal with them. This is not a hardware–software interface issue, but rather an environment–hardware–software interface issue. It is also well recognized that to build reliable and safe CPS requires multidisciplinary expertise and a new technical community that integrates experts from at least the following disciplines: computer science, wireless communications, signal processing, embedded systems, real-time systems, and control theory. CPS solutions often rely on some combination of ideas or results from two or more of these areas.

As WSNs proliferate, they are beginning to appear in most areas of daily life. These networks are open in the sense that they dynamically interact with the environment, the Internet, other systems, and people.

This openness gives rise to several other key technical problems of dealing with systems of systems and humans in the loop.

3.2 System Description and Operational Scenarios

Over the past 10 years, many advances in WSNs have been made on many fronts. WSNs provide an important platform upon which to build CPS. Thus, in this section, some of the key functions of WSN are described, but with an emphasis on how the physical world affects those functions. In particular, medium access control, routing, localization, time synchronization, and power management are discussed. For all of these cases, it is assumed that a WSN consists of many nodes organized into a multi-hop wireless network. Each node contains one or more sensors, possibly actuators, one or more wireless communication transceivers that it uses to communicate with neighboring nodes, computation capability (typically a microcontroller), memory, and a power source (batteries, energy-scavenging hardware, or sometimes a wall outlet).

CPS built upon WSN are used in many ways. To provide an overarching view of how such systems are used, operational scenarios are presented from the home health care, surveillance and tracking, and environmental science application areas. In presenting these scenarios, we consider the sensing, networking, and actuation aspects of the system. Of course, for each area many variations are possible.

- *Home health care:* Motion and temperature sensors are placed in each room of a home; contact sensors are placed on doors, cabinets, and appliances; accelerometers and pressure pads are placed on the beds; infrared sensors are placed on entrance and exit doors; medication sensors and radio-frequency identification (RFID) tags and readers are selectively placed; and a visual display, alarm horn, and lights act as actuators. Data are initially collected to learn about the person's normal activities of daily living, such as eating, sleeping, and toileting. Once the system has learned a person's normal behaviors, it monitors for anomalies. If an anomaly is detected, the system sends a message to the display as an intervention (actuation). The system might also activate lights to help avoid falls or improve mood. If an unsafe condition is detected, the horn sounds an alarm. In small homes or apartments, the wireless communications may be single hop; in a larger setting, a multi-hop network may be necessary. Figure 3.1 depicts this kind of "smart home" for health care.

- *Surveillance and tracking:* Motion, proximity, acoustic, magnetic, temperature, and camera sensors might be quickly deployed over a very large area (e.g., a port, a country's border, a forest, or a remote valley). Once deployed, the nodes determine their location via an appropriate localization protocol and self-organize into a wireless multi-hop network. Subsequent multimodal sensor processing detects moving objects, tracks them, and classifies them. If the classified objects are of interest, such as a truck in the wrong place, people crossing a border, a fire in a forest, or a tank entering a valley, then actuation is initiated. Actuation might include dispatching police, fire fighters, or soldiers. Figure 3.2 illustrates this kind of energy-efficient surveillance system.
- *Environmental science:* Water height sensors are installed at a (reservoir) dam and flow rate sensors are installed in each of the rivers and streams feeding into the reservoir. A controller at a base station node predicts how much water should be released based on the current water height and the readings from all the inflows. The rate flow sensors transmit data to the base station via a fixed multi-hop network.

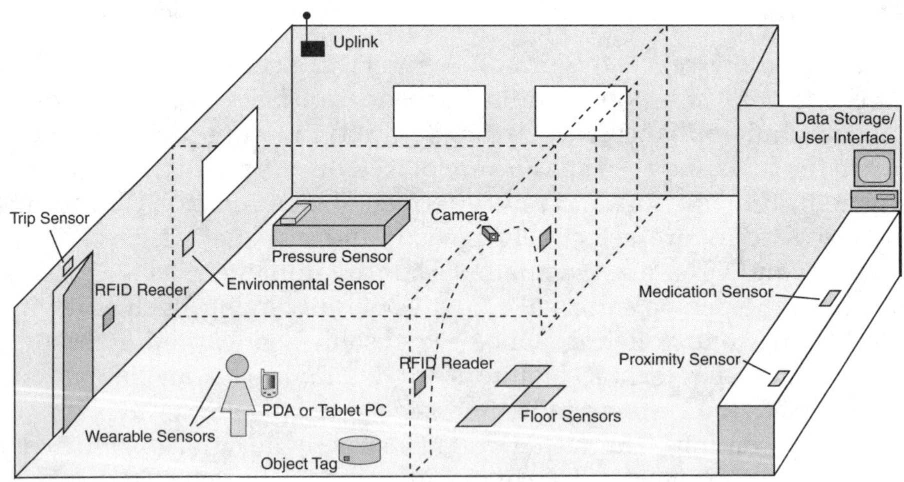

Figure 3.1: *Smart home for health care*

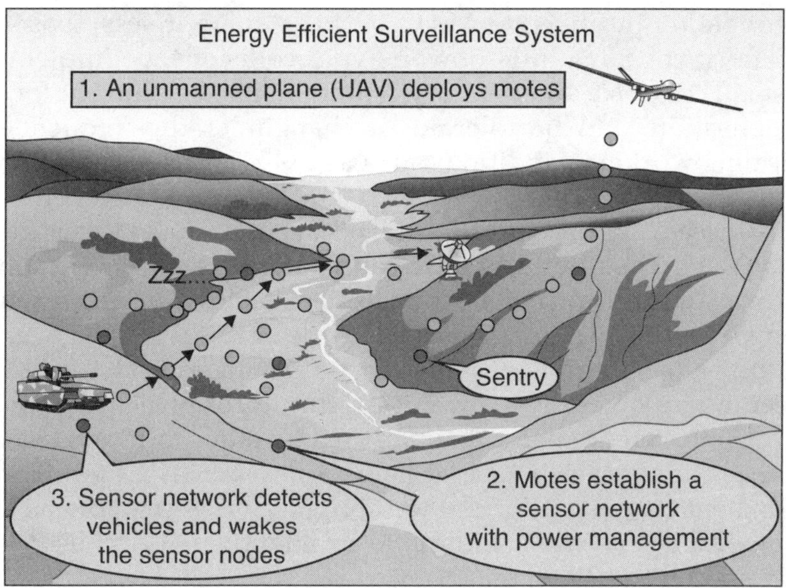

Figure 3.2: *Surveillance, tracking, and actuation*

3.2.1 Medium Access Control

A medium access control (MAC) protocol coordinates the actions of nearby nodes over a shared channel. The most commonly found MAC solutions are contention based. A general approach for a contention-based MAC is for a node that has a message to transmit to test the channel to see if it is busy: if it is not busy, then the node transmits its data; if it is busy, the node waits and tries again later. If two or more nodes transmit data and their messages collide, then those nodes wait random amounts of time before resending their packets, thereby attempting to avoid a subsequent collision. To conserve energy, some wireless MAC protocols also implement a *doze* mode, in which nodes not involved with sending or receiving a packet in a given time frame go into sleep mode. Many variations on this basic MAC scheme exist.

Many MAC protocols—but especially those for wired and ad hoc networks—are optimized for the general case and for arbitrary communication patterns and workloads. A WSN, however, typically has more focused requirements: It includes a local unicast or broadcast, traffic is generally from widely dispersed nodes to one or a few sinks (most traffic is then in one direction), it features periodic or rare event-based communication, and energy consumption is a major factor.

Consequently, an effective MAC protocol for wireless sensor networks must consume little power, avoid collisions, be implemented with small code size and memory requirements, be efficient for a single application, and be tolerant of changing radio-frequency and physical networking conditions such as noise levels and interference patterns.

It is these physical world issues, in fact, that make this a difficult CPS problem. The physical world causes wireless communications to perform in an uncertain manner due to many factors, including whether the device is indoors or outdoors, where the device is located (e.g., high in the air, on the ground, on a human body), which obstacles intervene between nodes (e.g., people, vehicles, furniture), which dynamically changing interference patterns are caused by other co-located systems, and what the weather is like, to name a few salient issues. Since the environment is dynamic and uncertain, it is necessary for the cyber-technologies in the MAC solution to be adaptive and robust to these conditions.

An example of a good MAC protocol for wireless sensor networks is B-MAC [Polastre04]. B-MAC is highly configurable and can be implemented with a few lines of code with a small memory footprint. Its interface allows dynamic choices of various functions as needed by a particular application. B-MAC consists of four main parts: clear channel assessment (CCA), packet backoff, link layer acknowledgments, and low-power listening. For CCA, B-MAC uses a weighted moving average of samples when the channel is idle to assess the background noise and efficiently detect valid packets and collisions. This is a clear example of the cyber world being implemented to deal with a physical world property, so that the system can adapt to noise levels as they change. The packet backoff time is configurable and is chosen from a linear range, rather than relying on the exponential backoff scheme typically used in other distributed systems. This approach reduces delays and works because of the typical communication patterns found in a wireless sensor network. B-MAC also supports a packet-by-packet link layer acknowledgment, so that only important packets need to pay the extra acknowledgment (ACK) cost. A low-power listening scheme is employed in which a node cycles between awake and sleep states. While awake, it listens for a long enough preamble to assess if it needs to stay awake or can return to sleep mode. This scheme saves significant amounts of energy.

Many non-WSN MAC protocols use a request to send (RTS) and clear to send (CTS) style of interaction. This approach works well for ad hoc mesh networks where packet sizes are large (thousands of bytes). However, the overhead of RTS-CTS packets needed to set up a packet transmission is not acceptable in wireless sensor networks, where packet sizes are on the order of 50 bytes. B-MAC, therefore, does not use a RTS-CTS scheme.

Means to support multichannel wireless sensor networks have also drawn researchers' attention. In these systems, it is necessary to extend MAC protocols to multichannel MACs. One such protocol is multifrequency media access control for wireless sensor networks (MMSN) [Zhou06]; the details of MMSN are complicated and are not described here. Multichannel protocols must support all of the features found in protocols such as B-MAC, but must also assign frequencies for each transmission. Consequently, multifrequency MAC protocols typically consist of two phases: channel assignment and access control. In the future, an increasing number of wireless sensor networks, and hence CPSs, are likely to employ multiple channels (frequencies). The advantages of multichannel MAC protocols include providing greater packet throughput and being able to transmit even in the presence of a crowded spectrum, perhaps arising from systems of systems interactions in the same physical area.

3.2.2 Routing

Multi-hop routing is a critical service required for WSNs. Consequently, a large amount of work has addressed this topic. First, consider two other systems that have developed many routing algorithms: the Internet and wireless mobile ad hoc network (MANET) systems. Internet routing techniques do not perform well in WSN. Such routing assumes the availability of highly reliable wired connections so that packet errors are rare. When wireless communication occurs in open environments (the physical world), the assumption of rare packet errors does not hold true for WSNs. For wireless MANET systems, many routing solutions depend on symmetric links (i.e., if node A can reliably reach node B, then node B can reach node A) between neighbors; this, too, is often not true for WSN. Asymmetry is yet another physical world constraint that must be addressed in CPS routing protocol software.

For WSN, which are often deployed in an ad hoc fashion, routing typically begins with neighbor discovery. Nodes send rounds of messages (packets) to each other and build local neighbor tables. At a minimum, these tables include each neighbor's ID and location. Nodes, therefore, must know their geographic location (a physical property) prior to neighbor discovery. Other information typically found in these tables includes nodes' remaining energy (a physical property), delay via that node (partly dependent on the physical environment, and which might necessitate retries), and an estimate of link quality (a function of the environment, power levels, and other factors). Once the routing tables are set up, in many WSN routing algorithms, messages are directed from a source location to a destination based on geographic coordinates, not IDs.

A typical WSN routing algorithm is geographic forwarding (GF) [Karp00]. In GF, a node is aware of its location, and a message contains the destination (geographic) address. A receiving node then computes which neighbor node makes the most progress toward the destination by using the distance formula from geometry. It then forwards the message to this next hop. Many variants of GF exist. For example, a node choosing a next hop could also take into account time delays, reliability of the link, and remaining energy.

Another important routing paradigm for WSN is directed diffusion [Intanagonwiwat00]. This solution integrates routing, queries, and data aggregation. With this approach, a query is disseminated throughout the network indicating an interest in data from remote nodes. A node with the appropriate requested data responds with an attribute–value pair. This attribute–value pair is drawn toward the requestor based on gradients, which are set up and updated during query dissemination and response. Along the path from the source to the destination, data can be aggregated to reduce communication costs. Data may also travel over multiple paths, thereby increasing the robustness of routing. Since directed diffusion is a generic framework, it can be used to implement software that addresses many CPS issues caused by the environment, such as dynamic link quality, asymmetric links, and voids.

Routing in WSNs often simultaneously addresses multiple issues such as reliability; integration with wake/sleep schedules; support for unicast, multicast, and anycast semantics; meeting real-time requirements; handling node mobility; and dealing with voids. Most CPS built on WSN require routing solutions that treat all of these issues in a holistic manner. Here, we briefly consider reliability and void issues, as they illustrate several critical CPS issues:

- *Reliability:* Since messages travel multiple hops it is important to have high reliability on each link; otherwise, the probability of end-to-end reception would be unacceptably low. Significant work has been done to identify reliable links using metrics such as received signal strength, link quality index (based on "errors"), and packet delivery ratio. Then, next hops are chosen using high-quality links. If necessary, a node might increase its power or even wait until the environment changes in a way such that a previously poor link becomes good. Retries also improve hop-by-hop reliability. However, being cognizant of the physical phenomenon of burst losses, some solutions wait for some time before retrying; in other words, they wait out the condition that is causing the burst losses.
- *Voids:* Since WSN nodes have a limited transmission range, for some nodes in the routing path there might potentially be no forwarding nodes in the direction a message is supposed to travel. This kind of void may occur due to failures, temporary obstacles that block communication, or nodes that are in the sleep state. Protocols such as greedy perimeter stateless routing (GPSR) [Karp00a] solve this problem by choosing some other node *not* in the correct direction in an effort to find a path around the void. For example, GPSR uses a left-hand rule to try to find a new route. Note that a void is a physical real-world situation; if it occurs and the routing protocol (the cyber element) does not account for voids, then the system will fail. In this failure case, it can be said that the routing protocol (the cyber) is not aware of the physical property of voids.

3.2.3 Node Localization

In WSNs, node localization is the problem of determining the geographical location of each node in the system. Localization is one of the most fundamental and difficult problems that must be solved for WSNs. It is a function of many parameters and requirements, potentially making its resolution very complex. For example, the following issues should be considered as part of node localization: What is the cost of extra localization hardware? Do beacons (nodes that know their locations) exist and if so, how many and what are their communication ranges? Which degree of location accuracy is required? Is the system indoors or outdoors? Is there a line of sight among the nodes? Is it a 2D or 3D localization problem? What is the energy budget (number of messages)? How long should it take to localize? Are clocks synchronized? Does the system reside in hostile or friendly territory? Which error assumptions are being made? Is the system subject to security attacks?

For some combinations of requirements and issues, the localization problem is easily solved. For example, if cost and form factor are not major concerns and accuracy of a few meters is acceptable, then for most outdoor systems, equipping each node with global positioning system (GPS) capabilities is a simple answer. If the system is manually deployed, one node at a time, then a simple GPS node carried by the deployer can automatically localize each node in turn, via a solution called walking GPS [Stoleru04]. While simple, the walking GPS solution is elegant and avoids any manual keying of the location for each node. To handle the robustness requirements of CPS, the walking GPS solution addresses missing satellite signals by nearby node cooperation when needed.

Most solutions for localization in WSN are either range based or range free. Range-based schemes use various techniques to first determine the distances between nodes (range) and then compute node locations using simple geometric principles [Stoleru05]. To determine the distances, extra hardware is usually employed—for example, hardware to detect the time differences in the arrivals of sound and radio waves. This difference can then be converted to a distance measurement. Physical world realities, however, often lead to poor distance estimates—a problem that must be addressed in some manner.

In range-free schemes, distances are not determined directly; instead, hop counts are used [Bulusu00]. Once the hop counts are determined, the distances between nodes are estimated using an average distance per hop. Based on this average distance value, simple geometric principles are used to compute locations. Range-free solutions are not as accurate as range-based solutions and often require more messages, thereby consuming more energy. However, they do not require extra hardware or a clear line of sight for every node.

The locations of nodes are used in a variety of system functions (e.g., routing, group sensor fusion among nearby nodes, sleep scheduling) and application semantics (e.g., tracking and location of an event); thus, node localization is a central service in WSN. Many localization solutions rely on accurately dealing with physical world properties such as the relationship between signal strength and distance.

3.2.4 Clock Synchronization

The clocks of each node in a WSN should read the same time within epsilon and remain that way. Since the accuracy of clocks tends to drift over time, they must be periodically resynchronized. In some instances when very high accuracy is required, it is even important for nodes to account for clock drift between synchronization periods. Clock synchronization is important for many reasons. When an event occurs in a WSN, it is often necessary to know where and when it occurred. Clocks are also used for many system and application tasks. For example, sleep/wake schedules, some localization algorithms, and sensor fusion are some of the services that often depend on synchronized clocks, as do application tasks such as tracking and computing velocity.

The Network Time Protocol (NTP) [Mills94], which is used to synchronize clocks on the Internet, is too heavyweight for WSN. Likewise, giving GPS capabilities to every node is often too costly. As alternatives, representative clock synchronization protocols have been developed specifically for WSN, including Reference Broadcast Synchronization (RBS) [Elson02], Timing-Sync Protocol for System Networks (TPSN) [Ganeriwal03], and Flooding Time Synchronization Protocol (FTSP) [Maroti04].

In RBS, a reference time message is broadcast to neighbors. Receivers record the time when the message is received. Nodes then exchange their recorded times and adjust their clocks to synchronize. This protocol does not suffer from transmitter-side nondeterminism, as timestamps are implemented only on the receiver side. Accuracies are approximately 30 microseconds for 1 hop. RBS does not address the possibility of multi-hop systems, but could be extended to cover such systems.

In TPSN, a spanning tree is created for the entire network. This solution assumes that all links in the spanning tree are symmetric. Pairwise synchronization is then performed along the edges of the tree starting at the root. Since there is no broadcasting as in RBS, TPSN is expensive. A key attribute of this protocol is that the timestamps are inserted into outgoing messages in the MAC layer, thereby reducing nondeterminism. Accuracy is in the range of 17 microseconds.

In FTSP, there are radio-layer timestamps, skew compensation with linear regression, and periodic flooding to make the protocol robust to failures and topology changes. Both transmission and reception of messages are timestamped in the radio layer, and differences in these times are used to compute and adjust clock offsets. Accuracy is in the range of 1–2 microseconds.

Determining the amount of clock drift and deciding whether the cyber-technologies must handle drift between synchronization times are CPS issues. The accuracy of these solutions versus the cost tradeoffs necessary when they are implemented, as well as the robustness of the solutions, are key CPS research questions.

3.2.5 Power Management

Many sensor devices that are used in WSN run on two AA batteries. Depending on the activity level of a node, its lifetime may be only a few days if no power management schemes are used. Since most systems require much longer lifetimes, significant research has been undertaken in an effort to increase the nodes' lifetime while still meeting functional requirements. At the hardware level, it is possible to add solar cells or scavenge energy in many ways, such as from motion or wind. Batteries are also improving. If form factor is not a problem, then it is also possible to add even more batteries to extend a node's lifetime. The capabilities and reliability of low-power circuits and microcontrollers are improving. Most hardware platforms allow multiple power-saving states (off, idle, on) for each component of the device (each sensor, the radio, the microcontroller). When such options exist, only the components required at a particular time need to be active. At the software level, power management solutions are targeted at minimizing communications, given that transmitting and listening for messages are energy-expensive operations, and at creating sleep/wake schedules for nodes or particular components of nodes.

Minimizing the number of messages is a cross-cutting problem. For example, a good MAC protocol will lead to fewer collisions and retries. With good routing, short paths on high-quality links, and congestion avoidance, collisions can be reduced, which in turn minimizes the number of messages sent. Efficient neighbor discovery, time synchronization, localization, query dissemination, and flooding can all reduce the number of messages, thereby increasing nodes' lifetime.

The proposed solutions for creating sleep/wake schedules vary considerably. Many solutions attempt to keep awake the minimum number of nodes, called sentries, which provide the required sensing coverage while permitting all the others to sleep. To balance energy consumption, a rotation is performed periodically, whereby new sentries are selected for the next period of time. Another commonly used sleep/wake schedule technique relies on duty-cycle nodes. As an example, a node may be awake for 200 milliseconds out of each second, leading to a 20% duty cycle. The duty cycle percentage chosen depends on the application's requirements, but the end result is usually a very significant energy savings. Note that duty-cycle and sentry solutions can be combined, as was done in the VigilNet military surveillance system [He06, He06a].

The cyber aspects of the power management solutions should also consider the physical realities of measuring the amount of energy remaining, the minimum battery energy required to operate, and uncertainties of energy consumption.

3.3 Key Design Drivers and Quality Attributes

A number of key design issues act as drivers for CPS solutions in sensor networks. Associated with each driving issue is a set of quality attributes that allows the designer to quantify the degree of satisfaction achieved by a particular design. This section describes several of the important driving factors. We refer to these as "aware" concerns. To construct CPS on WSN in open environments with humans in the loop, and where the CPS also evolve their functionality and operate for long periods of time, requires a comprehensive cyber solution that has many properties. For example, the software must be physically aware, real-time aware, validate aware, and security aware (to name a few of the key properties). There must be support for runtime adaptation and explicit support for detecting and resolving dependencies. This is a tall order.

3.3.1 Physically Aware

Probably the core CPS issue is developing cyber-technologies that are physically aware. "Physically aware" means having awareness of the environment (not hardware). CPUs, sensors, and actuators are hardware components and, of course, the cyber-technologies must be aware of these components. But this hardware awareness is well known in the embedded systems research arena. The impact of the environment on the cyber aspects of the system is more complicated, is open ended, and has received somewhat less attention. This impact must be a focal point for new CPS research.

To illustrate this point, let us reconsider two examples mentioned earlier. The cyber environment for some MANET multi-hop routing protocols works well in many ad hoc mesh networks. In many cases, however, the cyber world assumes that the environment supports symmetric communication between any two nodes: If node A can send a message to node B, then the reverse is also true. Symmetry cannot be assumed in many WSNs, so it can be said that the MANET (cyber) solution is unaware of this physical property. As a second example, consider a case in which the cyber element is aware of a physical world property. Notably, GPSR is a multi-hop routing protocol whose cyber component is *aware* of potential voids (areas where there is no current next-hop node in the direction of the destination) and GPSR has code that solves this physical world problem. Thus it can be said that GPSR is physically aware of voids. CPS must be physically aware of all the important physical world properties that can affect correctness and performance of the combined cyber-physical system.

Beyond these examples, since many CPS will employ wireless communications, sensing, and actuation, the physical world aspects of each must be addressed in the cyber. For wireless communication, the impact of the environment on communication is fully appreciated, including the impact of distance, power, weather, interference, reflection, diffraction, and so on. For sensing and actuation, the problem is more open ended. In all cases, a methodology is needed that can identify the environmental issues and support cyber-based solutions that can deal with them. Such solutions will require modeling the environment and making those models first-class entities in the system design and implementation. In addition, it is important to define correctness for open systems, which continuously evolve in terms of their functionality, numbers and types of components (hardware and software components may often be entering and leaving the system), and current state.

3.3.2 Real-Time Aware

Complex CPS in open environments often exhibit a wide variety of timing constraints, ranging from soft real-time to safety-critical.

The uncertain environments, the changing conditions in those environments and in the systems themselves, and the significant impact of humans in the loop all generate the need for new notions of real-time guarantees. Most previous real-time research has assumed a static, a priori, known task set with strict assumptions regarding their invocations, worst-case execution times, and (limited) interactions. While such real-time scheduling results are very useful for well-controlled, small, and closed systems, they are largely inadequate for CPS built on WSN.

Some initial work has begun on ways to deal with real-time issues for CPS. RAP [Lu02] uses a velocity monotonic algorithm that takes into account both time and distance and outperforms previous protocols in terms of the average end-to-end deadline miss ratio. It is suitable for soft real-time data streams having a single sink, though no guarantee is provided to individual streams. Another real-time protocol, called SPEED [He05], maintains a desired delivery speed across a sensor network by combining feedback control and nondeterministic geographic forwarding. It outperforms ad hoc, on-demand distance vector routing (AODV) [Perkins99], greedy perimeter stateless routing (GPSR) [Karp00], and dynamic source routing (DSR) [Johnson96] in terms of the end-to-end deadline miss ratio. Nevertheless, SPEED is not suitable for arbitrary streams; rather, it supports streams having a single source or single destination.

Robust implicit earliest deadline first (RI-EDF) [Crenshaw07] is a MAC layer protocol that provides a real-time guarantee by utilizing the rules of EDF to derive a network schedule. In this case, however, the network has to be fully linked; that is, every node must be within transmission range of every other node—something that is not practical in real-world systems.

Another MAC layer protocol, described by [Bui07], provides soft real-time and bandwidth guarantees by avoiding packet collisions. It uses multiple channels, which may not be available with all sensors. This protocol also assumes that the wireless link is not affected by jamming or electromagnetic interference (EMI), such that the only way a transmission can fail is because of interference from other streams.

[Abdelzaher04] proposes a sufficient condition for the schedulability of real-time streams on a wireless sensor network. This work provides important insights into the limitations of achieving guaranteed delays in WSN. [Li05] provides timeliness guarantees for multi-hop streams by explicitly avoiding collisions and scheduling messages based on per-hop timeliness constraints in real-time robotic sensor applications; unfortunately, this solution is not suitable for large-scale networks. Another transmission scheduling algorithm, presented in [Chipara07], performs better than time division multiple access (TDMA)–based protocols in terms of both real-time performance and throughput. This algorithm is mainly appropriate for scheduling queries from the base station and their responses, where there is a single destination for the streams. In [He06b], a real-time analysis is performed for a complex multi-hop military surveillance application by using a hierarchical decomposition scheme.

Finally, many other solutions have been proposed that use feedback control ideas to monitor deadline misses and drive the percentage of deadline misses to a small (target) number. Since feedback control embraces robustness and sensitivity analysis, it appears to be a promising direction for CPS real-time processing.

3.3.3 Runtime Validation Aware

CPS technology is applicable to a wide range of mission-critical applications, including emergency response, infrastructure monitoring, military surveillance, transportation, and medical applications. These applications must operate reliably and continuously due to the high cost associated with system failure. Continuous and reliable operation of WSN is difficult to guarantee due to hardware degradation and environmental changes, some of which were impossible for the original system designers to foresee. This is particularly true for CPS operating in open environments.

A great array of fault tolerance and reliability techniques have been developed over the last 50 years, many of which have been applied to WSNs [Clouqueur04, Paradis07, Ruiz04, Yu07]. Any WSN or CPS that must operate with high confidence is likely to utilize many of these schemes. However, most of the existing approaches, such as eScan [Zhao02] and congestion detection and avoidance (CODA) [Wan03], aim to improve the robustness of individual system components. Consequently, it is difficult to use such methodologies to validate the high-level functionality of the application as a whole. Similarly, self-healing applications, although attempting to provide continuous system operation, are not capable of—or, at least, have not yet been used to demonstrate—adherence of the system to key high-level functional requirements.

In WSN research, health-monitoring systems such as LiveNet [Chen08], Memento [Rost06], and MANNA [Ruiz03] employ sniffers or specific embedded code to monitor the status of a system. These applications monitor low-level components of the system, however—not the high-level application requirements. While they are helpful and necessary, they fall short of determining whether the application requirements are still being met or are not being met over the lifetime of the system.

As future CPS built upon WSN will operate in open environments, many changes will occur to these systems over time as their environment and functionality evolve. Keeping up with these changes will require more than low-level system monitoring; that is, it will necessitate a continuous or periodic runtime validation approach.

In WSN research, considerable work has been done on runtime assurance (RTA) [Wu10]. In this approach, high-level validation is performed at runtime. The validation attempts to show that a WSN will function correctly in terms of meeting its high-level application requirements, irrespective of any changes to the operating conditions and system state that have occurred since it was originally designed and deployed. The basic approach is to use program analysis and compiler techniques to facilitate automated testing of a WSN at runtime. The developer describes the requirements of the application using a high-level specification, which is compiled into both the code that will execute the application on the WSN and a set of input/output tests that can be used to repeatedly verify correct operation of the application over time. The test inputs are then supplied to the WSN at runtime, either periodically or by request. The WSN performs all computations, message passing, and other distributed operations required to produce output values and actions, which are compared to the expected outputs. This testing process produces a partial, but acceptable, level of end-to-end validation of only the essential system functions.

RTA differs from network health monitoring, which detects and reports low-level hardware faults, such as node or route failures. The end-to-end application-level tests used for RTA have two key advantages over the tests of individual hardware components used for health monitoring:

- *Fewer false positives:* RTA does not test nodes, logic, or wireless links that are not necessary for correct system operation, so it produces fewer maintenance dispatches than health monitoring systems.
- *Fewer false negatives:* A network health monitoring system will validate only that all nodes are alive and have a route to a base station, but does not test for more subtle causes of failure such as

topological changes, clock drift, or new obstacles that might have appeared in the environment and block communication or sensing. In contrast, the RTA approach tests for the many different ways that an application may fail to meet its high-level requirements because it uses end-to-end tests.

The goal of RTA is to provide a positive affirmation of correct application-level operation. For CPS, RTA must be further improved so that it more directly addresses system safety and operates in the presence of security attacks.

3.3.4 Security Aware

As CPS become pervasive and wirelessly accessible in open environments, they also become more vulnerable to security attacks. Since many CPS involve critical operations and safety, it is necessary to support the standard security properties, including confidentiality, integrity, authenticity, identification, authorization, access control, availability, auditability, tamper resistance, and non-repudiation. Achieving adequate security is extremely difficult in all systems, but especially so in CPS: These devices often have limited power, memory and execution capabilities, and real-time requirements, and they may even be physically vulnerable. As a consequence of these challenges, developers have had only limited success in applying security solutions to WSNs. Specifically, piecemeal solutions have been proposed at the physical, networking, and middleware layers of WSNs.

At the physical layer, the simplest type of attack seeks to destroy or disable the devices entirely, by creating a denial of service. Such destruction can often be mitigated using fault-tolerant protocols. For example, a mesh network can continue to operate despite some fraction of the devices being destroyed, by having the remaining connected nodes take over routing. Probing of the physical device to deconstruct its internal properties is another type of attack. By reading the contents of memory cells, secret keys may be recovered and then programmed into another device, which can fully masquerade as the original—yet is under the attacker's control. The messages originated by this clone are fully authentic, and the device can actively participate in formerly inaccessible transactions.

In yet another type of attack at the physical layer, since data-dependent computation affects the power consumption and timing of circuits, this aspect of operation can be analyzed statistically over many trials to determine the bit patterns of a key [Ravi04]. Electromagnetic emissions may be inspected similarly to power consumption. Proposed solutions include tamper-resistant packaging [Anderson96], better attack detection, fault recovery mechanisms, and reducing trust in external components [Suh03]. For example, if a device can detect a tampering attempt, it might erase its memory to prevent disclosure of sensitive data. Circuits may be shielded by distributing logical components across the die, bus traffic may be encrypted, and data values and timing can be randomized or "blinded" to thwart side-channel attacks.

Devices' use of wireless communication also leaves them vulnerable to denial-of-service attacks by radio jamming, which can be perpetrated at large distances and unobtrusively. [Xu05] proposes channel hopping and "retreats" to physically move away from the jammer. This approach is most appropriate for ad hoc networks, as it may consume too much energy to be practical for sensor devices. [Law05] proposes data blurting and changing transmission schedules as countermeasures. Another approach, when the jamming cannot be avoided, is for nodes to determine the extent of the jammed area in a wide-scale network by collaboratively mapping and avoiding the region [Wood03].

Many security issues also exist above the physical layer. For example, all communication and middleware protocols can be attacked. Incorporating security solutions for each protocol is often not feasible. An alternative approach is to use dynamic adaptation based on the type of attack that is currently under way. As an example, consider the following approach for routing.

Secure Implicit Geographic Forwarding (SIGF) [Wood06] is a family of routing protocols for WSNs that allows very lightweight operation when no attacks are occurring, and stronger defenses—at the cost of overhead and delay in reacting—when attacks are detected. The basic solution of SIGF has no routing tables; thus, attacks based on changes to the routing tables are not possible. However, if other types of attacks occur, such as a Sybil attack, then the SIGF solution relies on its ability to detect this attack (not always a good assumption), and dynamically invokes another form of SIGF that is resilient to that type of attack. Any damage that occurs during the time required to the system to detect and react to the threat is not prevented.

In long-lived CPS, reprogramming the system will be necessary as the system evolves. Over-the-network reprogramming presents significant security concerns. All other hardware and software defenses may be subverted by a flaw that allows an attacker to replace nodes' programs with custom code. [Deng06] proposes related schemes for securely distributing code in WSN. The first scheme uses a chain of hashes, where each message contains segment i of code and a hash of segment $i + 1$. Upon receipt of a message, the previous code segment can be immediately and efficiently verified. To bootstrap the chain, an error-correcting code (ECC) signature of the first hash value is computed using the base station's private key. This method is suitable when there is little message loss and packets are received mostly in order. The second scheme uses a hash tree to distribute all the hashes in advance, so that out-of-order packets can be quickly checked. Resistance to denial-of-service attacks is improved with this strategy, because packets need not be stored if they are corrupt.

While many security issues are exacerbated for CPS, it may also be possible to exploit the system's physical properties to aid security solutions. For example, mobile nodes can travel only at realistic speeds, so an attack that moves a node too far too fast by artificially changing its location can be readily detected. An accelerometer can be used to detect if a node is physically moved when it should not be. The electronic fingerprints of radio-frequency (RF) transmissions can provide confidence that a node is what it claims to be. CPS has the opportunity to develop these and other techniques that will not work in cyber-space alone.

3.4 Practitioners' Implications

Particular features of sensor networks affect multiple aspects of their development, but perhaps among the most significant ones are the programming abstractions.

Programming WSN and subsequently CPS in an efficient and correct manner is an important open research issue. Many research projects have addressed ways to facilitate WSN programming, with many of the resultant solutions being based on group abstractions. However, these group abstractions have several shortcomings that limit their applicability in CPS. For example, Hood [Whitehouse04] is a neighborhood programming abstraction that allows a given node to share data with a subset of nodes around it, specified using parameters such as the physical distance or number of wireless hops. Hood cannot group nodes that belong to different networks or that use heterogeneous communication platforms. If a mobile group member moves to another network, then it no longer belongs to that group. Additionally, all nodes must share the same code, actuators are not supported, group specification is fixed at compile time, and each instance of a Hood requires specific code compiled and deployed on the targeted nodes.

As a second example, an abstract region [Welsh04] is an abstraction similar to Hood: It allows the definition of a group of nodes according to geographic location or radio connectivity, and permits the sharing and reduction of neighborhood data. Abstract regions provide tuning parameters to obtain various levels of energy consumption, bandwidth consumption, and accuracy. But each definition of a region requires a dedicated implementation; therefore, each region is somehow separated from others and cannot be combined. As with the Hood solution, abstract regions cannot group sensors from different networks, nor can they address actuators, heterogeneous, or mobile devices.

For CPS, the programming abstraction called a Bundle [Vicaire10, Vicaire12] has been designed to overcome some of these shortcomings. Similar to other programming abstractions, a Bundle creates logical collections of sensing devices. Previous abstractions were focused on WSN and did not address key aspects of CPS. Bundles elevate the programming domain from a single WSN to complex systems of systems by allowing the programming of applications involving multiple CPS that are controlled by different administrative domains; they also support mobility both within and across CPS. Bundles can seamlessly group not only sensors, but also actuators, which constitute an important part of CPS. Bundles enable programming in a multiuser environment with fine-grained access right control and conflict resolution mechanisms. Bundles support heterogeneous devices, such as motes, personal digital assistants (PDAs), laptops, and actuators, according to the applications' requirements. They allow different applications to simultaneously use the same sensors and actuators. Conflicts among different users are managed with device-specific programs called resolvers. Bundles also facilitate feedback control mechanisms through dynamic membership updates and requirements reconfiguration based on feedback from the current members. Bundles are implemented in Java, which contributes to the ease and conciseness of programming with this abstraction.

An important feature of the Bundle abstraction is its dynamic aspect. The Bundle membership is updated periodically, so as to respect the membership specification. This feature supports the need for adaptation inherent in many CPS. Bundles use the concept of state for each actuator. Manipulating actuators using states is very different from manipulating actuators using remote method invocation (RMI). As an example, consider the case of turning a light on. A remote method call through RMI directly connects the application to the remote light actuator and turns it on. By contrast, in Bundles, the application simply generates a requirement for the light to be on and sends it to the negotiator by RMI. The negotiator of the light actuator then tries to fulfill this requirement by turning the remote light actuator on. If the actuator's next periodic update reveals to the negotiator that the requirement is not yet fulfilled, then it retries the remote method call until the action is successful. It is also possible to specify a timeout interval from the application level so that the negotiator continues to retry the remote method calls only until that interval expires. Note that the state of the actuator does not change in the negotiator until the requirement is actually fulfilled. The negotiator stores this requirement as long as the application does not cancel it (or terminate). Also, the negotiator may store several such requirements and decide, according to rules specified by the node owner, which requirement should be satisfied. Programmers can check at any time whether their requirements are being satisfied and take appropriate action.

Bundles address some of the key issues for CPS programming. Unfortunately, they fall short in the areas of correctness semantics, dependency detection and resolution, and explicit support for real-time and environment abstraction.

Another promising approach is to use a model-driven design wherein code is automatically generated. As yet, however, it is not possible to generate code automatically from the top down in such a manner as to meet physical world realisms, constraints, uncertainties, and the heterogeneous and distributed nature of a CPS built upon a WSN.

3.5 Summary and Open Challenges

CPS based on WSN have great potential across many application domains. Their increasing pervasiveness, their intimate interactions with humans, their wireless nature, and their wireless access to the Internet all combine to make them open systems. Openness provides significant benefits, but is not without problems—for example, ensuring privacy and security and demonstrating correctness. New cyber-physical solutions are required that will address the critical issues that the uncontrolled environment, the humans-in-the-loop element, and systems of systems bring to the forefront. Today's closed embedded systems technology, fixed real-time systems theory, and feedback control based on electromechanical laws fall short in providing such solutions. A new, exciting, multidisciplinary research field is needed to address these shortcomings and, indeed, is already emerging.

Seeing that the potential in this area achieves its full fruition requires solving many open challenges:

- Effective sensing, decision, and control architectures that can easily interoperate as systems of systems
- New multichannel MAC protocols with real-time guarantees
- Routing that deals with the realities of wireless and mobile nodes
- Localization solutions that operate indoors, with high precision and accuracy, low power requirements, and small accuracy variance among the many nodes of a network
- Efficient and highly accurate clock synchronization that can operate in large-scale networks
- Techniques for identifying the impact of the physical world on the cyber world to avoid fragile cyber (software) that tend to work correctly only under very limiting and often unrealistic assumptions
- Solutions that guarantee their ability to meet the timing and safety-critical constraints found in many CPS applications
- New runtime validation solutions for long-lived CPS to enable their continuous or periodic recertification
- Security solutions for resource-limited systems, as many CPS devices have limited sensing, computation, memory, and power
- High-level programming abstractions that address the need for ease of programming and hiding details while preserving the information needed to show correctness with respect to performance and semantics of the applications

References

[Abdelzaher04]. T. Abdelzaher, S. Prabh, and R. Kiran. "On Real-Time Capacity Limits of Multihop Wireless Sensor Networks." IEEE Real-Time Systems Symposium (RTSS), December 2004.

[Anderson96]. R. Anderson and M. Kuhn. "Tamper Resistance: A Cautionary Note." Usenix Workshop on Electronic Commerce, pages 1–11, 1996.

[Bui07]. B. D. Bui, R. Pellizzoni, M. Caccamo, C. F. Cheah, and A. Tzakis. "Soft Real-Time Chains for Multi-Hop Wireless Ad-Hoc Networks." Real-Time and Embedded Technology and Applications Symposium (RTAS), pages 69–80, April 2007.

[Bulusu00]. N. Bulusu, J. Heidemann, and D. Estrin. "GPS-less Low Cost Outdoor Localization for Very Small Devices." *IEEE Personal Communications Magazine*, October 2000.

[Chen08]. B. Chen, G. Peterson, G. Mainland, and M. Welsh. "LiveNet: Using Passive Monitoring to Reconstruct Sensor Network Dynamics." In *Distributed Computing in Sensor Systems*. Springer, 2008.

[Chipara07]. O. Chipara, C. Lu, and G.-C. Roman. "Real-Time Query Scheduling for Wireless Sensor Networks." IEEE Real-Time Systems Symposium (RTSS), pages 389–399, December 2007.

[Clouqueur04]. T. Clouqueur, K. K. Saluja, and P. Ramanathan. "Fault Tolerance in Collaborative Sensor Networks for Target Detection." *IEEE Transactions on Computers*, 2004.

[Crenshaw07]. T. L. Crenshaw, S. Hoke, A. Tirumala, and M. Caccamo. "Robust Implicit EDF: A Wireless MAC Protocol for Collaborative Real-Time Systems." *ACM Transactions on Embedded Computing Systems*, vol. 6, no. 4, page 28, September 2007.

[Deng06]. J. Deng, R. Han, and S. Mishra. "Secure Code Distribution in Dynamically Programmable Wireless Sensor Networks." *Proceedings of ACM/IEEE International Conference on Information Processing in Sensor Networks (IPSN)*, pages 292–300, 2006.

[Elson02]. J. Elson, L. Girod, and D. Estrin. "Fine-Grained Network Time Synchronization Using Reference Broadcasts." Symposium on Operating Systems Design and Implementation (OSDI), December 2002.

[Ganeriwal03]. S. Ganeriwal, R. Kumar, and M. Srivastava. "Timing-Sync Protocol for Sensor Networks." ACM Conference on Embedded Networked Sensor Systems (SenSys), November 2003.

[He06]. T. He, S. Krishnamurthy, J. Stankovic, T. Abdelzaher, L. Luo, T. Yan, R. Stoleru, L. Gu, G. Zhou, J. Hui, and B. Krogh. "VigilNet: An Integrated Sensor Network System for Energy Efficient Surveillance." *ACM Transactions on Sensor Networks*, vol. 2, no. 1, pages 1–38, February 2006.

[He05]. T. He, J. Stankovic, C. Lu, and T. Abdelzaher. "A Spatiotemporal Communication Protocol for Wireless Sensor Networks." *IEEE Transactions on Parallel and Distributed Systems*, vol. 16, no. 10, pages 995–1006, October 2005.

[He06a]. T. He, P. Vicaire, T. Yan, Q. Cao, L. Luo, R. Stoleru, L. Gu, G. Zhou, J. Stankovic, and T. Abdelzaher. "Achieving Long Term Surveillance in VigilNet." INFOCOM, April 2006.

[He06b]. T. He, P. Vicaire, T. Yan, L. Luo, L. Gu, G. Zhou, R. Stoleru, Q. Cao, J. Stankovic, and T. Abdelzaher. "Achieving Real-Time Target Tracking Using Wireless Sensor Networks." IEEE Real-Time and Embedded Technology and Applications Symposium (RTAS), May 2006.

[Intanagonwiwat00]. C. Intanagonwiwat, R. Govindan, and D. Estrin. "Directed Diffusion: A Scalable Routing and Robust Communication Paradigm for Sensor Networks." ACM MobiCom, August 2000.

[Johnson96]. D. B. Johnson and D. A. Maltz. "Dynamic Source Routing in Adhoc Wireless Networks." In *Mobile Computing*, pages 153–181. Kluwer Academic, 1996.

[Karp00]. B. Karp. "Geographic Routing for Wireless Networks." PhD dissertation, Harvard University, October 2000.

[Karp00a]. B. Karp and H. T. Kung. "GPSR: Greedy Perimeter Stateless Routing for Wireless Sensor Networks." IEEE MobiCom, August 2000.

[Law05]. Y. W. Law, P. Hartel, J. den Hartog, and P. Havinga. "Link-Layer Jamming Attacks on S-MAC." *Proceedings of International Conference on Embedded Wireless Systems and Networks (EWSN)*, pages 217–225, 2005.

[Li05]. H. Li, P. Shenoy, and K. Ramamritham. "Scheduling Messages with Deadlines in Multi-Hop Real-Time Sensor Networks." Real-Time and Embedded Technology and Applications Symposium (RTAS), 2005.

[Lu02]. C. Lu, B. Blum, T. Abdelzaher, J. Stankovic, and T. He. "RAP: A Real-Time Communication Architecture for Large-Scale Wireless Sensor Networks." Real-Time and Embedded Technology and Applications Symposium (RTAS), June 2002.

[Maroti04]. M. Maroti, B. Kusy, G. Simon, and A. Ledeczi. "The Flooding Time Synchronization Protocol." ACM Conference on Embedded Networked Sensor Systems (SenSys), November 2004.

[Mills94]). D. Mills. "Internet Time Synchronization: The Network Time Protocol." In *Global States and Time in Distributed Systems*. IEEE Computer Society Press, 1994.

[Paradis07]. L. Paradis and Q. Han. "A Survey of Fault Management in Wireless Sensor Networks." *Journal of Network and Systems Management*, June 2007.

[Perkins99]. C. E. Perkins and E. M. Royer. "Ad-Hoc on Demand Distance Vector Routing." Workshop on Mobile Computer Systems and Applications (WMCSA), pages 90–100, February 1999.

[Polastre04]. J. Polastre, J. Hill, and D. Culler. "Versatile Low Power Media Access for Wireless Sensor Networks." ACM Conference on Embedded Networked Sensor Systems (SenSys), November 2004.

[Ravi04]. S. Ravi, A. Raghunathan, and S. Chakradhar. "Tamper Resistance Mechanisms for Secure, Embedded Systems." *Proceedings of the 17th International Conference on VLSI Design*, page 605, 2004.

[Rost06]. S. Rost and H. Balakrishnan. "Memento: A Health Monitoring System for Wireless Sensor Networks." IEEE SECON, September 2006.

[Ruiz03]. L. Ruiz, J. Nogueira, and A. Loureiro. "MANNA: A Management Architecture for Wireless Sensor Networks." *IEEE Communications Magazine*, February 2003.

[Ruiz04]. L. Ruiz, I. Siqueira, L. Oliveira, H. Wong, J. Nogueira, and A. Loureiro. "Fault Management in Event Driven Wireless Sensor Networks." Modeling, Analysis and Simulation of Wireless and Mobile Systems (MSWiM), October 2004.

[Stankovic05]. J. Stankovic, I. Lee, A. Mok, and R. Rajkumar. "Opportunities and Obligations for Physical Computing Systems." *IEEE Computer*, vol. 38, no. 11, pages 23–31, November 2005.

[Stoleru04]. R. Stoleru, T. He, and J. Stankovic. "Walking GPS: A Practical Localization System for Manually Deployed Wireless Sensor Networks." IEEE Workshop on Embedded Networked Sensors (EmNets), 2004.

[Stoleru05]. R. Stoleru, T. He, J. Stankovic, and D. Luebke. "A High Accuracy, Low-Cost Localization System for Wireless Sensor Networks." ACM Conference on Embedded Networked Sensor Systems (SenSys), November 2005.

[Suh03]. G. Suh, D. Clarke, B. Gassend, M. van Dijk, and S. Devadas. "AEGIS: Architecture for Tamper-Evident and Tamper-Resistant Processing." *Proceedings of ICS*, pages 168–177, 2003.

[Vicaire12]. P. Vicaire, E. Hoque, Z. Xie and J. Stankovic. "Bundle: A Group Based Programming Abstraction for Cyber Physical Systems." *IEEE Transactions on Industrial Informatics*, vol. 8, no.2, pages 379–392, May 2012.

[Vicaire10]. P. A. Vicaire, Z. Xie, E. Hoque, and J. Stankovic. "Physicalnet: A Generic Framework for Managing and Programming Across Pervasive Computing Networks." IEEE Real-Time and Embedded Technology and Applications Symposium (RTAS), 2010.

[Wan03]. C. Wan, S. Eisenman, and A. Campbell. "CODA: Congestion Detection and Avoidance in Sensor Networks." ACM Conference on Embedded Networked Sensor Systems (SenSys), November 2003.

[Welsh04]. M. Welsh and G. Mainland. "Programming Sensor Networks Using Abstract Regions." Symposium on Networked Systems Design and Implementation (NSDI), pages 29–42, 2004.

[Whitehouse04]. K. Whitehouse, C. Sharp, E. Brewer, and D. Culler. "Hood: A Neighborhood Abstraction for Sensor Networks." MobiSYS, pages 99–110, 2004.

[Wood06]. A. Wood, L. Fang, J. Stankovic, and T. He. "SIGF: A Family of Configurable, Secure Routing Protocols for Wireless Sensor Networks." ACM Security of Ad Hoc and Sensor Networks, October 31, 2006.

[Wood03]. A. Wood, J. Stankovic, and S. Son. "JAM: A Jammed-Area Mapping Service for Sensor Networks." *Proceedings of IEEE Real-Time Systems Symposium (RTSS)*, page 286, 2003.

[Wu10]. Y. Wu, J. Li, J. Stankovic, K. Whitehouse, S. Son and K. Kapitanova. "Run Time Assurance of Application-Level Requirements in Wireless Sensor Networks." International Conference on Information Processing in Sensor Networks (IPSN), SPOTS Track, April 2010.

[Xu05]. W. Xu, W. Trappe, Y. Zhang, and T. Wood. "The Feasibility of Launching and Detecting Jamming Attacks in Wireless Networks." *Proceedings of MobiHoc*, pages 46–57, 2005.

[Yu07]. M. Yu, H. Mokhtar, and M. Merabti. "Fault Management in Wireless Sensor Networks." IEEE Wireless Communications, December 2007.

[Zhao02]. Y. Zhao, R. Govindan, and D. Estrin. "Residual Energy Scan for Monitoring Sensor Networks." IEEE Wireless Communications and Networking Conference (WCNC), March 2002.

[Zhou06]. G. Zhou, C. Huang, T. Yan, T. He, and J. Stankovic. "MMSN: Multi-Frequency Media Access Control for Wireless Sensor Networks." INFOCOM, April 2006.

Part II

Foundations

Chapter 4

Symbolic Synthesis for Cyber-Physical Systems

Matthias Rungger, Antoine Girard, and Paulo Tabuada

Cyber-physical systems (CPS) consist of cyber and physical components interacting through sensors and actuators. Consequently, the behavior of a CPS is described by the combination of discrete dynamics accounting for the behavior of cyber components (e.g., software and digital hardware) and continuous dynamics accounting for the behavior of physical components (e.g., temporal evolution of temperatures, positions, and velocities). The terms *discrete* and *continuous* refer to the domains where the quantities of interest live. For example, the quantities used to model cyber components typically live in finite sets, while the quantities used to model physical components are typically real values and, therefore, live in infinite sets. This hybrid nature is one of the main reasons why the modeling, analysis, and design of CPS are so difficult.

In this chapter we take the view that it is easier to design correct CPS than to verify the correctness of CPS. The key insight is that at design time we can make judicious choices that lead to predictable and correct systems. However, when the time comes to perform verification, such choices have already been made—and if incorrectly made, they may lead to undecidable or computationally hard verification problems. The approach we follow relies on abstracting the physical components so that they can be described by the same models used to describe cyber components. Once this is done, we can leverage the existing work on reactive synthesis of cyber systems to synthesize controllers enforcing the desired specifications. Throughout this chapter we will present this approach along with its assumptions and limitations.

4.1 Introduction and Motivation

The synthesis method presented in this chapter proceeds roughly in three steps. In the first step, a finite abstraction of the physical components, often referred to as a symbolic model or discrete abstraction, is computed. In the second step, existing reactive synthesis algorithms are employed to synthesize a controller for the finite model of the CPS obtained by composing finite models for the cyber components with finite abstractions for the physical components. In the third and final step, the synthesized controller is refined to a controller that acts on the physical components to enforce the desired specification on the CPS.

Early ideas on the construction of symbolic models of continuous systems that are adequate for controller design can be found in [Caines98, Förstner02, Koutsoukos00, Moor02]. For example, in [Koutsoukos00, Moor02], supervisory control techniques are employed on the abstract level to design the controller for the symbolic model. The correctness of the design procedure is ensured by exploiting certain properties of the symbolic model. A more modern approach to ensure the correctness of the synthesis method is based on so-called simulation or alternating simulation relations [Alur98, Milner89]. Instead of proving the correctness directly, one shows that the symbolic model and the original CPS are related by an alternating simulation relation, from which the correctness of the controller refinement follows [Tabuada09]. We follow this line of reasoning in this chapter.

4.2 Basic Techniques

In this section, we first introduce the basic modeling techniques to describe the behavior of a system as well as the specification of properties to be verified. We follow this with the definition of the synthesis problem, along with some examples.

4.2.1 Preliminaries

We use $\mathbb{N}$, $\mathbb{Z}$, and $\mathbb{R}$ to denote the natural (including zero), integer, and real numbers, respectively. Given a set A, we use A^n and 2^A to denote the n-fold Cartesian product and the set of all subsets of A, respectively. We denote the Euclidean norm of $x \in \mathbb{R}^n$ by $|x|$.

We denote the closed, open, and half-open intervals in $\mathbb{R}$ by $[a,b]$, $]a,b[$, $[a,b[$, and $]a,b]$, respectively. Intervals in $\mathbb{Z}$ are respectively denoted by $[a;b]$, $]a;b[$, $[a;b[$, and $]a;b]$.

Given a function $f: A \to B$ and $A' \subseteq A$, we define $f(A') := \{f(a) \in B \mid a \in A'\}$ as the image of A' under f. A set-valued function or mapping f from A to B is denoted by $f: A \rightrightarrows B$. Each set-valued map $f: A \rightrightarrows B$ defines a binary relation on $A \times B$; that is, $(a,b) \in f$ if $b \in f(a)$. The inverse mapping f^{-1} and the domain *domf* of a set-valued map f are defined by $f^{-1}(b) = \{a \in A \mid b \in f(a)\}$ and $\mathit{domf} = \{a \in A \mid f(a) \neq \varnothing\}$, respectively. We use *id* to denote the identity function; that is, $id: A \to A$ with $id(a) = a$ for all $a \in A$.

Given a subset $A \subseteq X$ of a metric space (X,d) with metric d, we use $A + \varepsilon\mathbb{B}$ to denote the inflated set $\{x \in X \mid \exists_{a \in A} : d(a,x) \leq \varepsilon\}$. The boundary of a set A is denoted by ∂A and the set difference of two sets A and B is denoted, as usual, by $A \backslash B = \{a \in A \mid a \notin B\}$.

Given a sequence $a: [0;T] \to A$, $T \in \mathbb{N} \cup \{\infty\}$ in some set A, we use a_t to denote its t-th element and $a_{[0;t]}$ to denote its restriction to the interval $[0;t]$. The set of all finite sequences is denoted by A^*. The set of all infinite sequences is denoted by A^ω. If A is equipped with a metric d, and B is a subset of A^ω, then we use $B + \varepsilon\mathbb{B}$ to denote the set $\{a \in A^\omega \mid \exists_{b \in B}, \forall_{t \in \mathbb{N}} : d(a_t, b_t) \leq \varepsilon\}$.

4.2.2 Problem Definition

In this section, we introduce a mathematical model of CPS that underlies our analysis and describe the synthesis problem with respect to linear temporal logic specifications.

4.2.2.1 Modeling Systems

To be able to capture the rich behavior of a CPS, we use a general notion of system.

Definition 1: A system S is a tuple $S = (X, X_0, U, \rightarrow, Y, H)$ consisting of the following:

- Set of states X
- Set of initial states $X_0 \subseteq X$
- Set of inputs U
- Transition relation $\rightarrow \subseteq X \times U \times X$
- Set of outputs Y
- Output map $H : X \rightarrow Y$

An internal behavior ξ of S is an infinite sequence $\xi \in X^\omega$ for which there exists an infinite sequence $\nu \in U^\omega$, such that $\xi_0 \in X_0$ and $(\xi_t, \nu_t, \xi_{t+1}) \in \rightarrow$ (also denoted by $\xi_t \xrightarrow{\nu_t} \xi_{t+1}$) holds for all $t \in \mathbb{N}$. Every internal behavior ξ of S induces an external behavior ζ of S—that is, an infinite sequence in Y that satisfies $\zeta_t = H(\xi_t)$ for all $t \in \mathbb{N}$. We use $B(S)$ to denote the set of all external behaviors of S.

Definition 2: Alternatively, we say that a system S is metric if Y is a metric space, or the system S is finite if X and U are finite sets. Otherwise, we say that S is an infinite system.

Following are two examples that illustrate the usefulness of the introduced notion of a system. We start with a finite state system, as given in Section 1.3 of [Tabuada09].

Example: Communication Protocol

In this example we model a communication protocol as a system according to Definition 1. We consider a sender and a receiver that exchange some messages over a faulty channel. The sender processes the data from a buffer and sends it to the receiver. After a message has been sent, the sender waits for a confirmation message from the receiver acknowledging the received message. If the confirmation message shows that the message was transmitted incorrectly, the sender retransmits the last message. In case of a correct transmission, the sender proceeds with the next message from the buffer.

This behavior is illustrated by the finite state automaton shown in Figure 4.1. The circles denote the states and the arrows denote the transitions. The upper part of each circle is labeled by the state and the lower part is labeled by the output. The transitions are labeled by the inputs.

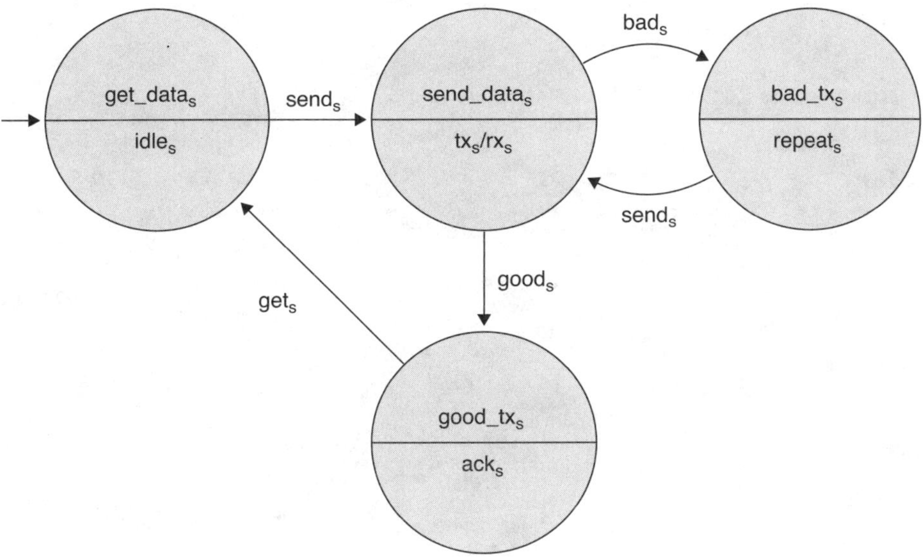

Figure 4.1: *Graphical representation of the finite-state system modeling the sender*

A message can be transmitted either successfully or unsuccessfully over the communication channel. If the received message is correct, the receiver sends a confirmation message to acknowledge the error-free message; otherwise, a repeat request is sent to the sender. The dynamic behavior of the receiver is illustrated in Figure 4.2.

The overall system S is given by the composition of the two systems. We do not provide details on how to compose systems, but rather refer the interested reader to [Tabuada09]. It is clear that S is a finite system. Some possible external behaviors of the sender are given by the following:

$$idle_s \rightarrow tx_s/rx_s \rightarrow ack_s \rightarrow get_s \rightarrow \ldots$$
$$idle_s \rightarrow tx_s/rx_s \rightarrow repeat_s \rightarrow tx_s/rx_s \rightarrow \ldots$$

Now we show that our mathematical system model is also adequate for continuous control systems, such as switched systems.

Example: DC-DC Boost Converter

In this example, we model a DC-DC boost converter, which is depicted in Figure 4.3 as a system. The DC-DC boost converter is a step-up converter that is used to generate a higher voltage on the load side compared to the source voltage. The converter operates in two modes.

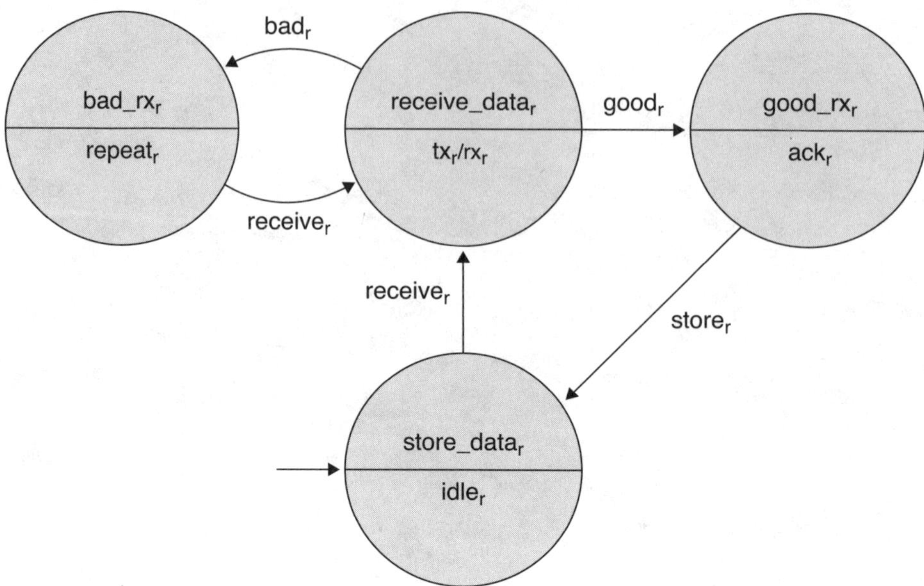

Figure 4.2: *Graphical representation of the finite-state system modeling the receiver*

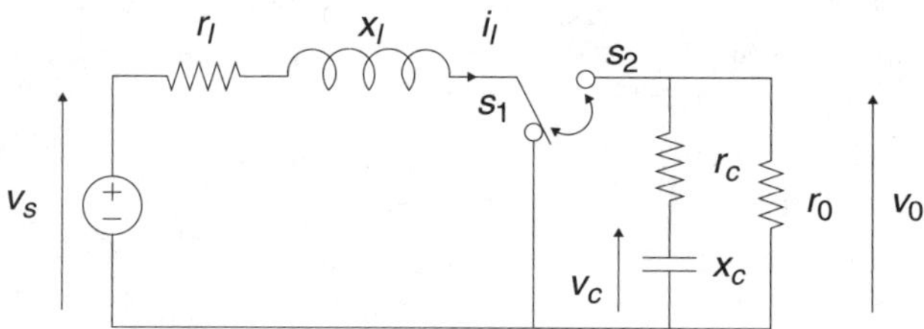

Figure 4.3: *The DC-DC boost converter*

When the switch is closed, electrical energy is stored in terms of a magnetic field in the inductor. When the switch is open, the inductor acts as a second source and its voltage adds up to the source voltage on the load side. DC-DC boost converters have been extensively studied in the literature, often from a hybrid control perspective (see, for example, [Beccuti06, Buisson05, Senesky03]), and also in the context of abstraction-based controller design [Girard10, Girard12, Reißig13].

The state of the system is given by $x(t) = [i_l(t), v_c(t)]^T$, where $i_l(t)$ and $v_c(t)$ denote the current through the inductor and the voltage across the capacitor at time t, respectively. The dynamic behavior is described by the switched linear system:

$$\dot{x}(t) = A_{g(t)}x(t) + b \tag{4.1}$$

In this system $A_i \in \mathbb{R}^{2\times 2}$, $i \in [1;2]$, and b are given as follows:

$$A_1 = \begin{bmatrix} -\frac{r_l}{x_l} & 0 \\ 0 & -\frac{1}{x_c}\frac{1}{r_0 + r_c} \end{bmatrix}, \quad A_2 = \begin{bmatrix} -\frac{1}{x_l}(r_l + \frac{r_c r_0}{r_c + r_0}) & -\frac{1}{x_l}\frac{r_0}{r_0 + r_c} \\ \frac{1}{x_c}\frac{r_0}{r_0 + r_c} & -\frac{1}{x_c}\frac{1}{r_0 + r_c} \end{bmatrix}, \quad b = \begin{bmatrix} \frac{v_s}{x_l} \\ 0 \end{bmatrix}$$

The system matrices A_i are used to describe the evolution of the state, depending on the position of the switch $g(t) \in [1;2]$.

In Section 4.2.4, we design a state-dependent controller to regulate the voltage across the load; it is implemented in a sample-and-hold manner with sampling time $\tau \in \mathbb{R}_{>0}$. Moreover, we restrict our attention to a subset $K \subseteq \mathbb{R}^2$ of the state space around the desired setpoint. We cast the sampled dynamics of Equation (4.1) on K as the following system:

$$S_\tau = (K, K, [1;2], \underset{\tau}{\rightarrow}, \mathbb{R}^2, id)$$

The transition relation is defined as follows:

$$x \overset{g}{\underset{\tau}{\rightarrow}} x' :\Leftrightarrow x' = \xi_{x,g}(\tau)$$

Here $\xi_{x,g}(\tau)$ is the solution of the differential equation (4.1) at time τ for the switch position $g \in [1;2]$. The system S_τ is a metric system. The external behavior coincides with the internal behavior and corresponds to the sampled continuous-time behavior defined by Equation (4.1).

We do not provide further examples, but refer the interested reader to [Tabuada09] and some recent articles in which the notion of a system according to Definition 1 is employed to describe networked control systems [Borri12] and stochastic control systems [Zamani14].

4.2.2.2 Linear Temporal Logic

In this section, we recall how linear temporal logic (LTL) can be used to specify the desired system behavior. The logic is defined over a set of atomic propositions $\mathcal{P}$, which is given by a finite set of subsets P_i, $i \in [1;p]$ of the output set Y; that is, the following is used for all $i \in [1;p]$:

$$P_i \subseteq Y \tag{4.2}$$

Intuitively, the atomic propositions represent the sets that are relevant for the particular desired system behavior.

Formally, a specification of the desired system behavior is given in terms of an LTL[1] formula φ, which is composed of propositional connectives (and "$\wedge$", negation "$\neg$") with the addition of the temporal operator's next "X", always "G", and eventually "F" over the set of atomic propositions $\mathcal{P}$. An LTL formula is evaluated over infinite sequences π in $2^{\mathcal{P}}$. A sequence π is said to satisfy $X\varphi$, $G\varphi$, or $F\varphi$ if φ holds true in the next time step, in every time step, or at some future time step(s), respectively. A precise definition of a sequence π satisfying a given LTL formula φ is given in Chapter 6, "Logical Correctness for Hybrid Systems," and can be found in several references, including [Baier08, Vardi96].

A sequence ζ in Y is said to satisfy a formula φ, denoted by $\zeta \models \varphi$, if the sequence $h \circ \zeta$, with $h(y) = \{P \in \mathcal{P} \mid y \in P\}$, satisfies φ. The set of all sequences in Y that satisfy an LTL formula is denoted as follows:

$$B(\varphi) = \{\zeta : \mathbb{N} \to Y \mid \zeta \models \varphi\} \tag{4.3}$$

We say that a system S satisfies an LTL formula φ, denoted by $S \models \varphi$, if every external behavior of S satisfies φ:

$$B(S) \subseteq B(\varphi) \tag{4.4}$$

Next we illustrate some widely used specifications and interpret their meaning in terms of the external behavior of a given system.

Safety

Safety properties are among the simplest and most widely used properties. They express the fact that certain bad system states should never be reached. Such a desired behavior is, for example, given by the following LTL formula:

$$\varphi = GP$$

In this formula, $P \subseteq Y$ is the set of safe outputs. The bad system states are given by $Y \backslash P$. Given that a system S satisfies GP, the always G operator implies that every external behavior $\zeta \in B(S)$ remains in the good states for all times; that is, $\zeta_t \in P$ for all $t \in \mathbb{N}$.

Let us revisit the previously introduced example of the communication protocol. We can use the following safety formula:

$$\varphi = G\neg(ack_s \wedge repeat_r)$$

1. For the sake of brevity of the presentation, we omit the until operator U and focus on a fragment of full LTL.

This expresses the requirement that the sender should never "believe" that a message was transmitted successfully while the receiver responded with a repeat request; that is, the sender and the receiver should never be in the states with outputs ack_s and $repeat_r$ simultaneously.

Reachability and Termination

Another fundamental property of a system S is expressed by the eventually operator:

$$\varphi = FP$$

This expression indicates that the set $P \subseteq Y$ should be reached after a finite number of steps; that is, every external behavior $\zeta \in B(S)$ satisfies $\zeta_t \in P$ for some $t \in \mathbb{N}$. For example, if S represents a computer program and $P \subseteq Y$ corresponds to the termination condition—that is, $H(x) \in P$ means that the program terminated—then $S \models FP$ implies that S terminates eventually.

Attractiveness (Under Constraints)

In the context of control systems, asymptotic stability is one of the most common specifications. Since stability is a topological property, we assume for the following discussion that S is a metric system with $Y = X$ and $H = id$, from which it follows that X is a metric space. Asymptotic stability with respect to an equilibrium $x^* \in X$ is expressed by two properties: stability and attractiveness. Stability implies that for every neighborhood N of the equilibrium, we can find a neighborhood N' of the equilibrium such that if we restrict the set of initial states to N'—that is, $X_0 \subseteq N'$—all behaviors remain within N for all times: $\xi_t \in N$ for all $t \in \mathbb{N}$. Attractiveness says that every behavior eventually converges to the equilibrium: $\lim_{t\to\infty} \xi_t = x^*$. We cannot express stability in LTL. However, it is possible to formulate a practical variant of attractiveness by the following formula:

$$\varphi = FGQ$$

This implies that every behavior ξ of a system S satisfying φ eventually reaches Q and stays in Q afterward; that is, there exists $t \in \mathbb{N}$ such that for all $t' \geq t$ we have $\xi_{t'} \in Q$. In addition to this property, we can easily impose certain constraints $P \subseteq Y$ during the transient system behavior, implying that $\xi_t \in P$ for all $t \in \mathbb{N}$, by the following expression:

$$\varphi = FGQ \wedge GP$$

Example: Regulating the State in the DC-DC Boost Converter
Recall the DC-DC boost converter introduced in the previous section. As mentioned earlier, the goal of a controller for the DC-DC converter is to regulate the output voltage across the load. This is usually done by regulating the state $[i_l(t), v_c(t)]^T$ around a reference state $x_{ref} = [i_{ref}, v_{ref}]^T$. To express the desired behavior in LTL we introduce the atomic proposition $Q = [1.1, 1.6] \times [1.08, 1.18]$. We use the LTL formula FGQ to express the notion that the system should eventually reach a neighborhood of the reference state and stay there afterward. In addition, we impose a safety constraint GP with $P = [0.65, 1.65] \times [0.99, 1.19]$ resulting in the final specification:

$$\varphi = FGQ \wedge GP$$

The safety constraint serves two purposes. First, using the safety property, we ensure that certain physical limitations of the DC-DC converter are never exceeded during the transient behavior. For example, with GP we ensure that the current always stays within $[0.65, 1.65]$. The second purpose of the safety constraint is to restrict the problem domain to a bounded subset of the state space, which facilitates the computation of finite symbolic models.

4.2.2.3 The Synthesis Problem

Now we proceed with the definition of the synthesis problem. We start by introducing the notion of a controller. In general, a controller for a system S is again a system S_C and the closed-loop system $S_C \times_I S$ is obtained by the composition of the system S_C and S with respect to an interconnection relation I [Tabuada09]. However, in this text we focus on rather simple safety and reachability problems of the following form:

$$\varphi = GP \text{ and } \varphi = FQ \wedge GP \tag{4.5}$$

Here, P and Q are subsets of Y. For those specifications, it is well known that a memoryless or state-feedback controller exists [Tabuada09]. We define a controller C for a system S as the set-valued map:

$$C : X \rightrightarrows U$$

This controller is used to restrict the possible inputs of the system S. The closed-loop system S/C follows from the composition of S with C by the following expression:

$$S/C = (X_C, X_{0,C}, U, \underset{C}{\rightarrow}, H, Y)$$

The modified state space $X_C = X \cap domC$, $X_{0,C} = X_0 \cap domC$ and the modified transition relation are given by the following expression:

$$(x,u,x') \in \underset{C}{\rightarrow} :\Leftrightarrow ((x,u,x') \in \rightarrow \wedge u \in C(x))$$

Example: Sampled-Data Dynamics for the DC-DC Boost Converter
A controller C for the sampled-data dynamics of DC-DC boost converter S_τ is given by the following expression:

$$C : \mathbb{R}^2 \rightrightarrows \{1,2\}$$

The closed-loop system is illustrated in Figure 4.4. At sampling times $t\tau$, $t \in \mathbb{N}$, the controller provides the valid inputs $\nu_t \in C(\xi_t)$ based on the current state ξ_t of the system. The inputs ν_t are kept constant over the sampling interval $[t\tau, (t+1)\tau[$.

In the implementation of the controller C on a digital device, we can pick any input $\nu_t \in C(\xi_t)$ according to some selection criterion.

With the precise notion of a closed-loop system, we formulate the synthesis problem as follows.

Problem 1: Given a system S and LTL specification φ of the form shown in Equation (4.5), we can find a controller C (if it exists) so that $\mathcal{B}(S/C) \neq \emptyset$ and S/C satisfies the specification; that is, $S/C \models \varphi$. (With specifications of the form in Equation (4.5), we might also be able to solve more general problems, such as attractiveness specifications.)

Remark 1: Suppose that the output map H of S is injective. We consider two sets $Q, P \subseteq Y$ with $Q \subseteq P$. Let C_1 and C_2 be two controllers for S:

$$S/C_1 \models GQ$$
$$S/C_2 \models FQ' \wedge GP$$

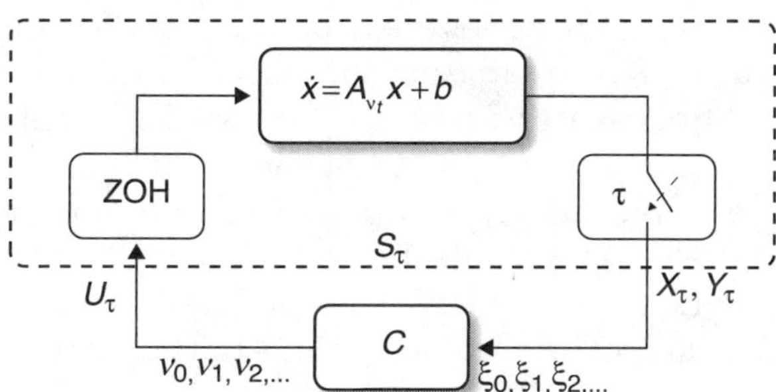

Figure 4.4: *The closed-loop system S_τ / C*

Here $Q' = H(dom\ C_1)$. Then we define the controller:

$$C(x) := \begin{cases} C_1(x) & \text{if } x \in dom\ C_1 \\ C_2(x), & \text{otherwise} \end{cases}$$

We claim that $S/C \models FGQ \wedge GP$. If $x \in dom\ C_1$, every external behavior ζ of S with initial state x stays in Q (and P) forever and, therefore, ζ satisfies $FGQ \wedge GP$. If $x \in dom\ C_2 \setminus dom\ C_1$, every external behavior with initial state x reaches Q' in finite time. Let $t \in \mathbb{N}$ be the first time for which $\zeta_t \in Q'$. Since H is injective, this implies that the corresponding internal behavior satisfies $\xi_t \in dom\ C_1$, after which controller C_1 is used to keep the behavior inside Q, which shows that ζ satisfies $FGQ \wedge GP$.

4.2.3 Solving the Synthesis Problem

As outlined earlier in this chapter, CPS and control systems are generally infinite systems, and the usual algorithms for controller synthesis (see, for example, [Pnueli89, Vardi95]) are not directly applicable. Nevertheless, we can employ those algorithms to synthesize controllers by using a finite-state system that abstracts the infinite system on a bounded domain. The correctness of this approach is ensured by establishing a so-called approximate simulation relation [Girard07] between the original system and the finite approximation—that is, the symbolic model. In this section, we introduce the precise definition of approximate simulation relation and show how this notion facilitates the transfer (or refinement) of a controller that is found for the symbolic model to a controller for the original system. We follow here the approach developed in [Tabuada08] and [Girard12].

4.2.3.1 Approximate Simulation Relations

Approximate simulation relations are usually defined with respect to two metric systems that have the same output space [Girard07] but are allowed to have different input spaces. To keep the discussion simple, we consider a more restrictive situation and assume not only that both systems are metric with the same output space, but also have the same input space. For this case, the definition of approximate (bi)simulation relation is as follows.

Definition 3: Let S and $\hat{S}$ be two metric systems with $U = \hat{U}$, $Y = \hat{Y}$, and metric d. Consider the parameter $\varepsilon \in \mathbb{R}_{\geq 0}$. The relation

$$R \subseteq X \times \hat{X}$$

is said to be an ε-approximate simulation relation (ε-*aSR*) from S to $\hat{S}$ if the following is true:

- For all $x_0 \in X_0$, there exists $\hat{x}_0 \in \hat{X}_0$ so that $(x_0, \hat{x}_0) \in R$ and for any $(x, \hat{x}) \in R$:
  - $d(H(x), \hat{H}(\hat{x})) \le \varepsilon$
  - $x \xrightarrow{u} x'$ implies the existence of $\hat{x} \xrightarrow{u} \hat{x}'$ such that $(x', \hat{x}') \in R$

We say that $\hat{S}$ approximately simulates S or that S is approximately simulated by $\hat{S}$, denoted by the following if there exists an ε-*aSR* from S to $\hat{S}$:

$$S \preccurlyeq_\varepsilon \hat{S}$$

If R is an ε-*aSR* from S to $\hat{S}$ and R^{-1} is an ε-*aSR* from $\hat{S}$ to S, then R is called an ε-approximate bisimulation relation (ε-*aBSR*) between S and $\hat{S}$. We say that S is ε-approximately bisimilar to $\hat{S}$, denoted by the following if there exists an ε-*aBSR* between S and $\hat{S}$:

$$S \cong_\varepsilon \hat{S}$$

The next theorem shows that approximate (bi)similarity implies approximate behavioral (equivalence) inclusion. The theorem comes from Proposition 9.4 in [Tabuada09].

Theorem 1: Let S and $\hat{S}$ be two metric systems with $U = \hat{U}$, $Y = \hat{Y}$, and metric d. Suppose that there exists an ε-*aSR* from S to $\hat{S}$. Then, for every external behavior ζ of S, there exists an external behavior $\hat{\zeta}$ of $\hat{S}$ such that

$$d\left(\zeta_t, \hat{\zeta}_t\right) \le \varepsilon, \ \forall t \in \mathbb{N} \tag{4.6}$$

Equivalently we have the following:

$$S \preccurlyeq_\varepsilon \hat{S} \Rightarrow B(S) \subseteq B\left(\hat{S}\right) + \varepsilon\mathbb{B} \tag{4.7}$$

If S and $\hat{S}$ are ε-approximately bisimilar, we have the following:

$$B(S) \subseteq B\left(\hat{S}\right) + \varepsilon\mathbb{B} \text{ and } B\left(\hat{S}\right) \subseteq B(S) + \varepsilon\mathbb{B} \tag{4.8}$$

In the following discussion, let us assume that we are given a system S and a finite system $\hat{S}$, that ε-approximately simulates S. Theorem 1 allows us to (approximately) answer verification problems for S using the abstraction $\hat{S}$. Suppose that we would like to know if S satisfies a safety property $\varphi = GP$. By utilizing well-known algorithms from automata theory [Vardi96], we can check whether $\hat{S}$ satisfies $\check{\varphi}_\varepsilon = GP_\varepsilon$ for $P_\varepsilon := P \setminus (\partial P + \varepsilon\mathbb{B})$. The formula $\check{\varphi}_\varepsilon$ represents a more restricted variant of φ and accounts for the approximate behavioral inclusion in Equation (4.7) in the sense that $B(\check{\varphi}_\varepsilon) \subseteq B(\varphi)$ holds. If $\hat{S}$ satisfies $\check{\varphi}_\varepsilon$ we deduce from Equation (4.7) that S satisfies φ. We express this fact in symbols as follows:

$$S \preccurlyeq_\varepsilon \hat{S} \wedge \hat{S} \models \varphi_\varepsilon \Rightarrow S \models \varphi$$

Note, however, that $S \preccurlyeq_\varepsilon \hat{S}$ does not allow us to make the reverse conclusion; that is, $\hat{S} \not\models \check{\varphi}_\varepsilon$ does not imply $S \not\models \varphi$. To reason about the reverse direction, we need the stronger notion of bisimilarity. Given that $S \cong_\varepsilon \hat{S}$, we deduce from Equation (4.8) that $\hat{S} \not\models \hat{\varphi}_\varepsilon$, with $\hat{\varphi}_\varepsilon = G(P + \varepsilon\mathbb{B})$, implies $\hat{S} \not\models \varphi$ and we obtain both directions as follows:

$$S \cong_\varepsilon \hat{S} \wedge \hat{S} \models \check{\varphi}_\varepsilon \Rightarrow S \models \varphi$$

$$S \cong_\varepsilon \hat{S} \wedge \hat{S} \not\models \hat{\varphi}_\varepsilon \Rightarrow S \not\models \varphi$$

In contrast to $\check{\varphi}_\varepsilon$ the formula $\hat{\varphi}_\varepsilon$ represents a relaxed variant of the formula φ and there exists a "gap" in the decision procedure based on approximately bisimilar models. However, as we will see in Section 4.2.4, under suitable stability assumptions this gap can be made arbitrarily small.

4.2.3.2 Controller Refinement

Let us now consider the following situation. Suppose we are given a system S, a specification φ, and a finite model $\hat{S}$ that is approximate bisimilar to S. Moreover, we obtained a controller $\hat{C}$ for $\hat{S}$ so that the closed-loop system $\hat{S}/\hat{C}$ satisfies $\check{\varphi}_\varepsilon$, where $\check{\varphi}_\varepsilon$ represents the specification for $\hat{S}$, which is derived from φ. Such a controller can be computed, for example, by the algorithms described in [Pnueli89, Vardi95] or [Tabuada09].

In this section we show how to refine the controller $\hat{C}$ for $\hat{S}$ to a controller C for S. Depending on the particular specification at hand, we will employ different refinement strategies. The controller refinement for general specifications based on bisimulation relations can be found in [Tabuada08]. The specialized case considered in this chapter is reported in [Girard12].

We start with the rather straightforward refinement strategy for safety specifications $\varphi = GP$. Let R be an ε-*aBSR* between S and $\hat{S}$. We then refine a controller $\hat{C}_s$ for $\hat{S}$ to a controller C_s for S:

$$C_s(x) := \left(\hat{C}_s(R(x)\right) \tag{4.9}$$

The control inputs available at a state x of the original system are given by the union of control inputs that are available to the symbolic model at the related states $R(x) \subseteq \hat{X}$.

The refinement strategy for reachability specifications $\varphi = GP \wedge FQ$ is formulated in terms of the minimal time function $T_{\hat{C}_r} : \hat{X} \to \mathbb{N} \cup \{\infty\}$ associated with a controller $\hat{C}_r$ for $\hat{S}$. The function is defined as the least upper bound on the entry time of all external behaviors of $\hat{S}/\hat{C}_r$ with initial state $\hat{x} \in \hat{X}$ in the set Q while staying in P:

$$T_{\hat{C}_r}(\hat{x}) := inf\{t \in \mathbb{N} \mid \forall_{\zeta \in B(\hat{S}/\hat{C}_r)} : \zeta_0 = \hat{H}(\hat{x}), \zeta_t \in Q, \forall_{t' \in [0;t]} \zeta_{t'} \in P\} \quad (4.10)$$

Since $\hat{S}/\hat{C}_r$ is finite, the function can be computed iteratively by a shortest-path algorithm. We obtain the refinement strategy for reachability specifications as follows. Let R be an ε-$aBSR$ between S and $\hat{S}$ and let $\hat{C}_r$ be a controller for $\hat{S}$. Let $T_{\hat{C}_r}$ be given according to Equation (4.10). Then we define the controller C_r' for S:

$$C_r(x) := \hat{C}_r\left(\underset{\hat{x} \in R(x)}{argmin}\, T_{\hat{C}_r}(\hat{x})\right) \quad (4.11)$$

Here, we use the convention that $argmin_{x \in A} f(x) = \varnothing$ for $inf_{x \in A} f(x) = \infty$. The control inputs available at a state x of the original system are given by the control inputs that are associated with a related state $\hat{x} \in R(x)$ that attains the minimum in $T_{\hat{C}_r}(\hat{x})$. Basically, we pick the inputs associated with a symbolic state $\hat{x}$ from which the symbolic model $\hat{S}/\hat{C}_r$ reaches the set Q the fastest.

In the following theorem, we summarize Theorem 1 and Theorem 3 of [Girard12], which ensure the correctness of the presented refinement procedures. Moreover, we provide a statement for the case when the specification for the symbolic model is unrealizable—that is, when there exists no controller such that the closed-loop system satisfies the specification. This statement follows immediately by the symmetry of the presented refinement strategies; that is, a controller for the original system can be refined to a controller for the symbolic model by exchanging the roles of the system and the symbolic model.

Theorem 2: Let S and $\hat{S}$ be two metric systems with $U = \hat{U}$, $Y = \hat{Y}$, and metric d. Let R be an ε-$aBSR$ from S to $\hat{S}$. Consider the sets $P, Q \subseteq Y$ and their approximations:

$$\check{P}_\varepsilon := P \setminus (\partial P + \varepsilon \mathbb{B}),\ \hat{P}_\varepsilon := P + \varepsilon \mathbb{B},$$

$$\check{Q}_\varepsilon := Q \setminus (\partial Q + \varepsilon \mathbb{B}),\ \hat{Q}_\varepsilon := Q + \varepsilon \mathbb{B}$$

The following implications hold:

$$\hat{S}/\hat{C}_s \models G\check{P}_\varepsilon \Rightarrow S/C_s \models GP$$

$$\hat{S}/\hat{C}_r \models G\check{P}_\varepsilon \wedge F\check{Q}_\varepsilon \Rightarrow S/C_r \models GP \wedge FQ$$

Here the controllers C_i, $i \in \{s,r\}$ for S are refined from the controllers $\hat{C}_i$ for $\hat{S}$ according to Equations (4.9) and (4.10), respectively.

Moreover, if $G\hat{P}_\varepsilon$ (or $G\hat{P}_\varepsilon \wedge F\hat{Q}_\varepsilon$) is unrealizable with respect to $\hat{S}$, then GP (or $GP \wedge FQ$) is unrealizable with respect to S.

4.2.4 Construction of Symbolic Models

In this section, we show how to compute symbolic models for control systems. We follow the approach presented in [Girard10]. Because our intention is to provide a rather easy-to-follow introduction to these methods, we focus on a special case of the general scheme outlined in [Girard10]. In particular, we consider switched affine systems $\Sigma = (K, G, A_g, b_g)$ of the following form:

$$\dot{\xi}(t) = A_g \xi(t) + b_g \tag{4.12}$$

Here control input $g \in G = [1;k]$, $k \in \mathbb{N}$, system state $\xi(t) \in K \subseteq \mathbb{R}^n$, and system matrices and affine terms are given by $A_g \in \mathbb{R}^{n \times n}$ and $b_g \in \mathbb{R}^n$, respectively.

A trajectory of Σ associated with an initial state $x \in \mathbb{R}^n$ and a constant input $g \in G$ is given by the following expression:

$$\xi_{x,g}(t) = e^{A_g t} x + \int_0^t e^{A_g (t-s)} b_g \, ds$$

where e^A denotes the matrix exponential defined by the series $\sum_{k=0}^{\infty} \frac{1}{k!} A^k$. Let us define the system that captures the sampled behavior of Σ.

Given a sampling time $\tau \in \mathbb{R}_{>0}$, we define $S_\tau(\Sigma)$ as follows:

$$S_\tau(\Sigma) = (K, K, G, \underset{\tau}{\rightarrow}, id, \mathbb{R}^n) \tag{4.13}$$

The transition relation is defined by the following:

$$(x, g, x') \in \underset{\tau}{\rightarrow} :\Leftrightarrow x' = \xi_{x,g}(\tau)$$

In the following sections, we construct a symbolic model for $S_\tau(\Sigma)$.

4.2.4.1 Stability Assumptions

The construction of symbolic models is based on certain incremental stability assumptions [Angeli02]. One variant of those assumptions is the existence of a common Lyapunov function that implies the switched system Σ (for piecewise constant switching signals) is incrementally globally uniformly asymptotically stable (δ-GUAS). Intuitively, δ-GUAS implies that trajectories associated with the same switching signal, but with different initial states, approach each other as time progresses.

We assume in the present discussion that there exist a symmetric, positive definite matrix $M \in \mathbb{R}^{n \times n}$ and a constant $\kappa \in \mathbb{R}_{>0}$ that satisfy the following inequality:

$$x^\top \left(A_g^\top M + MA_g\right)x \leq -2\kappa x^\top Mx, \ \forall x \in \mathbb{R}^n, \ g \in G \tag{4.14}$$

Conditions on the matrices A_g, $g \in G$ that imply the existence of such a matrix M and κ are given in [Shorten98] and [Liberzon99]. Given M, we define the following function:

$$V(x,y) = \sqrt{(x-y)^\top M(x-y)}$$

We argue that V is actually a δ-GUAS Lyapunov function for Σ [Girard13, Definition 2]. Let λ^- and λ^+ be the smallest and largest eigenvalues of M, respectively. Then we can verify that V satisfies the following inequalities for all $x, y \in \mathbb{R}^n$ and $g \in G$:

$$\sqrt{\lambda^-}\left|x-y\right| \leq V(x,y) \leq \sqrt{\lambda^+}\left|x-y\right| \tag{4.15}$$

$$D_1V(x,y)\left(A_g x + b_g\right) + D_2V(x,y)\left(A_g y + b_g\right) \leq -\kappa V(x,y) \tag{4.16}$$

Here D_iV denotes the partial derivative of V with respect to i-th argument. Observe that the inequalities in Equations (4.15) and (4.16) are sufficient for V being a δ-GUAS Lyapunov function and the following inequality holds:

$$V\left(\xi_{x,g}(t), \xi_{y,g}(t)\right) \leq e^{-\kappa t}V(x,y) \tag{4.17}$$

Moreover, it is easy to show that $W(x) := V(x,0)$ is a norm on $\mathbb{R}^n$ and, therefore, satisfies the triangle inequality. Then we use $V(x,y) = V(x-y,0)$ and obtain the following inequality for all $x, y, z \in \mathbb{R}^n$:

$$V(x,y) \leq V(x,y) + V(y,z) \tag{4.18}$$

4.2.4.2 The Symbolic Model

The construction of the symbolic model is based on a uniform discretization of the state space $K \subseteq \mathbb{R}^n$. We use the following notation. Given a set $K \subseteq \mathbb{R}^n$ and $\eta \in \mathbb{R}_{>0}$, the following expression denotes a uniform grid in K:

$$[K]_\eta := \{x \in K \mid \exists k \in \mathbb{Z}^n : x = k\eta 2/\sqrt{n}\}$$

Based on the definition of the grid, balls of radius η centered at the grid points $[K]_\eta$ cover the set K:

$$K \subseteq \bigcup_{\hat{x} \in [K]_\eta} \hat{x} + \eta\mathbb{B} \tag{4.19}$$

With the discretization parameter $\eta \in \mathbb{R}_{>0}$ at hand, we define the system:

$$S_{\tau,\eta}(\Sigma) = ([K]_\eta, [K]_\eta, G, \underset{\tau,\eta}{\rightarrow}, id, \mathbb{R}^n) \tag{4.20}$$

The transition relation is given by the following:

$$(\hat{x}, g, \hat{x}') \in \underset{\tau,\eta}{\rightarrow} :\Leftrightarrow \left|\hat{x}' - \xi_{x,g}(t)\right| \leq \eta \wedge \xi_{x,g}(\tau) \in K$$

Note that $S_\tau(\Sigma)$ and $S_{\tau,\eta}(\Sigma)$ are metric systems with the same input and output space. The metric is simply given by $d(x,y) = |x - y|$. Moreover, given that G is finite and K is bounded, the system $S_{\tau,\eta}(\Sigma)$ is finite.

The following theorem is from Theorem 4.1 in [Girard10].

Theorem 3: Consider a control system Σ, discretization parameters $\tau, \eta \in \mathbb{R}_{>0}$, and a desired precision $\varepsilon \in \mathbb{R}_{>0}$. Suppose a symmetric, positive definite matrix M and a constant $\kappa \in \mathbb{R}_{>0}$ satisfy Equation (4.14). Let $V(x,y) = \sqrt{(x-y)^\top M(x-y)}$ and λ^+, λ^- be the largest and smallest eigenvalues of M, respectively. If the relation

$$\sqrt{\lambda^-} e^{-\kappa\tau} \varepsilon + \sqrt{\lambda^+} \eta \leq \sqrt{\lambda^-} \varepsilon \tag{4.21}$$

holds, then the following is an ε-*aBSR* between $S_\tau(\Sigma)$ and $S_{\tau,\eta}(\Sigma)$ with respect to the metric $d(x,y) = |x - y|$:

$$R = \left\{(x, \hat{x}) \in K \times [K]_\eta \mid V(x, \hat{x}) \leq \sqrt{\lambda^-} \varepsilon\right\} \tag{4.22}$$

Let us provide the proof of this theorem, as it nicely illustrates the relationship between Equation (4.21) and R being an ε-*aBSR* between S and $\hat{S}$.

We show only that R is an ε-*aSR* from S to $\hat{S}$. The reverse direction follows by the symmetry of the arguments. Recall Definition 3 of an ε-*aSR*. Let us check the first condition. From Equation (4.19), it follows that for every $x \in K$, there exists $\hat{x} \in [K]_\eta$ such that $|x - \hat{x}| \leq \eta$. We use Equation (4.15) to see that $V(x, \hat{x}) \leq \sqrt{\lambda^+}\eta$. From Equation (4.21), it follows that $\sqrt{\lambda^+}\eta \leq \sqrt{\lambda^-}\varepsilon$, which implies that $(x, \hat{x}) \in R$.

To show the second condition, let $(x, \hat{x}) \in R$. From Equation (4.15), we get the inequality $|x - \hat{x}| \leq 1/\sqrt{\lambda^-}V(x, \hat{x})$; it follows from the definition of R that $|x - \hat{x}| \leq \varepsilon$.

We proceed with the third condition in Definition 3. Let $(x, \hat{x}) \in R$ and $(x, g, x') \in \underset{\tau}{\rightarrow}$. Then, by definition of the transition relation, we have $x' = \xi_{x,g}(\tau)$. We pick $\hat{x}' \in [K]_\eta$ so that $(\hat{x}, g, \hat{x}') \in \underset{\tau,\eta}{\rightarrow}$ with the following:

$$\left|\hat{x}' - \xi_{\hat{x},g}(\tau)\right| \leq \eta \tag{4.23}$$

By the definition of the transition relation of $S_{\tau,\eta}(\Sigma)$, such an $\hat{x}'$ always exists. Now we deduce by the following series of inequalities that $(x', \hat{x}') \in R$:

$$\begin{aligned}
V(x', \hat{x}') &\overset{(18)}{\leq} V\left(\xi_{x,g}(\tau), \xi_{\hat{x},g}(\tau)\right) + V\left(\xi_{\hat{x},g}(\tau), \hat{x}'\right) \\
&\overset{(17\ \&\ 23)}{\leq} e^{-\kappa\tau}V(x, \hat{x}) + V(\eta, 0) \\
&\overset{(22\ \&\ 15)}{\leq} e^{-\kappa\tau}\sqrt{\lambda^-}\,\varepsilon + \sqrt{\lambda^+}\eta \\
&\overset{(21)}{\leq} \sqrt{\lambda^-}\varepsilon
\end{aligned}$$

Example: Symbolic Model for the DC-DC Boost Converter

Continuing our running example of the DC-DC boost converter, we construct a symbolic model $S_{\tau,\eta}(\Sigma)$ of $S_\tau(\Sigma)$ with the switched dynamics $\Sigma = (K, [1;2], A_g, b)$, where A_g and b are given in Equation (4.1). We report here the numerical results obtained in [Girard11].

Recall that the desired specification for the converter is given by the LTL formula $\varphi = FGQ \wedge GP$, where Q represents a neighborhood of the reference state and P are our safety constraints. In the discussion that follows, we restrict the computation of the symbolic model to P and, therefore, set $K = P$. Moreover, we focus on the following specification instead of φ:

$$\varphi' = FQ \wedge GP$$

Nevertheless, as stated in Remark 1, it is a straightforward process to also synthesize a controller for φ with the presented methods.

As a first step in the construction of $S_{\tau,\eta}(\Sigma)$, we introduce a state transformation to provide better numerical conditioning. We rescale the voltage, which results in the new state $x = [i_l, 5v_c]^{\top}$. There exist both a symmetric positive definite matrix M and a constant $\kappa \in \mathbb{R}_{>0}$ satisfying Equation (4.14), which are given by $\kappa = 0.014$ and the following:

$$M = \begin{bmatrix} 1.0224 & 0.0084 \\ 0.0084 & 1.0031 \end{bmatrix}$$

The eigenvalues of M are $\lambda^- = 1$ and $\lambda^+ = 1.02554$. We fix the desired accuracy to $\varepsilon = 0.1$ and the sampling time to $\tau = 0.5$. We choose the state-space discretization parameter as $\eta = 97 \cdot 10^{-5}$ so that Equation (4.21) holds, and we compute $S_{\tau,\eta}(\Sigma)$, which results in a symbolic model with approximately $7 \cdot 10^5$ states.

The atomic propositions in the new state space, which are illustrated in Figure 4.5, are given by $Q' = [1.1, 1.6] \times [5.4, 5.9]$ and $P' = [0.65, 1.65] \times [4.95, 5.95]$. We use $\check{Q}_\varepsilon = [1.2, 1.5] \times [5.5, 5.8]$ and $\check{P}_\varepsilon = [0.75, 1.55] \times [5.05, 5.85]$ to synthesize a controller $\hat{C}$ so that $\hat{S}_{\tau,\eta}/\hat{C} \models F\check{Q}_\varepsilon \wedge G\check{P}_\varepsilon$. The controller C for S follows by Equation (4.11).

The domain of the controller and some sample trajectories are illustrated in Figure 4.5. The atomic propositions Q' and P' are illustrated by the small and large dashed rectangle, respectively. The black region shows the target set for the symbolic model enlarged by ε balls; that is, $\check{P}_\varepsilon + \varepsilon\mathbb{B}$. The dark gray shaded regions show the domain of the controller *domC*. In particular, the light gray shaded region and the dark gray shaded region show $C^{-1}(1)$ and $C^{-1}(2)$, respectively. Some example trajectories of the closed-loop system $S_\tau(\Sigma)/C$ are shown by the dotted lines.

4.3 Advanced Techniques

Throughout this chapter, we have simplified the presentation on various occasions with the aim of providing easy-to-follow arguments while still conveying the fundamental ideas underlying the symbolic controller synthesis for CPS. For example, in the definition of the synthesis problem, we restricted the LTL formulas to simple safety and

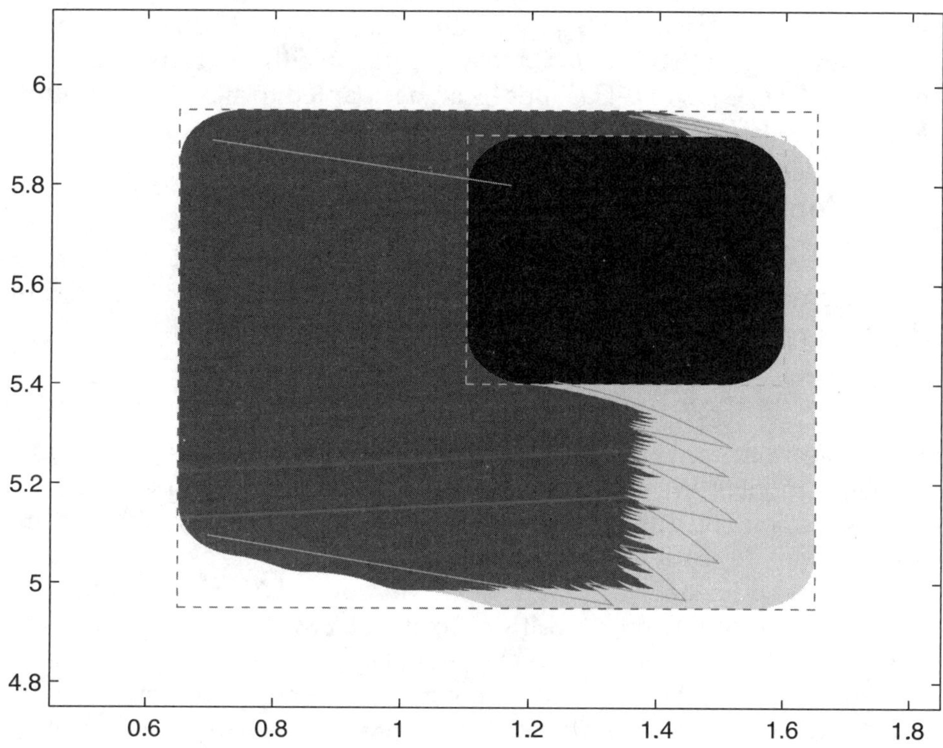

Figure 4.5: *Atomic propositions in the new state space*

reachability specifications. In Section 4.2.3, we assumed that the concrete system and the symbolic model have the same set of inputs, and thereby avoided the more involved notion of ε-approximated alternating simulation relations. In Section 4.2.4, we presented the computation of symbolic models only for the sampled behavior of switched linear systems and neglected the inter-sample behavior.

In the following subsections, we discuss some generalizations (related methods) of the presented scheme and provide follow-up references for the interested reader. We explain the curse of dimensionality and refer to existing methods that address this issue. We describe two possibilities to account for the continuous-time behavior of a system and comment on practical aspects of the implementation of the synthesized controllers on embedded devices. We conclude this section by listing software tools that can be used to compute symbolic models for the design of controllers for CPS.

Note that we do not address extensions with respect to more complex specifications. It is conceptually clear that once we have a method to compute symbolic models (together with the appropriate simulation relations) for a particular class of CPS, any controller synthesis method developed for a finite system that is tailored to a specific specification language—for example, LTL, CTL, or μ-Calculus—can be adapted to synthesize controllers for this class of CPS, while enforcing the respective specification.

4.3.1 Construction of Symbolic Models

We now consider the various ways to construct symbolic models. In this discussion, we start with some basic algorithms and then present the most recent innovations.

4.3.1.1 Basic Algorithms

In this section, we make reference to various approaches to compute symbolic models. We restrict our attention to CPS with complex continuous dynamics: The differential equation defining the continuous dynamics is state dependent and therefore excludes, for example, timed automata and linear hybrid automata. Moreover, we focus on symbolic models that facilitate the algorithmic controller synthesis, as opposed to symbolic models for verification [Alur00, Henzinger00, Tiwari08].

Symbolic models of two-dimensional continuous-time single-integrator and double-integrator systems are analyzed in [Kress-Gazit09] and [Fainekos09a], respectively. General discrete-time linear control systems are treated in [Kloetzer08, Tabuada06, Wongpiromsarn12]. [Yordanov12] presents an algorithm to compute symbolic models for piecewise affine systems.

Methods to construct symbolic models for switched polynomial systems are presented in [Ozay13]. General nonlinear systems (with time delays), switched nonlinear systems, and networked control systems under incremental stability assumptions, similar to the one presented here, are considered in [Pola08] ([Pola10]), [Girard10], and [Borri12], respectively. [Reißig11], [Zamani12], and [Tazaki12] provide algorithms to compute symbolic models of nonlinear systems without stability assumptions.

Symbolic models that are adequate for synthesis of hybrid control systems with state-dependent autonomous transitions are not yet fully understood. Two approaches for the special cases of hybrid systems with discrete-time linear dynamics and order-minimal hybrid systems are described in [Mari10] and [Bouyer06], respectively.

Algorithms to compute symbolic models that lead to controllers that are robust against model uncertainties, actuation errors, or uncontrollable environment dynamics are rather rare. In contrast to the unperturbed case, the symbolic model for perturbed systems needs to be related to the concrete system by an alternating simulation relation instead of a simulation relation [Tabuada09]. Alternating simulation relations are used to account for the uncontrollable disturbances and ensure the correctness of the controller refinement under all possible perturbations. [Wongpiromsarn12] considers discrete-time linear systems with bounded disturbances. Symbolic models of perturbed nonlinear systems under stability assumptions are considered in [Pola09]; they are examined without stability assumptions in [Liu14].

4.3.1.2 Advanced Algorithms

The most fundamental problem that prevents widespread application of the symbolic approach to controller synthesis is the so-called curse of dimensionality (CoD)—that is, the exponential increase in the number of states in the symbolic model with respect to the state-space dimension of the continuous dynamics. The algorithms mentioned to this point in the chapter have all been based on a uniform discretization or partitioning of the state space, so that the CoD is unavoidable. A number of other approaches, however, are specialized for on particular systems or employ search heuristics with the objective of reducing the computational burden.

For example, symbolic models for flat systems are developed in [Tabuada05], [Colombo11], and [Colombo13]. Flat systems have the nice property that the external behavior of the system—that is, the output trajectories—are reproducible (using an input transformation) by a chain of integrators. Therefore, any synthesis problem with respect to the nonlinear system is reducible to a (often lower-dimensional) linear system. By exploiting this property and reducing the chain of integrators further to a first-order system [Girard09], a safety synthesis problem involving an eight-dimensional nonlinear system has been solved [Colombo13]. Another particular system class—monotone systems—is analyzed in [Moor02a].

Rather than focusing on particular system properties, various methods try to reduce the computational burden, by adaptively improving the quality of the symbolic model while solving the synthesis problem on the abstract domain. For example in [Camara11] and [Mouelhi13], symbolic models on multiple time scales are used to locally refine the symbolic model where the controller synthesis fails. In [Rungger12], a heuristic is presented with the objective of reducing one of the most expensive operations—that is, the reachable set computations—in the construction of symbolic models. [Gol12] and [Rungger13] present techniques to adapt the computation of the symbolic model of a discrete-time linear system to a given specification, thereby reducing the domain on which the symbolic model needs to be computed. The aforementioned schemes vary with respect to the considered system dynamics, ranging from linear systems [Gol12, Rungger13], to incrementally stable switched nonlinear systems [Camara11, Mouelhi13], to general nonlinear systems [Rungger12]. Various methods have been used to compute symbolic models of CPS, including those with as many as five continuous variables.

Another interesting approach is described in [Le Corronc13]. These authors propose a quite different scheme to compute symbolic models of incrementally stable switched nonlinear systems. Instead of following the well-known methods in which the states of the symbolic model correspond to grid points or partition elements in the state space, they use bounded sequences of the switching signal as symbolic states. This approach has the potential to yield symbolic models that are constructed independently of the state-space dimension. For example, the authors were able to construct a symbolic model of a five-dimensional switched linear system with three modes.

Yet another advanced approach to compute symbolic models with the objective of reducing the computational burden induced by CoD is based on so-called sampling-based techniques [Bhatia11, Karaman11, Maly13]. Rather than computing the symbolic model on a grid or partition of the continuous state space, the symbolic model is constructed incrementally. New transitions are added to the abstract system by sampling the continuous input space and state space. Several different heuristics are used to guide the growth of the symbolic model. In [Maly13], a symbolic model of a hybrid system with a five-dimensional continuous state space is constructed to solve a synthesis problem with respect to a co-safety LTL specification.

4.3.2 Continuous-Time Controllers

The symbolic models of continuous-time control systems, obtained by methods such as the one presented in Section 4.3.1 and similar approaches [Girard10, Reißig11, Zamani12] account only for the sampled behavior of the original system. Nevertheless, it is possible to use some estimates of the inter-sample time behavior of the continuous system to provide guarantees of the continuous-time behavior. Such an analysis is presented in [Fainekos09]. An alternative event-triggered scheme, in which the LTL specification is directly defined based on the continuous-time behavior of the system, is implemented in [Kloetzer08] and [Liu13]. A drawback of this method becomes evident with the implementation of the synthesized controller, as an elaborate event-detection mechanism is needed to realize the controller on an embedded device. [Liu14] provides a discussion of that matter.

Another issue that relates to the implementation of a controller that is obtained using the symbolic synthesis methods is the required state information. For example, the controller synthesis approach outlined in Section 4.2.3 assumes that complete state information is available at the controller. A more realistic assumption is that only quantized state information is available. Controller refinement schemes that address this issue and require only quantized state information are developed in [Mari10], [Girard13], and [Reißig14].

4.3.3 Software Tools

We conclude this section with a list of available software tools to compute symbolic models of CPS. One of the earliest developed tools for controller synthesis based on symbolic models is LTLCon [Kloetzer08]. This tool supports linear systems and general LTL specifications (without the "next" operator) with atomic propositions defined as conjuncts of strict linear inequalities. Another tool with similar capabilities, focusing on linear systems and finite horizon GR(1) formulas, is TuLip [Wongpiromsarn11]. GR(1) formulas represent a fragment of full LTL formulas for which controllers can be synthesized efficiently. Even though restricted, GR(1) formulas are still sufficiently expressive to formulate interesting properties and are widely used in the robotics community to specify robot mission statements. A tool that supports GR(1) formulas with respect to simple single-integrator dynamics is Linear Temporal Logic Mission Planning (LTLMoP) [Kress-Gazit09]. A nice feature of LTLMoP is its ability to formulate the specification in structured English.

More complex continuous dynamics—in particular, discrete-time piecewise affine systems in connection with full LTL specifications—are supported by conPAS2 [Yordanov12].

Code System from Maria (CoSyMA) [Mouelhi13] is a tool that supports incrementally stable switched systems. CoSyMA uses an advanced multiscale abstraction technique to compute the symbolic models and handles time-bounded reachability and safety specifications.

A versatile tool that provides algorithms to compute symbolic models of various dynamical systems is Pessoa [Mazo10]. It offers synthesis procedures for safety and reachability specifications and is specifically tailored to linear systems. However, Pessoa also supports the construction of symbolic models of nonlinear systems.

4.4 Summary and Open Challenges

This chapter presented a particular variant of the symbolic approach to controller synthesis for CPS. We attempted to strike a balance between the complexity of the problem instance and the clarity of presentation. Often we preferred a simpler problem statement because it led to an easier-to-follow presentation. Specifically, we restricted the specification language to simple safety and reachability requirements. Moreover, we focused on approximate (bi)simulation relations rather than on the more powerful but also more involved notion of approximate alternating (bi) simulation relations. Similarly, we presented a particularly straightforward method for constructing symbolic models for incrementally asymptotically stable switched linear systems. Nevertheless, we provided several pointers to references that treat the more general and complex cases. Examples illustrated the various definitions and results.

The interdisciplinary field of automated controller design for CPS is a rather young research area. The pioneering work reported in [Koutsoukos00], [Moor02], [Förstner02], and [Caines98] has clarified, in principle, how we might leverage discrete algorithmic synthesis methods for CPS, yet numerous questions remain to be answered and challenges are as yet unresolved. The most pressing challenge is perhaps the need to decrease the computational complexity of the symbolic approach. The CoD needs to be addressed for some practically interesting classes of CPS before we can see a widespread application of this approach in practice. Furthermore, when it comes to the deployment of synthesized controllers on an embedded device, the memory size of the controller might represent an obstacle.

In addition to complexity issues, several theoretical questions need to be answered, such as those dealing with the completeness of the symbolic approach. As we have seen, if the control system satisfies certain incremental stability assumptions, it is possible to construct approximate bisimilar abstractions, which implies approximate completeness. However, completeness results for control systems with unstable dynamics or more complex dynamics—for example, hybrid systems—remain an open question.

References

[Alur98]. R. Alur, T. A. Henzinger, O. Kupferman, and M. Y. Vardi. "Alternating Refinement Relations." In *Concurrency Theory*, pages 163–178. Springer, 1998.

[Alur00]. R. Alur, T. A. Henzinger, G. Lafferriere, and G. J. Pappas. "Discrete Abstractions of Hybrid Systems." *Proceedings of the IEEE*, vol. 88, no. 7, pages 971–984, 2000.

[Angeli02]. D. Angeli. "A Lyapunov Approach to Incremental Stability Properties." *IEEE Transactions on Automatic Control*, vol. 47, no. 3, pages 410–421, 2002.

[Baier08]. C. Baier and J. P. Katoen. *Principles of Model Checking*. MIT Press, Cambridge, MA, 2008.

[Beccuti06]. G. A. Beccuti, G. Papafotiou, and M. Morari. "Explicit Model Predictive Control of the Boost DC-DC Converter." *Analysis and Design of Hybrid Systems*, vol. 2, pages 315–320, 2006.

[Bhatia11]. A. Bhatia, M. R. Maly, L. E. Kavraki, and M. Y. Vardi. "Motion Planning with Complex Goals." *IEEE Robotics and Automation Magazine*, vol. 18, no. 3, pages 55–64, 2011.

[Borri12]. A. Borri, G. Pola, and M. Di Benedetto. "A Symbolic Approach to the Design of Nonlinear Networked Control Systems." *Proceedings of the ACM International Conference on Hybrid Systems: Computation and Control*, pages 255–264, 2012.

[Bouyer06]. P. Bouyer, T. Brihaye, and F. Chevalier. "Control in O-Minimal Hybrid Systems." *Proceedings of the 21st Annual IEEE Symposium on Logic in Computer Science*, pages 367–378, 2006.

[Buisson05]. J. Buisson, P. Richard, and H. Cormerais. "On the Stabilisation of Switching Electrical Power Converters." In *Hybrid Systems: Computation and Control*, pages 184–197. Springer, 2005.

[Caines98]. P. E. Caines and Y. J. Wei. "Hierarchical Hybrid Control Systems: A Lattice Theoretic Formulation." *IEEE Transactions on Automatic Control*, vol. 43, no. 4, pages 501–508, 1998.

[Camara11]. J. Camara, A. Girard, and G. Gössler. "Safety Controller Synthesis for Switched Systems Using Multi-Scale Symbolic Models." *Proceedings of the Joint 50th IEEE Conference on Decision and Control and European Control Conference*, pages 520–525, 2011.

[Colombo11]. A. Colombo and D. Del Vecchio. "Supervisory Control of Differentially Flat Systems Based on Abstraction." *Proceedings of the Joint 50th IEEE Conference on Decision and Control and European Control Conference*, pages 6134–6139, 2011.

[Colombo13]. A. Colombo and A. Girard. "An Approximate Abstraction Approach to Safety Control of Differentially Flat Systems." European Control Conference, pages 4226–4231, 2013.

[Fainekos09a]. G. E. Fainekos, A. Girard, H. Kress-Gazit, and G. J. Pappas. "Temporal Logic Motion Planning for Dynamic Robots." *Automatica*, vol. 45, no. 2, pages 343–352, 2009.

[Fainekos09]. G. E. Fainekos and G. J. Pappas. "Robustness of Temporal Logic Specifications for Continuous-Time Signals." *Theoretical Computer Science*, vol. 410, no. 42, pages 4262–4291, 2009.

[Förstner02]. D. Förstner, M. Jung, and J. Lunze. "A Discrete-Event Model of Asynchronous Quantized Systems." *Automatica*, vol. 38, no. 8, pages 1277–1286, 2002.

[Girard11]. A. Girard. "Controller Synthesis for Safety and Reachability via Approximate Bisimulation." Technical report, Université de Grenoble, 2011. http://arxiv.org/abs/1010.4672v1.

[Girard12]. A. Girard. "Controller Synthesis for Safety and Reachability via Approximate Bisimulation." *Automatica*, vol. 48, no. 5, pages 947–953, 2012.

[Girard13]. A. Girard. "Low-Complexity Quantized Switching Controllers Using Approximate Bisimulation." *Nonlinear Analysis: Hybrid Systems*, vol. 10, pages 34–44, 2013.

[Girard07]. A. Girard and G. J. Pappas. "Approximation Metrics for Discrete and Continuous Systems." *IEEE Transactions on Automatic Control*, vol. 52, pages 782–798, 2007.

[Girard09]. A. Girard and G. J. Pappas. "Hierarchical Control System Design Using Approximate Simulation." *Automatica*, vol. 45, no. 2, pages 566–571, 2009.

[Girard10]. A. Girard, G. Pola, and P. Tabuada. "Approximately Bisimilar Symbolic Models for Incrementally Stable Switched Systems." *IEEE Transactions on Automatic Control*, vol. 55, no. 1, pages 116–126, 2010.

[Gol12]. E. A. Gol, M. Lazar, and C. Belta. "Language-Guided Controller Synthesis For Discrete-Time Linear Systems." *Proceedings of the ACM International Conference on Hybrid Systems: Computation and Control*, pages 95–104, 2012.

[Henzinger00]. T. A. Henzinger, B. Horowitz, R. Majumdar, and H. Wong-Toi. "Beyond HyTech: Hybrid Systems Analysis Using

Interval Numerical Methods." In *Hybrid Systems: Computation and Control*, pages 130–144. Springer, 2000.

[Karaman11]. S. Karaman and E. Frazzoli. "Sampling-Based Algorithms for Optimal Motion Planning." *International Journal of Robotics Research*, vol. 30, no. 7, pages 846–894, 2011.

[Kloetzer08]. M. Kloetzer and C. Belta. "A Fully Automated Framework for Control of Linear Systems from Temporal Logic Specifications." *IEEE Transactions on Automatic Control*, vol. 53, pages 287–297, 2008.

[Koutsoukos00]. X. D. Koutsoukos, P. J. Antsaklis, J. A. Stiver, and M. D. Lemmon. "Supervisory Control of Hybrid Systems." *Proceedings of the IEEE*, vol. 88, no. 7, pages 1026–1049, 2000.

[Kress-Gazit09]. H. Kress-Gazit, G. E. Fainekos, and G. J. Pappas. "Temporal-Logic–Based Reactive Mission and Motion Planning." *IEEE Transactions on Robotics*, vol. 25, no. 6, pages 1370–1381, 2009.

[Le Corronc13]. E. Le Corronc, A. Girard, and G. Gössler. "Mode Sequences as Symbolic States in Abstractions of Incrementally Stable Switched Systems." *Proceedings of the 52nd IEEE Conference on Decision and Control*, pages 3225–3230, 2013.

[Liberzon99]. D. Liberzon and A. S. Morse. "Basic Problems in Stability and Design of Switched Systems." *IEEE Control Systems Magazine*, vol. 19, no. 5, pages 59–70, 1999.

[Liu14]. J. Liu and N. Ozay. "Abstraction, Discretization, and Robustness in Temporal Logic Control of Dynamical Systems." *Proceedings of the ACM International Conference on Hybrid Systems: Computation and Control*, pages 293–302, 2014.

[Liu13]. J. Liu, N. Ozay, U. Topcu, and R. M. Murray. "Synthesis of Reactive Switching Protocols from Temporal Logic Specifications." *IEEE Transactions on Automatic Control*, vol. 58, no. 7, pages 1771–1785, 2013.

[Maly13]. M. R. Maly, M. Lahijanian, L. E. Kavraki, H. Kress-Gazit, and M. Y. Vardi. "Iterative Temporal Motion Planning for Hybrid Systems in Partially Unknown Environments." *Proceedings of the 16th International Conference on HSCC*, pages 353–362, 2013.

[Mari10]. F. Mari, I. Melatti, I. Salvo, and E. Tronci. "Synthesis of Quantized Feedback Control Software for Discrete Time Linear Hybrid Systems." In *Computer Aided Verification*, pages 180–195. Springer, 2010.

[Mazo10]. M. Mazo, Jr., A. Davitian, and P. Tabuada. "Pessoa: A Tool for Embedded Controller Synthesis." In *Computer Aided Verification*, pages 566–569. Springer, 2010.

[Milner89]. R. Milner. *Communication and Concurrency*. Prentice Hall, 1995.

[Moor02a]. T. Moor and J. Raisch. "Abstraction Based Supervisory Controller Synthesis for High Order Monotone Continuous Systems." In *Modelling, Analysis, and Design of Hybrid Systems*, pages 247–265. Springer, 2002.

[Moor02]. T. Moor, J. Raisch, and S. O'Young. "Discrete Supervisory Control of Hybrid Systems Based on l-Complete Approximations." *Discrete Event Dynamic Systems*, vol. 12, no. 1, pages 83–107, 2002.

[Mouelhi13]. S. Mouelhi, A. Girard, and G. Gössler. "Cosyma: A Tool for Controller Synthesis Using Multi-Scale Abstractions." *Proceedings of the ACM International Conference on Hybrid Systems: Computation and Control*, pages 83–88, 2013.

[Ozay13]. N. Ozay, J. Liu, P. Prabhakar, and R. M. Murray. "Computing Augmented Finite Transition Systems to Synthesize Switching Protocols for Polynomial Switched Systems." American Control Conference, pages 6237–6244, 2013.

[Pnueli89]. A. Pnueli and R. Rosner. "On the Synthesis of a Reactive Module." *Proceedings of the 16th ACM SIGPLAN-SIGACT Symposium on Principles of Programming Languages*, pages 179–190, 1989.

[Pola08]. G. Pola, A. Girard, and P. Tabuada. "Approximately Bisimilar Symbolic Models for Nonlinear Control Systems." *Automatica*, vol. 44, no. 10, pages 2508–2516, 2008.

[Pola10]. G. Pola, P. Pepe, M. Di Benedetto, and P. Tabuada. "Symbolic Models for Nonlinear Time-Delay Systems Using Approximate Bisimulations." *Systems and Control Letters*, vol. 59, no. 6, pages 365–373, 2010.

[Pola09]. G. Pola and P. Tabuada. "Symbolic Models for Nonlinear Control Systems: Alternating Approximate Bisimulations." *SIAM Journal on Control and Optimization*, vol. 48, no. 2, pages 719–733, 2009.

[Reißig11]. G. Reißig. "Computing Abstractions of Nonlinear Systems." *IEEE Transactions on Automatic Control*, vol. 56, no. 11, pages 2583–2598, 2011.

[Reißig13]. G. Reißig and M. Rungger. "Abstraction-Based Solution of Optimal Stopping Problems Under Uncertainty." *Proceedings of the 52nd IEEE Conference on Decision and Control*, pages 3190–3196, 2013.

[Reißig14]. G. Reißig and M. Rungger. "Feedback Refinement Relations for Symbolic Controller Synthesis." *Proceedings of the 53rd IEEE Conference on Decision and Control*, 2014.

[Rungger13]. M. Rungger, M. Mazo, and P. Tabuada. "Specification-Guided Controller Synthesis for Linear Systems and Safe Linear-Time Temporal Logic." *Proceedings of the ACM International Conference on Hybrid Systems: Computation and Control*, pages 333–342, 2013.

[Rungger12]. M. Rungger and O. Stursberg. "On-the-Fly Model Abstraction for Controller Synthesis." American Control Conference, pages 2645–2650, 2012.

[Senesky03]. M. Senesky, G. Eirea, and T. J. Koo. "Hybrid Modelling and Control of Power Electronics." In *Hybrid Systems: Computation and Control*, pages 450–465. Springer, 2003.

[Shorten98]. R. N. Shorten and K. S. Narendra. "On the Stability and Existence of Common Lyapunov Functions for Stable Linear Switching Systems." *Proceedings of the 37th IEEE Conference on Decision and Control*, vol. 4, pages 3723–3724, 1998.

[Tabuada08]. P. Tabuada. "An Approximate Simulation Approach to Symbolic Control." *IEEE Transactions on Automatic Control*, vol. 53, no. 6, pages 1406–1418, 2008.

[Tabuada09]. P. Tabuada. *Verification and Control of Hybrid Systems: A Symbolic Approach*. Springer, 2009.

[Tabuada05]. P. Tabuada and G. J. Pappas. "Hierarchical Trajectory Refinement for a Class of Nonlinear Systems." *Automatica*, vol. 41, no. 4, pages 701–708, 2005.

[Tabuada06]. P. Tabuada and G. J. Pappas. "Linear Time Logic Control of Discrete-Time Linear Systems." *IEEE Transactions on Automatic Control*, vol. 51, no. 12, pages 1862–1877, 2006.

[Tazaki12]. Y. Tazaki and J. Imura. "Discrete Abstractions of Nonlinear Systems Based on Error Propagation Analysis." *IEEE Transactions on Automatic Control*, vol. 57, pages 550–564, 2012.

[Tiwari08]. A. Tiwari. "Abstractions for Hybrid Systems." *Formal Methods in System Design*, vol. 32, no. 1, pages 57–83, 2008.

[Vardi95]. M. Y. Vardi. "An Automata-Theoretic Approach to Fair Realizability and Synthesis." In *Computer Aided Verification*, pages 267–278. Springer, 1995.

[Vardi96]. M. Y. Vardi. "An Automata-Theoretic Approach to Linear Temporal Logic." In *Logics for Concurrency*, pages 238–266. Springer, 1996.

[Wongpiromsarn12]. T. Wongpiromsarn, U. Topcu, and R. M. Murray. "Receding Horizon Temporal Logic Planning." *IEEE Transactions on Automatic Control*, vol. 57, no. 11, pages 2817–2830, 2012.

[Wongpiromsarn11]. T. Wongpiromsarn, U. Topcu, N. Ozay, H. Xu, and R. M. Murray. "TuLip: A Software Toolbox for Receding Horizon Temporal Logic Planning." *Proceedings of the ACM International Conference on Hybrid Systems: Computation and Control*, pages 313–314, 2011.

[Yordanov12]. B. Yordanov, J. Tumová, I. Černá, J. Barnat, and C. Belta. "Temporal Logic Control of Discrete-Time Piecewise Affine Systems." *IEEE Transactions on Automatic Control*, vol. 57, no. 6, pages 1491–1504, 2012.

[Zamani12]. M. Zamani, G. Pola, M. Mazo, and P. Tabuada. "Symbolic Models for Nonlinear Control Systems Without Stability Assumptions." *IEEE Transactions on Automatic Control*, vol. 57, pages 1804–1809, 2012.

[Zamani14]. M. Zamani, I. Tkachev, and A. Abate. "Bisimilar Symbolic Models for Stochastic Control Systems Without State-Space Discretization." *Proceedings of the ACM International Conference on Hybrid Systems: Computation and Control*, pages 41–50, 2014.

Chapter 5

Software and Platform Issues in Feedback Control Systems

Karl-Erik Årzén and Anton Cervin

Control theory is one of the key intellectual underpinnings that allows us to analyze the interaction of software with the physical world. However, control theory abstracts software actions as actions that take no time to execute. In reality, software execution takes time—and this execution time can be affected by a number of platform issues, including the speed of the processor, the scheduler used to share the processor between multiple tasks, and network delays that occur when computations go across networks, among others. In this chapter we discuss such software and platform effects, as well as techniques used to design control systems that take these effects into account.

5.1 Introduction and Motivation

Similar to cyber-physical systems (CPS), control is a discipline in which tight interaction with the physical environment is necessary. This statement is not meant to imply that all control systems should be labeled CPS. Rather, it is only in the following cases that the control can be considered as cyber-physical in nature:

- *Hybrid control:* In hybrid control, the process under control and/or the controller are modeled as hybrid dynamic systems—for example, using hybrid automata. This is the topic of Chapter 4.
- *Control of CPS applications:* Control of CPS applications covers the case in which control is used in one of the typical CPS applications—for example, the electric grid, urban transportation systems, data centers, and clean energy buildings.
- *Large-scale distributed control:* CPS applications are often large scale, distributed, and sparse in nature. Partly motivated by these characteristics, distributed control and optimization are currently attracting high interest in the control community. The background for this attention is the poor scalability of many centralized multivariable control design methods, such as linear quadratic Gaussian (LQG) control and H_∞ control. An alternative approach is possible for so-called positive systems, in which the sparsity of the system can be exploited both in performance and stability analysis and in the controller synthesis [Rantzer12]. Rather than solving a large-scale, ill-conditioned synthesis problem, a sequence of smaller, well-posed problems is solved. Alternatively, the smaller problems may be solved in parallel using, for example, multiple cores, or they may be solved in a decentralized way by local agents that are required to have only local process knowledge.
- *Resource-aware control:* In traditional control, the computing system implementing the controller is viewed as a machine that is able to realize the abstraction of a discrete-time difference equation in an ideal way. The fact that computations take time, which in many cases is not constant, along with the fact that the amount of computations that can be performed in parallel is limited by the number of processors available, is often disregarded. A similar situation holds for networked control, in which control loops are closed over a communication network that is often idealized as having a constant delay—even though for most network protocols this delay is far

from constant, and, in particular for wireless networks, it can even be unbounded due to lost packets. The result of these differences is temporal nondeterminism in most cases, caused by limited computing and communication resources in the implementation platform and by the sharing of these resources among multiple applications. This is a common situation in CPS. In resource-aware control, this nondeterminism and the effect it has on the control performance should be taken into account in the analysis, design, and implementation of the control system. That is the topic of this chapter.

5.2 Basic Techniques

In this section, we lay out the basic techniques that allow us to take into account the computation time in the control design.

5.2.1 Controller Timing

A controller performs three main operations: sampling, computation, and actuation. During sampling, the output of the process under control (i.e., the input to the controller) is obtained using, for example, a sensor connected to an A/D-converter. During computation, the output of the controller (i.e., the control signal) is calculated as a function of process output, the desired value, or the reference value for the process output and the internal state of the controller. Finally, during actuation, the control signal is effectuated using, for example, a D/A-converter in combination with an actuator. If the controller is implemented in a single processor, all the operations are implemented in this processor. If the control system is closed over a network—that is, in a networked controller—the sampling, computation, and actuation may be implemented in different computing nodes.

Traditionally, control assumes that sampling is performed periodically and that actuation is performed with as little latency as possible. This scenario is depicted in Figure 5.1, where I denotes the sampling and O denotes the actuation.

In contrast, in embedded and networked applications, the situation may be like that shown in Figure 5.2. In this scenario, it is assumed that the nominal sampling instants of the controller are given by $t_k = hk$, where h is the nominal sampling interval of the controller.

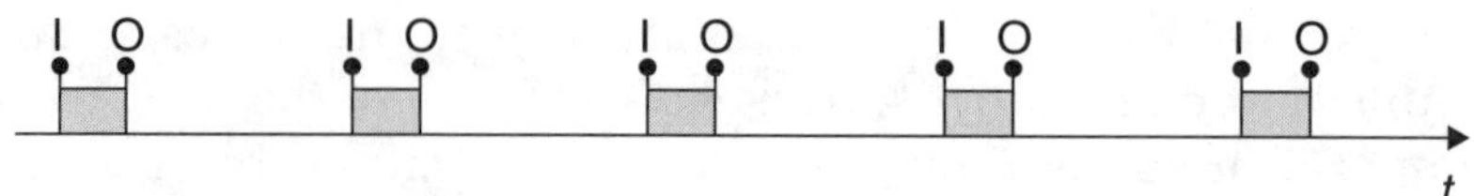

Figure 5.1: *Periodic controller timing*

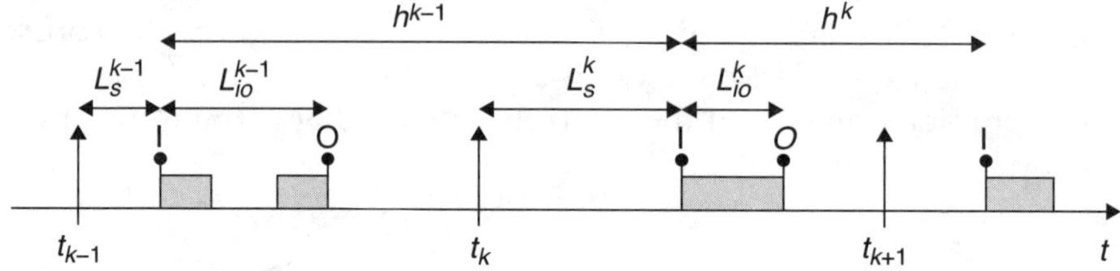

Figure 5.2: *Controller timing*

Due to the resource sharing, the sampling may be delayed for some time L_s—the sampling latency of the controller. A dynamic resource-sharing policy will introduce variations in this interval. The sampling jitter is quantified by the difference between the maximum and minimum sampling latencies in all executions of the controller:

$$J_s \stackrel{\text{def}}{=} \max_k L_s^k - \min_k L_s^k \tag{5.1}$$

Normally, it can be assumed that the minimum sampling latency of a controller is zero, in which case we have the following:

$$J_s = \max_k L_s^k$$

Jitter in the sampling latency will also introduce jitter in the sampling interval h. From Figure 5.2, it is seen that the actual sampling interval in period k is given by

$$h^k = h - L_s^{k-1} + L_s^k \tag{5.2}$$

The sampling interval jitter is quantified as follows:

$$J_h \stackrel{\text{def}}{=} \max_k h^k - \min_k h^k \tag{5.3}$$

We can see that the sampling interval jitter is upper bounded by the following:

$$J_h \leq 2J_s \tag{5.4}$$

After some computation time and possibly additional time spent waiting for access to the computing resources, the controller will actuate the control signal. The delay from the sampling to the actuation is called the input-output latency, denoted L_{io}. Varying execution times or delays caused by the resource scheduling will introduce variations in this interval. The input-output jitter is quantified as follows:

$$J_{io} \stackrel{\text{def}}{=} \max_k L_{io}^k - \min_k L_{io}^k \tag{5.5}$$

5.2.2 Control Design for Resource Efficiency

A discrete-time controller can be designed according to two main principles:

- Discretization of a controller designed in continuous time
- Design of a controller in discrete time based on a discretized process model

In both cases it is assumed that the process to be controlled is modeled by a continuous-time state space model. For the sake of simplicity, we will assume this model is linear and time invariant (LTI):

$$\begin{aligned} \frac{dx}{dt} &= Ax(t) + Bu(t) \\ y(t) &= Cx(t) + Du(t) \end{aligned} \tag{5.6}$$

Here, $y(t)$ is the process output, $x(t)$ is the process state, and $u(t)$ is the process input. The system has n state variables, r inputs, and p outputs, and A, B, C, and D are matrices of appropriate size.

In the first approach, the controller is designed using continuous-time control theory using some suitable design method—for example, PID control, state feedback, output feedback, and so on. The resulting controller will also be a continuous-time dynamic LTI system that can be described in state space form, as in the following example:

$$\begin{aligned} \frac{dx_c}{dt} &= Ex_c(t) + F_y y(t) + F_r r(t) \\ u(t) &= Gx_c(t) + H_y y(t) + H_r r(t) \end{aligned} \tag{5.7}$$

Here x_c is the state of the controller, $y(t)$ is the controller input (i.e., the output of process), $r(t)$ is the reference value for the controller, and $u(t)$ is the controller output. This controller is then discretized by approximating the continuous-time derivatives with discrete-time finite differences—for example, a forward difference approximation, a backward difference approximation, or a Tustin approximation. The resulting controller will then be a discrete-time state space equation:

$$\begin{aligned} x_c(kh+h) &= f(x_c(kh), y(kh), r(kh)) \\ u(kh) &= g(x_c(kh), y(kh), r(kh)) \end{aligned} \tag{5.8}$$

This controller is straightforward to translate into code. The approximation in most cases works well if the sampling period, h, is short.

In the second approach, the continuous-time process model is first discretized. This is done by solving the system equations, as shown in Equation (5.6), while looking at only how the state evolves from one sampling instant $x(kh)$ to the next sampling instant $x(kh+h)$ [Åström11]. This is normally done under the assumption that the process input is held constant between the sampling instants. For example, this case applies if the so-called zero-order-hold (ZOH) actuation is used. The resulting discrete-time process model can then be written as follows:

$$\begin{aligned} x(kh+h) &= \Phi x(kh) + \Gamma u(kh) \\ y(kh) &= Cx(kh) + Du(kh) \end{aligned} \tag{5.9}$$

We then have the following:

$$\begin{aligned} \Phi &= \Phi(kh + h, kh) = e^{Ah} \\ \Gamma &= \Gamma(kh+h, kh) = \int_0^h e^{As}\, ds B \end{aligned} \tag{5.10}$$

Hence, at the sampling times, the system is described by a linear time-invariant difference equation. Note that this simplification is obtained without any approximations. Since the output $y(kh)$ in most cases is measured before the control signal $u(kh)$ is applied to the system, it follows that $D = 0$. The ZOH sampling can also be done for processes containing time delays.

Given the discrete-time process model shown in Equation (5.9), the control design is then done using some discrete-time control design method—for example, PID control, state feedback, output feedback, and so on—in a manner completely analogous to that used in the continuous-time case. Controllers designed for discrete-time processes typically do not require as short a sampling period as controllers that have been obtained by discretizing a continuous-time controller.

To decrease the controller's resource consumption, two approaches may be used: (1) reduce the computation time or (2) sample less frequently. We will discuss these techniques in the next section.

5.3 Advanced Techniques

As the complexity of applications increases and more features need to be fitted into a smaller number of computers, an increasingly more efficient use of computation resources is required. In this section we discuss advanced techniques for efficient resource usage in control applications.

5.3.1 Reducing the Computation Time

In most cases, the amount of computation and its associated computation time depend on the dynamics of the controlled process and the requirements of the closed system. Hence, in general, it is difficult to influence the amount of computation that is required, though there are still some possibilities.

If the controller is implemented on a low-end microcontroller without hardware support for floating-point arithmetic, considerable savings in execution time can be obtained if fixed-point arithmetic is used instead of software-emulated floating-point arithmetic. For small controllers, for example, a savings of approximately one order of magnitude can be obtained with this technique.

If CPU resources can be traded against memory, then it is sometimes possible to precompute part of the control algorithm. This approach is, for example, used extensively in control of combustion engines in automotive systems, where nonlinear mappings are represented as lookup tables rather than approximated by nonlinear functions. Similar techniques are used in model-predictive control (MPC) [Maciejowski01]. In conventional MPC, a quadratic programming problem is solved at each sample *kh*, generating a sequence of control signals, $u(kh)+u(kh+h), u(kh+2h), \ldots$. Following the receding horizon principle, only the first of these control signals is applied to the process, and then in the next sample the entire procedure is repeated; that is, the optimization is performed anew. This sequence can be rather time-consuming. In contrast, in explicit MPC [Bemporad00], the control signal is precomputed, generating a set of piecewise affine functions. During execution, a lookup is performed based on the current operating point and the selected function is applied.

It is also possible to use an "anytime approach" in control. With this technique, a controller generates a control signal that gives better closed-loop performance as more computing resources are provided to it—that is, the longer it is allowed to run. Again, this approach may be applied with MPC. Under certain conditions, if the optimization is executed long enough to find a feasible solution—that is, if the solution guarantees that the constraints on the state and control signal are met—then the control loop will be stable, even if the optimal solution has not been found yet [Scokaert99]. It is also plausible to consider controllers that consist of a discrete set of control algorithms, each requiring a different amount of computing resources. Depending on the resources at hand, the overall controller then dynamically selects which algorithm to apply. Although these kinds of anytime control algorithms are interesting from a research point of view, they are currently quite far away from industrial adoption.

For networked control, the network bandwidth plays a similar role as the CPU time. Reducing the bandwidth consumed by a networked controller corresponds to sending less information over the network; that is, it uses higher levels of quantization. In the extreme case, only single symbols are sent—for example, if a signal should be increased or decreased. A great deal of research is being done in this area (see [Yüksel11] and [Heemels09]), but this approach also remains quite far from industrial use.

5.3.2 Less Frequent Sampling

An alternative to minimizing the amount of computations that are performed in each sample is to minimize the number of samples—that is, to sample less frequently. Two approaches to achieve this goal are possible:

- Reduce the sampling rate of a periodic controller.
- Sample aperiodically.

Reducing the sampling rate of a controller is often straightforward. In many cases, the control performance increases gradually with the sampling frequency, up to some limit where numerical effects reduce the performance. Acceptable control performance is often obtained for a large range of sampling periods. For example, following is an often-used rule of thumb for selecting the sampling interval:

$$h\omega_c \approx 0.05 \text{ to } 0.14 \tag{5.11}$$

Here ω_c is the crossover frequency (in radians per second) of the continuous-time system. With the longest suggested sampling interval, there is still a considerable margin up to the sampling interval where the closed-loop system becomes unstable.

The use of long sampling intervals is not without problems, however. Between the sampling instants, the process runs in an open loop, which means that any disturbances affecting the plant will not be seen and counteracted by the controller until the next sampling instant. Thus, the control performance may be acceptable from a theoretical standpoint, yet be unacceptable in practice unless all the disturbances are modeled and included in the performance analysis.

Since acceptable performance usually is obtained for a range of sampling periods, it is possible to use the sampling period as a gain-scheduling variable—that is, to dynamically decide how large the sampling period should be based on how much computing resource is available. Depending on the type of controller used, the controller parameters might have to be recalculated every time the sampling period is changed.

It is also possible to let the range of acceptable sampling periods depend on the state of the controller. This approach is motivated by the observation that the requirements for a high sampling rate are larger during transitional periods than while the controller is in a stable period. Examples of such transitions could be a change in the reference value or the arrival of a load disturbance.

A potential risk with all schemes that switch the controller dynamics in a deterministic way is switching-induced instabilities [Lin09]. Switching between two closed-loop controllers in a deterministic way may result in an unstable system, even if each of the controllers is stable on its own. However, this risk is mainly present in the case where the switches are very frequent—for example, at every sampling instant.

5.3.3 Event-Based Control

An alternative to periodic sampling is to sample the process aperiodically—for example, only when the error signal has changed sufficiently much. During recent years the interest in aperiodic, or event-based, control has increased substantially, mostly motivated by the communication constraints in network control; that is, communication can be viewed as expensive. However, event-based control also has other motivations, such as when the sensing operation, actuation, or computation is expensive—that is, when computing resources are scarce. Another motivation is that control in biological systems, including humans doing manual control, in general is event based rather than periodic.

Several event-based control settings have been defined. In aperiodic event-based control, no lower bound on the inter-arrival times of the sampling instants exist, such that the controller may execute infinitely often. In sporadic event-based control, a lower bound does exist. In that case, once the controller has performed an execution, it must wait for at least the amount of time specified as the lower bound before it is allowed to execute again. In essence, there is a price associated with executing too early.

Different event-detection mechanisms are also possible. One possibility is to sample only when the controller error crosses a certain threshold. Another possibility is to sample when the error has changed more than a certain value, an approach referred to as sample-on-delta. In both cases, the controller must be complemented by a sensing mechanism that decides when it is time to sample. This mechanism can sometimes be built into the sensor hardware.

The work on event-based control can be divided into model-free and model-based approaches. The model-free approach is often based on a PID scheme that is complemented by a mechanism that decides when the control execution should take place and controller parameters that depend on the actual time-varying sampling interval. Several examples of this setup have been reported that achieved similar control performance as in the periodic case but with less computing resource (see [Årzén99], [Beschi12], and [Durand09]).

With the model-based approach to event-based control, the overall aim is to develop a system theory for aperiodically sampled systems. The nonlinear effects caused by the aperiodic sampling make this a very challenging task. Although many results have been derived, they are still mostly applicable to low-order systems. For more information, see [Henningsson12] and [Heemels08].

An alternative version of event-based control is provided by the self-triggered control [Tabuada07, Wang08]. In addition to computing the control signal, a self-triggered controller calculates the next time instant when the controller should be invoked. A prerequisite for this approach to work is that the disturbances acting on the system must have been correctly modeled and taken into account in the control design.

5.3.4 Controller Software Structures

When implementing a controller in software, the ideal case is to have negligible sampling latency and a short input-output latency. If the input-output latency jitter is also negligible, then the control design becomes straightforward, as it is easy to compensate for a constant input-output latency at design time.

Deterministic sampling and actuation are easiest to achieve when static scheduling is used—that is, when the sampling and actuation are performed at predefined instants determined by the clock and the execution is performed in a dedicated time slot or execution window. If the controller is networked, then the communication also has to be statically scheduled to minimize the input-output latency. A drawback with static scheduling, however, is that it tends to under-utilize the computing resources. Typically the execution window is sized based on the longest (worst-case) execution time of the controller code. Due to data dependencies, code branching, effects caused by memory hierarchies, and pipelining, there is typically a large difference between the worst-case execution time and the average-case execution time. With static scheduling, there is no easy way to make use of the CPU time that remains unused within each time slot, which leads to resource under-utilization.

Another consequence of the time variation is that although the execution of the controller is completed and the actuation could be performed, the actuation will not be performed until the prescheduled actuation instant arrives. If the difference between the average case and the worst case is large, the control performance achieved with a varying input-output latency is often better than the performance achieved with a longer but constant latency, even if the constant latency is compensated for—that is, if it is included in the control design. Hence, from a control performance point of view, it can be advantageous to relax the constraint on deterministic actuation instants.

This last point is the motivation for more dynamic implementation techniques—for example, implementing the controllers as tasks in a real-time operating system (RTOS) using priority or deadline-based scheduling, and using event-triggered network protocols rather than time-triggered protocols. When implementing a controller as a task, it is still beneficial to use deterministic sampling. This can be achieved by performing the sampling in a dedicated high-priority task or by using hardware support. To reduce the input-output latency, a common practice is to split the controller code into two parts, *Calculate Output* and *Update State*, and to organize the controller code as follows: *Sample*, *Calculate Output*, *Actuate*, and *Update State*.

The *Calculate Output* segment contains only the part of the control algorithm that depends on the current input signal. All of the remaining work—the updates of the controller state and precalculations—takes place in the *Update State* segment—that is, after the actuation.

Consider a general linear controller in discrete-time state space form:

$$\begin{aligned} x_c(k+1) &= Ex_c(k) + F_y y(k) + F_r r(k) \\ u(k) &= Gx_c(k) + H_y y(k) + H_r r(k) \end{aligned} \tag{5.12}$$

For simplicity, we have assumed that $h = 1$. The pseudocode for this controller would then be the following:

```
y = getInput(); r = getReference();    // Sample
u = u1 + Hy*y + Hr*r;                  // Calculate Output
sendOutput(u);                         // Actuate
xc = E*xc + Fy*y + Fr*r; u1 = G*xc;    // Update State
```

5.3.5 Sharing of Computing Resources

When a controller is implemented as a task in a real-time operating system or when a controller is closed over a network, then the CPU and network resources are shared with other tasks and other communication links. This sharing gives rise to interference that affects the temporal determinism, and hence the control performance, negatively. The nature of the interference is determined by how the shared resource is scheduled.

If fixed-priority task scheduling is used, a task will be affected only by tasks of higher or equal priority. In contrast, in earliest-deadline-first (EDF) scheduling, all tasks may influence all other tasks. For networks, the protocol determines the specific nature of the interference. For example, when the Controller Area Network (CAN) protocol is used, a network transmission is nonpreemptive once it has successively passed the arbitration. Alternatively, if a shared Ethernet network is used, repeated collisions may occur, causing potentially infinite delays. The interference can be interpreted as a nonlinearity: A small change in a task parameter (i.e., the period, execution time, deadline, or priority) may give rise to difficult-to-predict changes in input-output latency and jitter.

One way to reduce or remove the interference is to use reservation-based resource scheduling. Reservation-based scheduling divides the physical resource into a number of virtual resources—that is, virtual processors or virtual networks—that are temporally isolated from each other. The virtual processor can be seen as a processor that executes at a speed given by the following bandwidth:

$$B_i = \frac{Q_i}{P_i}$$

Here Q_i is the budget of processor i and P_i is the period of processor i. The reservation mechanism guarantees the tasks associated with processor i access to the CPU for Q_i time units each P_i time units.

A number of reservation mechanisms have been proposed, both for fixed-priority and EDF task scheduling. One of the more popular ones is the Constant Bandwidth Server (CBS) for EDF [Abeni98]. A similar development is also taking place for networks, where bandwidth reservation mechanisms can be found at several protocol layers, providing the abstraction by which the sending node and the receiving node communicate using a dedicated slower network.

An alternative and more straightforward way to provide temporal isolation is to use time division, in which the resource is shared among the users according to a precalculated cyclic schedule. This schedule is is divided into time slots, and each time slot is assigned to a single task in the case of CPU scheduling, and to a single sending node in the case of network scheduling. Time-division CPU scheduling, which is also known as static (cyclic) scheduling, is the oldest scheduling model used for control applications. Its high level of determinism is the main reason why it remains the preferred method for safety-critical applications, such as flight control systems. In the network case, several time-division protocols are available, including TT-CAN, TTP, and TT-Ethernet. This approach is also becoming increasingly common with protocols where time-driven communication is combined with event-driven communication as, for example, in Flexray and Profinet-IO. To reduce the input-output latency in networked control, it is important that the task schedules and the network schedules are synchronized.

5.3.5.1 The Control Server

An example of a reservation-based scheduling model developed especially for control is the control server [Cervin03]. In this model, a hybrid scheme is used that combines time-triggered I/O and intertask communication with dynamic, reservation-based task scheduling. The model provides the following features:

- Isolation between unrelated tasks
- Short input-output latencies
- Minimal sampling jitter and input-output jitter
- Simple interface between the control design and the real-time design

- Predictable control and real-time behavior, including during overruns
- Ability to combine several tasks (components) into a new task (component) with predictable control and real-time behavior

Isolation between unrelated tasks is achieved through the use of constant-bandwidth servers. The servers make each task appear as if it is running on a dedicated CPU with a given fraction of the original CPU speed. To facilitate short latencies, a task may be divided into a number of segments, which are then scheduled individually. A task may, for example, read inputs (from the environment or from other tasks) only at the beginning of a segment and write outputs (to the environment or to other tasks) only at the end of a segment. All communication is handled by the kernel and, therefore, is not prone to jitter.

The last three features in the previously given list are addressed through the combination of bandwidth servers and statically scheduled communication points. For periodic tasks with constant execution times, the control server model creates the illusion of a perfect division of the CPU. The model makes it possible to analyze each task in isolation, from both the scheduling and the control points of view. As with ordinary EDF scheduling, the schedule for the task set is simply determined by the total CPU utilization (ignoring context switches and the I/O operations performed by the kernel). The performance of a controller can also be viewed as a function of its allotted CPU share. These properties make the control server model very suitable for feedback scheduling applications.

Furthermore, this model makes it possible to combine two or more communicating tasks into a new task. The new task will consume a fraction of the CPU equal to the sum of the utilization of the constituting tasks. It will have a predictable I/O pattern, and, therefore, demonstrate predictable control performance. Control tasks, in turn, may be treated as real-time components, which can be combined into new components.

5.3.6 Analysis and Simulation of Feedback Control Systems

As soon as event-driven and dynamic implementation techniques are used, it is important to be able to analyze how the implementation of the technique affects the control performance. This analysis consists of two steps. The first step consists of determining how the platform parameters—for example, scheduling methods and scheduling parameters such as priorities and deadlines—and the network protocol and protocol parameters affect the sampling latency and jitter as well as the input-output latency and jitter. This relationship is often difficult to evaluate analytically.

The second step consists of evaluating how the latency distributions affect the control performance—that is, how temporally robust the system is. Nevertheless, one must understand that the control performance is also influenced by a number of other factors:

- The dynamics of the open-loop system
- The desired dynamics of the closed-loop system
- The type of controller used and how it has been designed
- Disturbances acting on the system
- The means by which control performance is defined

Constant delays can be accounted for in all model-based discrete-time control design methods. However, some design methods (e.g., LQG control) can also take latency distributions into account. Control performance can be defined in many ways. One general approach is to measure the cost (i.e., the inverted performance) as a quadratic function of the state variables and the control signal.

This section profiles three approaches to temporal analysis of control systems presented: Jitterbug, the Jitter Margin, and TrueTime.

5.3.6.1 Jitterbug

Jitterbug [Cervin03a, Lincoln02] is a MATLAB-based toolbox that computes a quadratic performance criterion for a linear control system under various timing conditions. This tool can also compute the spectral density of the signals in the system. Using the toolbox, one can assert how sensitive a control system should be to delay, jitter, lost samples, and so on, without resorting to simulation. Jitterbug is quite general and can also be used to investigate jitter-compensating controllers, aperiodic controllers, and multirate controllers. The main contribution of this toolbox, which is built on well-known theory (LQG theory and jump linear systems), is to make it easy to apply this type of stochastic analysis to a wide range of problems.

Jitterbug offers a collection of MATLAB routines that allow the user to build and analyze simple timing models of computer-controlled systems. To build a control system, the user connects a number of continuous- and discrete-time systems. For each subsystem, optional noise and cost specifications may be given. In the simplest case, the discrete-time systems are assumed to be updated in order during the control period. For each discrete system, a random delay (described by a discrete probability density function) can be specified that must elapse before the next system is updated. The total cost of the system (summed over all subsystems) is computed algebraically if the timing model system is periodic or iteratively if the timing model is aperiodic.

To make the performance analysis feasible, Jitterbug can handle only a certain class of system. The control system is built from linear systems driven by white noise, and the performance criterion to be evaluated is specified as a quadratic, stationary cost function. The timing delays in one period are assumed to be independent from the delays in the previous period. Also, the delay probability density functions are discretized using a time-grain that is common to the whole model.

Although a quadratic cost function can hardly hope to capture all aspects of a control loop, it can still be useful during design space exploration or tradeoff analysis, when one wants to judge several possible controller implementations against each other. A higher value of the cost function typically indicates that the closed-loop system is less stable (i.e., more oscillatory), whereas an infinite cost means that the control loop is unstable. The cost function can easily be evaluated for a large set of design parameters and can be used as a basis for the control and real-time design.

Jitterbug Models

In Jitterbug, a control system is described by two parallel models: a signal model and a timing model. The signal model is given by a number of connected, linear, continuous- and discrete-time systems. The timing model consists of a number of timing nodes and describes when the different discrete-time systems should be updated during the control period.

An example is shown in Figure 5.3, where a computer-controlled system is modeled by four blocks. The plant is described by the continuous-time system G, and the controller is described by the three discrete-time systems H_1, H_2, and H_3. The system H_1 could represent a periodic sampler, H_2 could represent the computation of the control

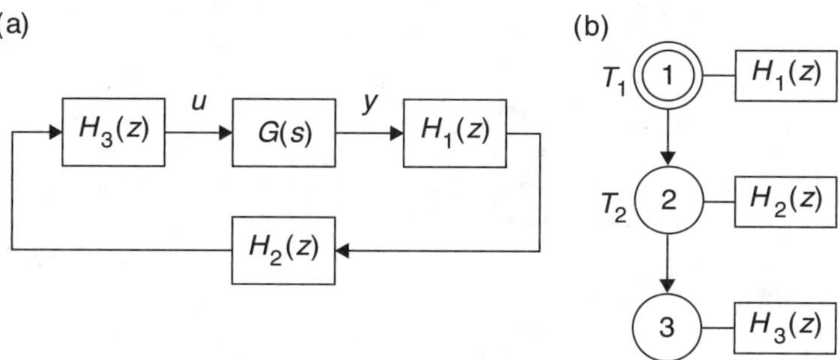

Figure 5.3: *A simple Jitterbug model of a computer-controlled system: (a) signal model and (b) timing model*

signal, and H_3 could represent the actuator. The associated timing model says that, at the beginning of each period, H_1 should first be executed (updated). Then there is a random delay τ_1 until H_2 is executed, and another random delay τ_2 until H_3 is executed. These delays could model computational delays, scheduling delays, or network transmission delays.

The same discrete-time system may be updated in several timing nodes. For each of the various cases, one can specify a different equation for the update operation. This approach can be used to model a filter, for example, where the update equations look different depending on whether a measurement value is available. One can also make the update equations depend on the time that has passed since the first node became active. Such an approach can be used to model jitter-compensating controllers, for example.

For some systems, it is desirable to specify alternative execution paths (and thereby multiple next nodes). In Jitterbug, two such cases can be modeled (Figure 5.4):

- A vector n of next nodes can be specified with a probability vector p. After the delay, execution node $n(i)$ will be activated with probability $p(i)$. This strategy can be used to model a sample being lost with some probability.
- A vector n of next nodes can be specified with a time vector t. If the total delay in the system since the node exceeds $t(i)$, node $n(i)$ will be activated next. This strategy can be used to model time-outs and various compensation schemes.

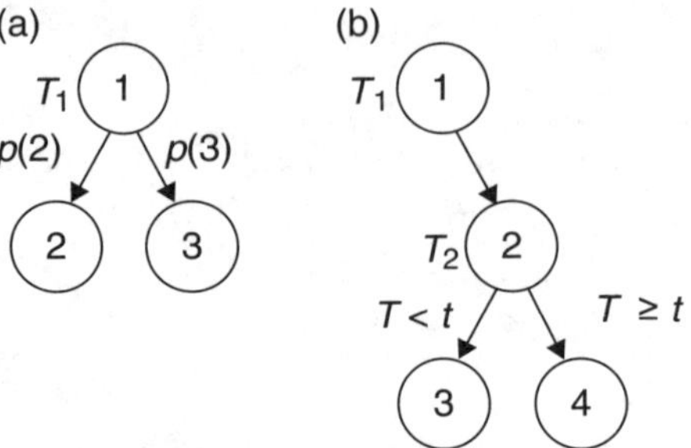

Figure 5.4: *Alternative execution paths in a Jitterbug execution model: (a) random choice of path and (b) choice of path depending on the total delay from the first node*

Jitterbug Example

This MATLAB script shows the commands needed to compute the performance index of the control system defined by the timing and signal models in Figure 5.3:

```
G = 1000/(s*(s+1));                  % Define the process
H1 = 1;                              % Define the sampler
H2 = -K*(1+Td/h*(z-1)/z);            % Define the controller
H3 = 1;                              % Define the actuator
Ptau1 = [ ... ];                     % Define delay prob distribution 1
Ptau2 = [ ... ];                     % Define delay prob distribution 2
N = initjitterbug(delta,h);          % Set time-grain and period
N = addtimingnode(N,1,Ptau1,2);      % Define timing node 1
N = addtimingnode(N,2,Ptau2,3);      % Define timing node 2
N = addtimingnode(N,3);              % Define timing node 3
N = addcontsys(N,1,G,4,Q,R1,R2);     % Add plant, specify cost and noise
N = adddiscsys(N,2,H1,1,1);          % Add sampler to node 1
N = adddiscsys(N,3,H2,2,2);          % Add controller to node 2
N = adddiscsys(N,4,H3,3,3);          % Add actuator to node 3
N = calcdynamics(N);                 % Calculate internal dynamics
J = calccost(N);                     % Calculate the total cost
```

The process is modeled by the following continuous-time system:

$$G(s) = \frac{1000}{s(s+1)}$$

The controller is a discrete-time PD-controller implemented as follows:

$$H_2(z) = -K\left(1 + \frac{T_d}{h}\frac{z-1}{z}\right)$$

The sampler and the actuator are described by the following trivial discrete-time systems:

$$H_1(z) = H_3(z) = 1$$

The delays in the computer system are modeled by the two (possibly random) variables τ_1 and τ_2. The total delay from sampling to actuation is thus given by $\tau_{tot} = \tau_1 + \tau_2$.

Using the defined Jitterbug model, it is straightforward to investigate, for example, how sensitive the control loop is to slow sampling and constant delays (by sweeping over suitable ranges for these parameters), and to random delays with jitter compensation. For more details and other illustrative examples (including multirate control, overrun handling, and notch filters subject to lost samples), see [Cervin10]. Jitterbug also contains commands for designing LQG controllers both for constant input delays and for input delays defined by a discrete-time probability density function.

When modeling the random choice of paths, it is straightforward to extend this example to also analyze the effect of lost samples caused by networked control over unreliable network links—for example, wireless links. The corresponding timing model is depicted in Figure 5.5, where for simplicity it is assumed that the sampler and controller are co-located in one node, $C(z)$, and that the controller sends the control signal to the actuator node, $A(z)$, with a probability of p of losing the signal.

5.3.6.2 The Jitter Margin

A drawback when using Jitterbug is that its timing model is rather simplistic; for example, it does not allow dependent delay distributions and the distributions may not change over time. Another issue is the means

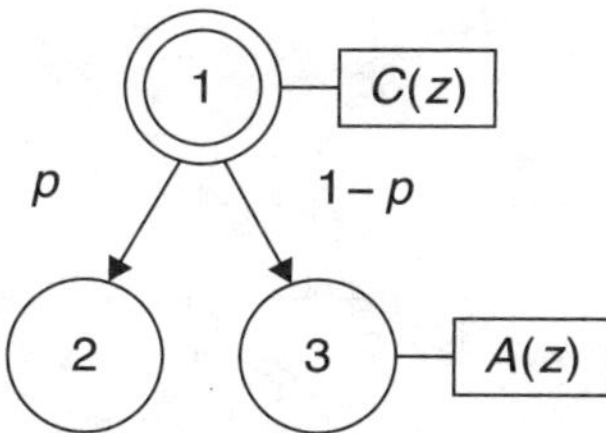

Figure 5.5: *Jitterbug timing model of a networked control loop with probability of losing the control signal*

used to obtain the distributions. Real-time scheduling theory focuses on worst-case delays. Although research in this area is increasing, the published theory and results for response-time distributions remain sparse.

Another drawback with Jitterbug is that the cost obtained with this model holds only for the particular timing distributions used. In many cases, one is instead interested in an analysis that holds for arbitrary timing variations, below some maximum upper bound. The Jitter Margin, for which two versions are available, meets this need. In [Kao04], the result was given for the output jitter-only case, assuming that the sampling jitter was zero—corresponding to deterministic sampling. This result was later extended in [Cervin12] to handle both input jitter and output jitter. Here, only the latter result will be discussed.

Jitter Margin Models

The nominal system model is a standard linear sampled-data control loop (Figure 5.6), consisting of a single-input/single-output, continuous-time plant $P(s)$, a periodic sampler with interval h, a discrete-time controller $K(z)$ (assuming positive feedback), and a zero-order-hold circuit. It is assumed that the nominal closed-loop system is stable. The performance of the system is measured by the $\mathcal{L}_2$-induced gain from the disturbance input d to the plant output y.

The time-varying delays introduced by the implementation are characterized by three non-negative parameters: the input jitter J_i, the output jitter J_o, and the nominal input-output delay L (Figure 5.7).

Ideally, the controller should execute at periodic time instances $t_k = kh$, $k = \{0, 1, 2,\ldots\}$. With jitter and delay, however, the controller input is sampled somewhere in the interval that follows:

$$\left[t_k - \frac{J_i}{2}, t_k + \frac{J_i}{2}\right]$$

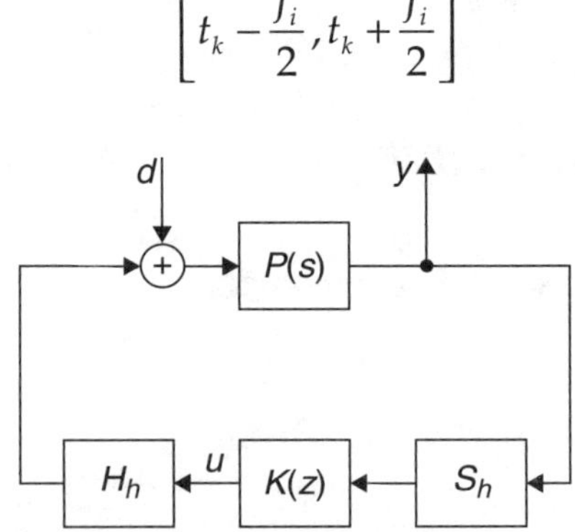

Figure 5.6: *Nominal standard linear sampled data system model*

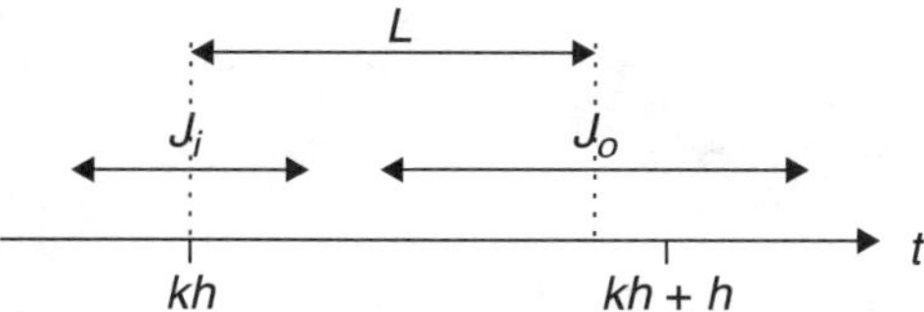

Figure 5.7: *Controller timing model with input jitter J_i, output jitter J_o, and nominal input-output delay L*

The output is updated somewhere in the following interval:

$$\left[t_k + L - \frac{J_o}{2}, t_k + L + \frac{J_o}{2}\right]$$

It follows that any valid set of timing parameters must respect the following relationship:

$$|J_i - J_o| \leq 2L \tag{5.13}$$

The timing model allows both $L > h$ and $J_i, J_o > h$. It does not care about how or if the actual controller timing varies from period to period, though.

The sampled-data system with delay and jitter is modeled in Figure 5.8, with continuous-time plant $P(s)$, time-varying delay operator Δ_i, periodic sampler S_h, discrete-time controller $K(z)$, zero-order-hold H_h, time-varying delay operator Δ_o, and nominal input-output delay e^{-sL}.

The nominal input-output delay is modeled as a delay at the plant input, while the jitters are modeled using a pair of time-varying delay operators:

$$\Delta_i(v) = v(t - \delta_i(t)), \quad -J_i/2 \leq \delta_i(t) \leq J_i/2 \tag{5.14}$$

$$\Delta_o(v) = v(t - \delta_o(t)), \quad -J_o/2 \leq \delta_o(t) \leq J_o/2 \tag{5.15}$$

Here, $\delta_i(t)$ and $\delta_o(t)$ may assume any values within their bounds at a given time t.

No relationship is assumed between $\delta_i(t)$ and $\delta_o(t)$. In reality, however, the delays are likely to be dependent. This is especially true if $J_i + J_o > 2L$, since the output operation cannot take place before the input operation in a real system. This may introduce some conservativeness in the analysis.

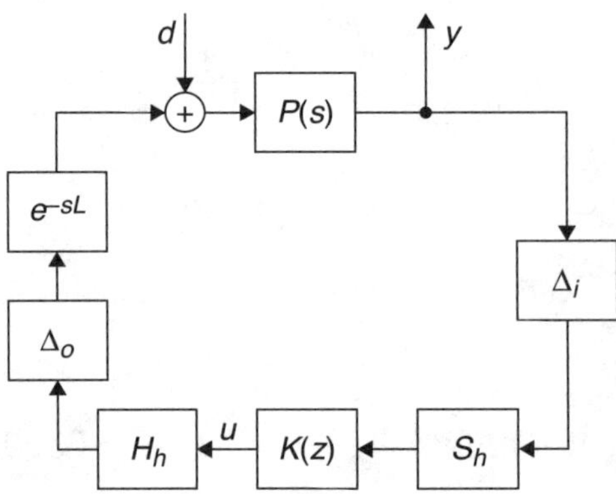

Figure 5.8: *Sampled-data system with delay and jitter*

The Jitter Margin analysis is based on the small-gain theorem [Zames66] (see [Cervin12] for details). Hence, only the stability criterion is sufficient—that is, a system for which the analysis predicts instability can, in reality, be stable. The actual analysis entails computing all input-output gains of a mixed continuous-time/discrete-time system, or a total of nine gains. The system will be stable if three inequality conditions on the gains are met. If the system is stable, then the analysis also provides an upper-bound on the gain from d to y (again see [Cervin12] for details).

Jitter Margin Example

As an example, consider the system given by the following:

$$P(s) = \frac{1}{s^2 - 0.01}$$

$$K(z) = \frac{-12.95z^2 + 10z}{z^2 - 0.2181z + 0.1081}, \qquad h = 0.2$$

We fix $L = 0.08$ and independently vary J_i and J_o between 0 and $2L$. Figure 5.9 reports the performance degradation of the system relative to the nominal case $L = J_i = J_o = 0$.

As can be seen in Figure 5.9, this system is slightly more sensitive to input jitter than to output jitter. The performance degradation level curves bend only very slightly inward, indicating that the combined analysis is not very conservative.

5.3.6.3 TrueTime

Both Jitterbug and Jitter Margin assume knowledge of the control loop's latency. To obtain this value, simulation is often the only option

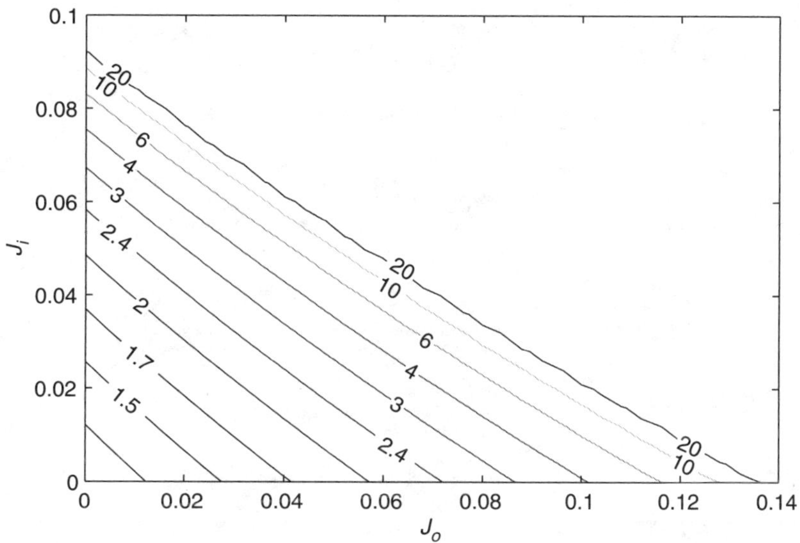

Figure 5.9: *Worst-case performance degradation as a function of the amount of input jitter and output jitter*

available. TrueTime [Cervin03a, Henriksson03, Henriksson02] is a MATLAB/Simulink-based tool that facilitates simulation of the temporal behavior of a multitasking real-time kernel executing controller tasks. The tasks are controlling processes that are modeled as ordinary continuous-time Simulink blocks. TrueTime also makes it possible to simulate simple models of communication networks and their influence on networked control loops.

In TrueTime, kernel and network Simulink blocks are introduced; the interfaces of these blocks are shown in Figure 5.10. The Schedule outputs display the allocation of common resources (CPU, network) during the simulation. The kernel blocks are event driven and execute code that models, for example, I/O tasks, control algorithms, and network interfaces. The scheduling policy of the individual kernel blocks is arbitrary and decided by the user. Likewise, in the network blocks, messages are sent and received according to the chosen network model. Two network blocks are available—one for wirebound networks and one for wireless networks. The library also contains a simple battery model to simulate battery-powered devices, an ultrasound propagation model for simulating ultrasound-based localization methods, and separate send and receive blocks that can be used in simulation models where network simulation functionality is required but kernel simulation is not needed.

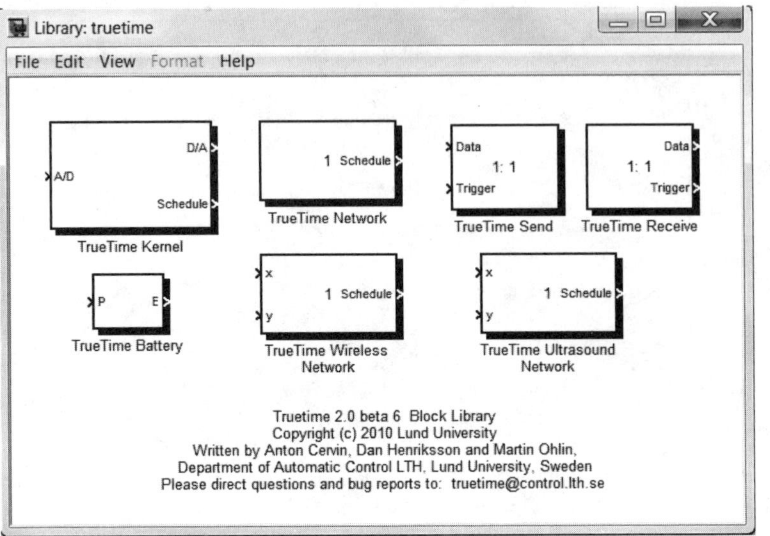

Figure 5.10: *The TrueTime block library*

The level of simulation detail is also chosen by the user: It is often neither necessary nor desirable to simulate code execution on an instruction level or network transmissions on a bit level. TrueTime allows the execution time of tasks and the transmission times of messages to be modeled as constant, random, or data dependent. Furthermore, this tool allows simulation of context switching and task synchronization using events, semaphores, or monitors.

TrueTime can be used as an experimental platform for research on dynamic real-time control systems. For instance, it is possible to study compensation schemes that adjust the control algorithm based on measurements of actual timing variations (i.e., to treat the temporal uncertainty as a disturbance and manage it with feedforward or gain scheduling). It is also easy to experiment with more flexible approaches to real-time scheduling of controllers, such as feedback scheduling (see [Cervin02]). The available CPU or network resources are dynamically distributed according to the current situation (e.g., CPU load, performance of the different loops) in the system.

The Kernel Block

The kernel block is a MATLAB S-function that simulates a computer with a simple but flexible real-time kernel, A/D and D/A converters, a network interface, and external interrupt channels. The kernel executes user-defined tasks and interrupts handlers. Internally, the kernel maintains several data structures that are commonly found in a real-time kernel: a ready queue, a time queue, and records for tasks, interrupt handlers, semaphores, monitors, and timers that have been created for the simulation. It also supports simulation of dynamic voltage and frequency scaling (DVFS) and simulation of constant bandwidth servers. The kernel and network S-functions are implemented as discrete event simulators that decide the next time instant when the block should be executed.

An arbitrary number of tasks can be created to run in the TrueTime kernel. Tasks may also be created dynamically as the simulation progresses. Tasks are used to simulate both periodic activities, such as controller and I/O tasks, and aperiodic activities, such as communication tasks and event-driven controllers. Aperiodic tasks are executed through the creation of task instances (jobs).

Each task is characterized by a number of static (e.g., relative deadline, period, and priority) and dynamic (e.g., absolute deadline and release time) attributes. In addition, the user can attach two overrun handlers to each task: a deadline overrun handler (triggered if the task misses its deadline) and an execution-time overrun handler (triggered if the task executes longer than its worst-case execution time).

Interrupts may be generated in two ways: externally (associated with the external interrupt channel of the kernel block) or internally (triggered by user-defined timers). When an external or internal interrupt occurs, a user-defined interrupt handler is scheduled to serve the interrupt.

The execution of tasks and interrupt handlers is defined by user-written code functions. These functions can be written either in C++ (for speed) or as MATLAB m-files (for ease of use). Control algorithms may also be defined graphically using ordinary discrete Simulink block diagrams.

Simulated execution occurs at three distinct priority levels: the interrupt level (highest priority), the kernel level, and the task level (lowest priority). The execution may be preemptive or nonpreemptive; this property can be specified individually for each task and interrupt handler.

At the interrupt level, interrupt handlers are scheduled according to fixed priorities. At the task level, dynamic-priority scheduling may be used. At each scheduling point, the priority of a task is given by a user-defined priority function, which reflects the task attributes. This approach makes it easy to simulate different scheduling policies. For instance, a priority function that returns a priority number implies fixed-priority scheduling, whereas a priority function that returns the absolute deadline implies earliest-deadline-first scheduling. Predefined priority functions exist for rate-monotonic, deadline-monotonic, fixed-priority, and earliest-deadline-first scheduling.

The Network Block

The network block is event driven and executes when messages enter or leave the network. When a node tries to transmit a message, a triggering signal is sent to the network block on the corresponding input channel. When the simulated transmission of the message ends, the network block sends a new triggering signal on the output channel corresponding to the receiving node. The transmitted message is put in a buffer at the receiving computer node.

A message contains information about the sending and receiving computer nodes, arbitrary user data (typically measurement signals or control signals), the length of the message, and optional real-time attributes such as a priority or a deadline.

The network block simulates medium access and packet transmission in a local area network. Eight simple models of wire-bound networks are currently supported: CSMA/CD (e.g., Ethernet), CSMA/AMP (e.g., CAN), Round Robin (e.g., Token Bus), FDMA, TDMA (e.g., TTP), Flexray, PROFINET IO, and Switched Ethernet. There are two models of wireless protocols: IEEE 802.11b (WLAN) and IEEE 802.15.3 (Zigbee).

The propagation delay is ignored, since it is typically very small in a local area network. Only packet-level simulation is supported; in other words, it is assumed that higher protocol levels in the kernel nodes have divided long messages into packets.

Configuring the network block involves specifying a number of general parameters, such as transmission rate, network model, and probability for packet loss. Protocol-specific parameters that need to be supplied include the time slot and the cyclic schedule in the case of TDMA.

Execution Model

The execution of tasks and interrupt handlers is defined by code functions. A code function is further divided into code segments according to the execution model shown in Figure 5.11. The execution of the code

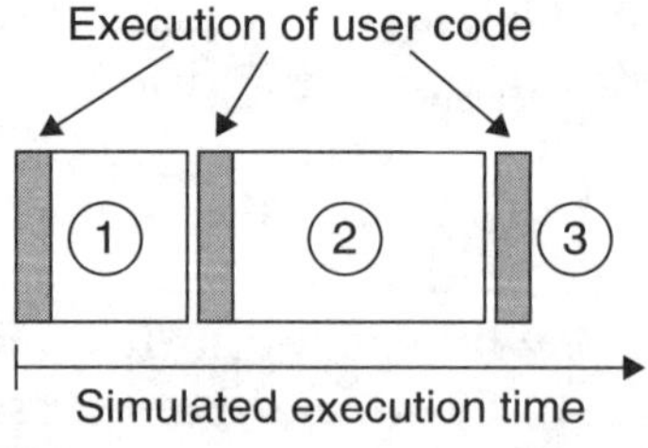

Figure 5.11: *Execution model with code divided in sections*

associated with tasks and interrupt handlers is modeled by a number of code segments with different execution times. Execution of user code occurs at the beginning of each code segment.

The code can interact with other tasks and with the environment at the beginning of each code segment. This execution model makes it possible to model input-output latencies, blocking when accessing shared resources, and so on. The number of segments can be chosen to simulate an arbitrary time granularity of the code execution. Technically, it would, for example, be possible to simulate very fine-grained details occurring at the machine instruction level, such as race conditions. Doing so, however, would require a large number of code segments.

The simulated execution time for each segment is returned by the code function, and can be modeled as constant, random, or even data dependent. The kernel keeps track of the current segment and calls the code functions with the proper arguments during the simulation. Execution resumes in the next segment when the task has been running for the time associated with the previous segment. As a consequence, preemption by higher-priority activities and interrupts may cause the actual delay between executions of segments to be longer than the execution time.

The following listing shows an example of a code function corresponding to the timeline in Figure 5.11. The function implements a simple PI-controller without anti-windup. In the first segment, the plant is sampled and the control signal is computed. In the second segment, the control signal is actuated and the controller state is updated. The third segment indicates the end of execution by returning a negative execution time.

```
function [exectime, data] = PIcontroller(segment, data)
 switch segment,
  case 1, data.r = ttAnalogIn(1);
    data.y = ttAnalogIn(2);
    data.e = data.r - data.y;
    data.u = data.K*data.e + data.I;
    exectime = 0.002;
  case 2,
    ttAnalogOut(1, data.u);
    data.I = data.I + (data.K*data.h/data.Ti)*data.e;
    exectime = 0.001;
  case 3,
    exectime = -1;
 end
```

The data structure represents the local memory of the task and is used to store the control signal and the measured variable between calls to the different segments. A/D and D/A conversion is performed using the kernel primitives `ttAnalogIn` and `ttAnalogOut`, respectively.

Note that the input-output latency of this controller will be at least 2 ms (i.e., the execution time of the first segment). If preemption by other high-priority tasks occurs, however, the actual input-output latency will be longer.

A limitation of the TrueTime kernel block is that it does not model any other resources than CPU time. For example, the user must explicitly model delays caused by memory accesses and cache misses and include these delays in the segment execution times.

5.4 Summary and Open Challenges

Truly resource-aware control systems should take the real-time software and implementation platform into account at all development stages. However, as we have seen in this chapter, the relationships among control performance, task timing, and resource sharing are far from trivial. This chapter presented a few tools that may help at the analysis stage—namely, Jitterbug, the Jitter Margin, and TrueTime. Unfortunately, adequate tools for the design, implementation, and operation stages are still lacking to a large extent.

Co-design that integrates the control, computing, and communication aspects of the system holds the promise of better resource utilization and higher control performance, but may ultimately prove to be too cumbersome. As technological systems become ever more complex, it is important to reduce the engineering efforts required for the integration of subsystems. The concept of components with well-defined interfaces has proved very successful in several engineering disciplines. To achieve composability, thereby reducing development times, some amount of performance is typically sacrificed. The control server presented in Section 5.3.5.1 represents one such proposal. To define good design interfaces between the control, computation, and communication elements of a system perhaps represents the most important challenge in the area of resource-aware control.

Even with proper design interfaces in place, more research is needed at each layer. Event-based control is a subarea within control theory that needs further development before it can be applied in general. Even standard design techniques such as LQG are not yet fully developed enough to deal with implementation effects such as sampling jitter and input-output jitter.

Computing and communication resources are rarely constant in CPS. Given this reality, a resource-aware control system should be able to adapt to changes in the availability of resources. Meeting this goal may require both online identification of system parameters and online redesign.

References

[Abeni98]. L. Abeni, G. Buttazzo, S. Superiore, and S. Anna. "Integrating Multimedia Applications in Hard Real-Time Systems." *Proceedings of the 19th IEEE Real-Time Systems Symposium*, pages 4–13, 1998.

[Årzén99]. K. Årzén. "A Simple Event-Based PID controller." Preprints 14th World Congress of IFAC, Beijing, P.R. China, January 1999.

[Åström11]. K. J. Åström and B. Wittenmark. *Computer-Controlled Systems: Theory and Design*. Dover Books, Electrical Engineering Series, 2011.

[Bemporad00]. A. Bemporad, M. Morari, V. Dua, and E. N. Pistikopoulos. "The Explicit Solution of Model Predictive Control via Multiparametric Quadratic Programming." American Control Conference, pages 872–876, Chicago, IL, June 2000.

[Beschi12]. M. Beschi, S. Dormido, J. Sánchez, and A. Visioli. "On the Stability of an Event-Based PI Controller for FOPDT Processes." IFAC Conference on Advances in PID Control, Brescia, Italy, July 2012.

[Cervin12]. A. Cervin. "Stability and Worst-Case Performance Analysis of Sampled-Data Control Systems with Input and Output Jitter." American Control Conference, Montreal, Canada, June 2012.

[Cervin03]. A. Cervin and J. Eker. "The Control Server: A Computational Model for Real-Time Control Tasks." *Proceedings of the 15th Euromicro Conference on Real-Time Systems*, Porto, Portugal, July 2003.

[Cervin02]. A. Cervin, J. Eker, B. Bernhardsson, and K. Årzén. "Feedback-Feedforward Scheduling of Control Tasks." *Real-Time Systems*, vol. 23, no. 1–2, pages 25–53, July 2002.

[Cervin03a]. A. Cervin, D. Henriksson, B. Lincoln, J. Eker, and K. Årzén. "How Does Control Timing Affect Performance? Analysis and Simulation of Timing Using Jitterbug and TrueTime." *IEEE Control Systems Magazine*, vol. 23, no. 3, pages 16–30, June 2003.

[Cervin10]. A. Cervin and B. Lincoln. *Jitterbug 1.23 Reference Manual.* Technical Report ISRN LUTFD2/TFRT-0.1em-7604-0.1em-SE, Department of Automatic Control, Lund Institute of Technology, Sweden, July 2010.

[Durand09]. S. Durand and N. Marchand. "Further Results on Event-Based PID Controller." *Proceedings of the European Control Conference*, pages 1979–1984, Budapest, Hungary, August 2009.

[Heemels09]. W. P. M. H. Heemels, D. Nesic, A. R. Teel, and N. van de Wouw. "Networked and Quantized Control Systems with Communication Delays." *Proceedings of the Joint 48th IEEE Conference on Decision and Control and 28th Chinese Control Conference*, 2009.

[Heemels08]. W. P. M. H. Heemels, J. H. Sandee, and P. P. J. Van Den Bosch. "Analysis of Event-Driven Controllers for Linear Systems." *International Journal of Control*, vol. 81, no. 4, pages 571–590, 2008.

[Henningsson12]. T. Henningsson. "Stochastic Event-Based Control and Estimation." PhD thesis, Department of Automatic Control, Lund University, Sweden, December 2012.

[Henriksson03]. D. Henriksson and A. Cervin. *TrueTime 1.1: Reference Manual.* Technical Report ISRN LUTFD2/TFRT-0.1em-7605-0.1em-SE, Department of Automatic Control, Lund Institute of Technology, October 2003.

[Henriksson02]. D. Henriksson, A. Cervin, and K. Årzén. "TrueTime: Simulation of Control Loops Under Shared Computer Resources." *Proceedings of the 15th IFAC World Congress on Automatic Control*, Barcelona, Spain, July 2002.

[Kao04]. C. Kao and B. Lincoln. "Simple Stability Criteria for Systems with Time-Varying Delays." *Automatica*, vol. 40, no. 8, pages 1429–1434, August 2004.

[Lin09]. H. Lin and P. J. Antsaklis. "Stability and Stabilizability of Switched Linear Systems: A Survey of Recent Results." *IEEE Transactions on Automatic Control*, vol. 54, no. 2, pages 308–322, February 2009.

[Lincoln02]. B. Lincoln and A. Cervin. "Jitterbug: A Tool for Analysis of Real-Time Control Performance." *Proceedings of the 41st IEEE Conference on Decision and Control*, Las Vegas, NV, December 2002.

[Maciejowski01]. J. M. Maciejowski. *Predictive Control: With Constraints.* Prentice Hall, 2001.

[Rantzer12]. A. Rantzer. "Distributed Control of Positive Systems." ArXiv e-prints, February 2012.

[Scokaert99]. P. O. M. Scokaert, D. Q. Mayne, and J. B. Rawlings. "Suboptimal Model Predictive Control (Feasibility Implies Stability)." *IEEE Transactions on Automatic Control*, vol. 44, no. 3, pages 648–654, March 1999.

[Tabuada07]. P. Tabuada. "Event-Triggered Real-Time Scheduling of Stabilizing Control Tasks." *IEEE Transactions on Automatic Control*, vol. 52, no. 9, pages 1680–1685, September 2007.

[Wang08]. X. Wang and M. D. Lemmon. "State Based Self-Triggered Feedback Control Systems with L2 Stability." *Proceedings of the 17th IFAC World Congress*, 2008.

[Yüksel11]. S. Yüksel. "A Tutorial on Quantizer Design for Networked Control Systems: Stabilization and Optimization." *Applied and Computational Mathematics*, vol. 10, no. 3, pages 365–401, 2011.

[Zames66]. G. Zames. "On the Input-Output Stability of Time-Varying Nonlinear Feedback Systems, Parts I and II." *IEEE Transactions on Automatic Control*, vol. 11, 1966.

Chapter 6

Logical Correctness for Hybrid Systems

Sagar Chaki, Edmund Clarke, Arie Gurfinkel, and John Hudak

A hybrid system is a dynamic system whose behavior changes both discretely and continuously. Hybrid systems constitute a powerful formalism for understanding, modeling, and reasoning about a wide range of devices that affect our day-to-day lives. These range from the simple, such as a room thermostat, to the critical, such as the airbag controller in a vehicle. Hybrid systems also enable rigorous analysis to be performed that can verify and predict correct behavior of such devices. In this chapter, we discuss different types of functional correctness problems that arise in the context of hybrid systems. For each category, we present sample problems and offer possible solutions using state-of-the-art tools and techniques. We round out the chapter with a discussion of open problems and future challenges in hybrid system verification.

6.1 Introduction and Motivation

It is fair to say that without the correct operation of a wide variety of devices, modern life would come to a standstill. In many cases, such devices operate behind the scenes, but are nevertheless critical to sustain a technology-reliant civilization. A key characteristic shared by many such devices is that they are built on digital computation technology, yet must operate in an inherently analog world. In this sense, they are canonical examples of hybrid systems [Alur95]—that is, discrete programs with analog environments. Yet another feature of many such hybrid systems (e.g., pacemakers, vehicle airbag controllers, nuclear plant controllers) is that their incorrect operation can lead to catastrophic consequences. The combination of these two facts implies that verifying "functional correctness"—that is, the safe and secure operation—of hybrid systems has fundamental importance. This topic is the focus of this chapter.

Informally, a hybrid system can be viewed as a state machine having two types of transitions: discrete jumps and continuous evolution. For example, a standard representation of a hybrid system is a hybrid automaton [Henzinger96], a type of finite-state machine in which states represent discrete modes, and transitions represent switching between the modes. Transitions are guarded by conditions on a set of continuous variables. Within each mode, the changes in the values of the continuous variables are specified via differential equations. In addition, the system stays within a mode only as long as a mode-specific invariant remains satisfied by the continuous variables.

Figure 6.1 shows the hybrid automaton (reproduced from [Alur95]) for a thermostat that attempts to maintain the temperature between m and M by controlling a heater. It has two modes: l_0, where a heater is turned off, and l_1, where a heater is turned on. The system begins in mode l_0 when the temperature x equals M. It remains in l_0 as long as the invariant $x > m$ for model l_0 holds. At any point in time when the transition guard $x = m$ holds, the system switches to mode l_1. It remains in l_1 as long as $x < M$ holds, and then switches back to mode l_0 whenever the transition guard $x = M$ holds. Note that this system is time deterministic since the mode invariants and transition guards are set up such that once the system enters a mode, the time of its departure from the mode is uniquely determined.

We focus on the class of functional correctness problems referred to as reachability. Informally, such a problem is specified by a formula Φ

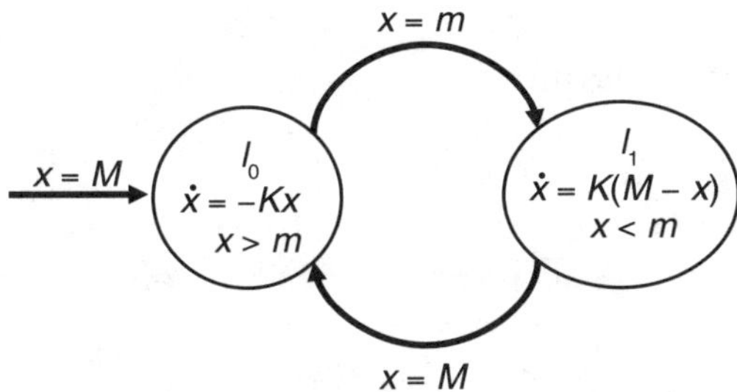

Figure 6.1: *A hybrid automaton for a thermostat*

describing a set of "good" states, where a state is an assignment of values to system variables. The system is functionally correct if it never reaches a state that does not belong to Φ. For example, a functional correctness property for the hybrid system in Figure 6.1 is expressed by the formula $m \leq x \leq M$, which asserts that the temperature of the hybrid system always stays in the range $[m, M]$. In practice, reachability is sufficient to capture the class of all functional correctness requirements—including many types of safety and security criteria—whose violations can be demonstrated via finite executions of the system, known as counterexamples.

The verification of the reachability of hybrid systems is a widely studied problem, involving decades of research and development by both the academic and practitioner communities. Presenting all of this work in its full detail is beyond the scope of this chapter. Instead, we limit our presentation here in three ways: by problem categorization, by solution strategy, and by target audience.

First, we categorize the reachability problem for hybrid systems into three classes: discrete, real time, and fully hybrid: Roughly speaking, a discrete reachability problem is one that can be solved by ignoring completely the continuous dynamics of the system. In contrast, a real-time reachability problem can be solved only by considering clocks and the passage of time. Finally, a fully hybrid problem requires reasoning about the dynamics of one or more continuous variables in addition to clocks.

Second, we restrict our attention to solutions for hybrid system reachability that are based on an exhaustive, automated, and algorithmic verification technique known as model checking [Clarke09]. While developed originally for verifying finite-state machines [Clarke83], model checking has since been extended to a wide variety of systems, including software [Jhala09] and hybrid systems [Frehse08, Henzinger97]. An important property of model checking is that it produces counterexamples as diagnostic feedback if the system being verified is found to violate the desired reachability criterion.

Third, this chapter is directed primarily toward practitioners who are interested in learning more about the state of the practice in applying model checking to verifying correctness of hybrid systems. It does not aim to be exhaustive, but rather tries to exemplify first how to classify problems into categories, and then how to select and use some current model checking tools to solve those problems effectively. At the same time, where appropriate, we point the reader to published literature that describes the fundamental concepts and algorithms underlying many of these tools.

The rest of the chapter is devoted to the three problem categories in increasing order of complexity: discrete, real time, and fully hybrid. For each category, we discuss which kinds of problems belong to the category, and demonstrate, via examples, which model checking tools apply to those categories and how to use them effectively. We conclude the chapter with a brief overview of some other approaches to verifying the functional correctness of hybrid systems, ongoing work, and open problems.

6.2 Basic Techniques

This section focuses on the fundamentals of hybrid systems verification. In particular, we discuss discrete verification and temporal logic, which has provided the foundation for the rest of the techniques that have provided multiple examples and the tools available to the practitioner.

6.2.1 Discrete Verification

A discrete event system (DES) is a dynamic system that behaves in accordance with an abrupt occurrence, at possibly unknown irregular intervals of physical events [Ramadge89]. It contains discrete states, whose state changes are event driven; that is, the state evolution depends entirely on the occurrence of asynchronous discrete events over time. For example, events may correspond to a change in a control setpoint during a process, messages arriving in a queue, a discretized continuous variable exceeding some threshold, or the failure of a component in a system (e.g., car, plane, manufacturing process). Discrete event systems are found in a number of domains, including manufacturing, robotics, logistics, computer operating systems, and network communication systems. They require some form of control to ensure an orderly control of events for the purpose of eventually achieving some (good) behavior or always avoiding some (bad) behavior. Eventually achieving a good behavior is termed a liveness property, whereas always avoiding a bad behavior is termed a safety property [Lamport77, Schneider85].

The capability to design and analyze discrete event systems is based on the fact that they can be modeled (represented) in an automata formalism. The representations are abstractions of the actual system. Formalisms utilize precise semantics to express behavior, with typical examples including state charts [Harel87, Harel87a, Harel90], Petri nets [Rozenburg98], Markov chains [Usatenko10], and so on. Given that this section focuses on model checking of discrete event systems, we will focus on state chart representations and provide a brief overview of applicable logic and model checking tools.

6.2.1.1 Model Checking Tools and Associated Logic

The models used in model checking are formal representations of system behavior. State charts are one formal representation that is widely used [Harel87, Harel87a, Harel90]. Similarly, claims are formal specifications of expected properties of the system that one is attempting to verify. The formal definition of models and claims enables automated tools to verify whether particular claims hold against a model. If a claim is well written and the model is a faithful representation of the system, the verification using model checking will indicate whether the system possesses or does not possess the expected property represented by the claim.

Depending on the kind of expected property and the type of model checker, different notations can be used to formalize the claim. For example, we might use classical propositional logic to define the various claims if we are concerned about static properties of the system.

Temporal logic is an extension to propositional logic that considers time; it is a formal approach for specifying and reasoning about the dynamic characteristics of a system. The formalism of temporal logic does not include time explicitly (i.e., counting or measuring time in the sense of a timing device). Rather, time is represented as a sequence of states (a state trace) in the behavior of a system. These sequences of states (traces) can be either finite $\langle s_0, s_1, s_2, \ldots, s_n \rangle$ or infinite $\langle s_0, s_1, s_2, \ldots \rangle$. States represent finite time intervals of fixed conditions for a system. For example, the state of "red" for a traffic light is when the color of the light is red.

Two prominent forms of temporal logic are distinguished [Clarke99]: linear temporal logic (LTL) and branching temporal logic, which includes computational tree logic (CTL). The difference between the two lies in the conceptualization of unfolding time. In linear temporal logic, the evolution of time is viewed as linear—that is, a single infinite sequence of states. In contrast, CTL views the evolution of time as possibly proceeding along a multiplicity of paths (i.e., branches). Each path is a linear sequence of states.

We now define the syntax and semantics of LTL briefly. Further details are available elsewhere [Clarke99].

6.2.1.2 Linear Temporal Logic

Let *AP* be a set of "atomic propositions." A LTL formula φ satisfies the following BNF grammar:

$$\varphi := \top \mid p \mid \neg\varphi \mid \varphi \wedge \varphi \mid \varphi \vee \varphi \mid X\ \varphi \mid \varphi\ U\ \varphi \mid G\ \varphi \mid F\ \varphi$$

Here $\top$ denotes "true", $p \in AP$ is an atomic proposition, $\neg$ denotes logical negation, $\wedge$ denotes conjunction, $\vee$ denotes disjunction, X denotes the next-state temporal operator, U denotes the "until" temporal operator, G denotes the "globally" temporal operators, and F denotes the "eventually" temporal operator.

For example, let $AP = \{p, q, r\}$. Then some possible LTL formulas are as follows:

$$\varphi_1 = G(p \Rightarrow XFq)$$

$$\varphi_2 = G\left(p \Rightarrow XF(q \vee r)\right)$$

As we will describe in the next section, every LTL formula denotes a property over paths (i.e., sequences of system states). Informally, φ_1 holds on a path if globally (i.e., at each state of the path) either the proposition p holds, or from the next state, q holds eventually. Similarly, φ_2 holds on a path if globally (i.e., at each state of the path) either the proposition p holds, or from the next state, q or r holds eventually.

An LTL formula is interpreted over Kripke structures. Informally, a Kripke structure is a finite-state machine whose states are labeled with atomic propositions. Formally, a Kripke structure M is a 4-tuple (S, I, R, L) where S is a finite set of states; $I \in S$ is the initial state; $R \subseteq S \times S$ is the transition relation; and $L : S \mapsto 2^{AP}$ is a mapping from states to atomic propositions. Informally, L maps each state to the set of atomic propositions that are true in that state.

Figure 6.2 shows a Kripke structure $M = (S, I, R, L)$ with three states: $S = \{s_0, s_1, s_2\}$, where s_0 is the initial state. The transition relation is shown by arrows; that is, $(s, s') \in R$ if there is an edge from s to s' in the figure. Each state s is labeled with $L(s)$; that is, $L(s_0) = \{p\}$, $L(s_1) = \{r\}$, and $L(s_2) = \{q\}$.

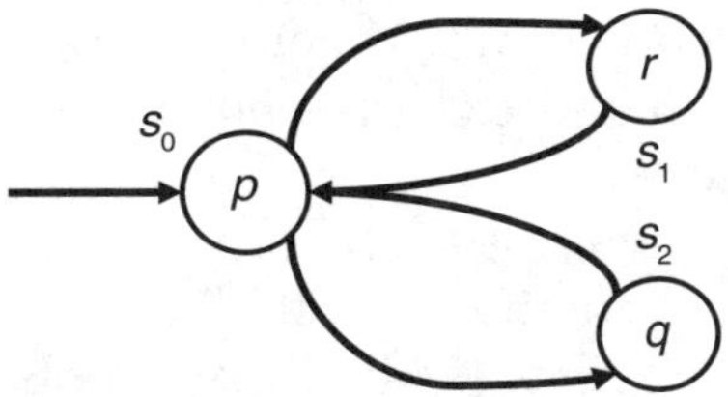

Figure 6.2: *An example Kripke structure with* $AP = \{p,q,r\}$

A trace of a Kripke structure $M = (S, I, R, L)$ is an infinite sequence of states $\langle s_0, s_1, s_2, \ldots \rangle$ such that

$$s_0 = I \land \forall i \geq 0.(s_i, s_{i+1}) \in L$$

In other words, a trace begins with the initial state and then follows the transition relation infinitely often. Given a trace $\pi = \langle s_0, s_1, s_2, \ldots \rangle$, we write π^i to mean the infinite suffix of π starting with s_i. Then we say that π satisfies a LTL formula φ, denoted $\pi \models \varphi$, if the following holds:

- $\varphi = \top$
- $\varphi \in \mathrm{AP}$, then $\varphi \in \mathrm{L}(s_0)$; that is, the first state of π is labeled with φ
- $\varphi = \neg\varphi'$, then $\pi \not\models \varphi'$
- $\varphi = \varphi_1 \land \varphi_2$, then $\pi \models \varphi_1$ and $\pi \models \varphi_2$
- $\varphi = \varphi_1 \; U \; \varphi_2$, then there exists a $k \geq 0$ such that $\forall 0 \leq i < k.\pi^i \models \varphi_1 \land \pi^k \models \varphi_2$

Note that $\lor$, F, and G can be expressed in terms of the other operators:

$$\varphi_1 \lor \varphi_2 \equiv \neg(\neg\varphi_1 \land \neg\varphi_2)$$

$$F\varphi \equiv \top \, U \, \varphi$$

$$G\varphi \equiv \neg(F\neg\varphi)$$

Finally, a Kripke structure M satisfies an LTL formula φ, denoted $M \models \varphi$, if for every trace π of M, $\pi \models \varphi$.

For example, let M be the Kripke structure from Figure 6.2. Let $\varphi_1 = G(p \Rightarrow XFq)$ and $\varphi_2 = G(p \Rightarrow XF(q \lor r))$. Then, $M \not\models \varphi_1$ since for the trace $\pi = \langle s_0, s_1, s_0, s_1, \ldots \rangle$, we have that $\pi \not\models \varphi_1$ (that is, we cannot verify that proposition φ_1 evaluates to true for trace π). On the other hand, $M \models \varphi_2$.

Given that state charts of the system to be verified are constructed and the claims of interest are expressed in the appropriate logic, an automated tool is necessary to apply the claims to the model. Some tools that have been widely used in both academia and industrial settings include NuSMV [Cimatti99], SPIN [Holzmann04], Cadance SMV [Jhala01], and LTSA [Magee06]. All of these tools use LTL as the property language; in some tools, CTL can also be used. A detailed discussion of model checking theory, logics, and tooling is beyond the scope of this section. The reader is referred to [Clarke99], [Huth04], and [Baier08] as foundational material and to [Peled01], who describes model checking in the context of software reliability.

6.2.1.3 Example: Helicopter Flight Control Verification

In this section, we outline how model checking can be used within the context of a real-world system by using a helicopter [FAA12, Gablehouse69] as an example. An overview of the system is presented to establish the context, and we then focus on a specific subsystem (the stabilizer) on which to verify a liveness property. We outline the use of a set of commercial tools—Simulink [Simulink13], Stateflow [Simulink13a], and the associated Simulink Design Verifier (SDV) [Simulink13b]—to show how properties can be verified.

A typical helicopter is composed of a number of computer-based subsystems, as shown in Figure 6.3.

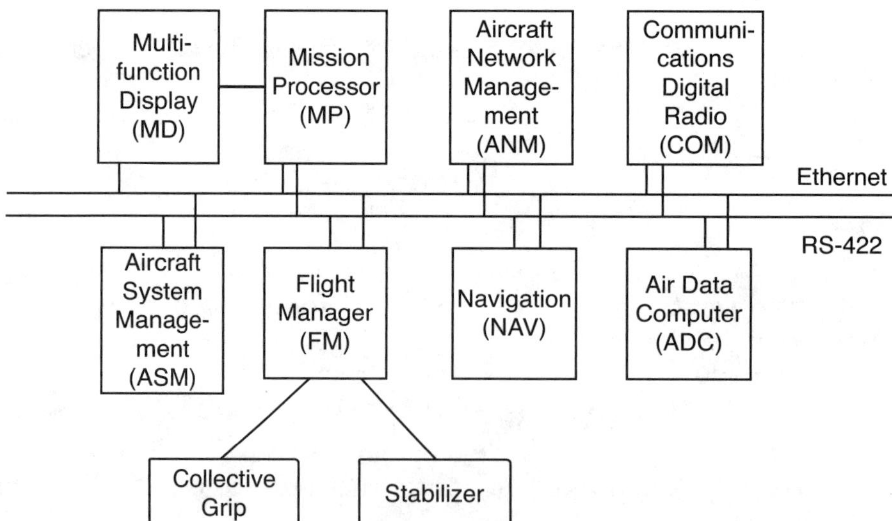

Figure 6.3: *Computer-based subsystems of a helicopter*

The basic subsystems of a typical helicopter include the aircraft systems management (ASM), mission processor (MP), multifunction display (MD), communications (COM), navigation (NAV), flight controls (FC), air data computer (ADC), and flight manager (FM). These subsystems communicate over redundant serial buses such as Ethernet, RS-422, or Mil-1550. From an operational perspective, the pilot controls the helicopter through controls in the cockpit (cyclic, collective flight grip). The controls are interfaced to the FM, and their states are communicated to their subsystems via the serial buses. The pilot receives information about the helicopter's status via the multifunction display. Using the hard and soft keys of the MD, the pilot can also initiate various flight modes, status displays, navigation displays, and so on. In many current helicopter designs, the majority of the controls are "fly by wire," meaning that the control devices contain electronic sensors that convert positional or force quantities into electronic signals that are read by the subsystem computer, and acted upon via the embedded control logic. The subsystem computer also interfaces to the appropriate actuators, which adjust such things as motor speed, rotor blade pitch, stabilizer angle, and so on.

Verification of Operational Requirements

The requirements of the system to be realized drive the design of that system. These requirements are specified as goals of the entire system, which are systematically decomposed and mapped onto the appropriate subsystems. The derived requirements specify the expected behavior of subsystems or collections of subsystems. Ultimately, they can be traced back to the top-level requirement. The requirements are derived in part from subject-matter experts (SMEs). One such group of SMEs for helicopters comprises the pilots who provide insight into the desired operation of a new helicopter.

The verification activity can follow the requirements decomposition. In this activity, each successive requirement decomposition is verified to ensure consistency with the higher-level requirement. The decomposed requirements are then mapped onto the appropriate subsystems and, in many cases, across subsystems—for example, the mechanical linkages/actuators and the associated computer control system of actuators.

The approach to model checking of an entire system is beyond the scope of this text, but suffice it to say that to establish a focus for model checking, one must determine the scope of the verification as well as the perspective [Comella_Dorda01, Hudak02]. In our example, we focus on the operation of the helicopter's stabilizer. The scope is the logic that controls the stabilizer over the entire nominal operational behavior of the helicopter, and the perspective taken is that of the pilot. Specifically, we look at questions regarding correct positioning of the stabilizer under certain operating (flight) conditions and pilot inputs.

Functionality of the Helicopter Stabilizer

Helicopters are inherently complex and unstable rotorcraft. The rotational moments generated by the main rotor can cause the body of the helicopter to rotate around the rotor axis. The tail rotor acts to counteract the horizontal rotational moments. During horizontal flight, a combination of the rotational moments in conjunction with thrust generated by the main rotor can make the helicopter unstable at certain airspeeds. This instability usually results in the rising of the helicopter's tail, which under certain situations can cause the helicopter to "nose over." A stabilizer airfoil ("stabilizer," for short) changes the direction of the airflow in the tail area and forces the tail down at higher air speeds.

From preliminary design documentation of helicopter systems, we are able to get some insight into how the stabilizer is controlled. Specifically, we learned that the stabilizer interfaces with the flight manager that controls, through a linear motor actuator, the horizontal stabilizer flight surface. The flight manager has a control algorithm that operates in two modes: automatic and manual. This suggests that the helicopter operation as selected by the pilot consists of an automatic and manual mode; in addition, a closed-loop control algorithm controls the positioning of the stabilizer.

Further review of both preliminary design documents and pilots' manuals reveals how the pilot interacts with the stabilizer. Specifically, the pilot interacts with the stabilizer in two ways: (1) by choosing the operational mode of the stabilizer position control system (auto or manual) or (2) in manual mode, by directly interacting with the stabilizer actuators. The interaction point is the collective flight grip, which incorporates a switch for control of the stabilizer. This three-position switch allows the stabilizer to be moved throughout its operational range and to be reset to the automatic mode. The nose up/nose down (NU/ND) switch on the flight grip enables the pilot to issue nose up/down commands as well as reset commands to the stabilizer. In addition, the pilot may request entry into nap of earth (NOE) mode by manipulating the multifunction pilot display. NOE mode is a flight mode that allows the helicopter to fly close to the earth—for example, during search operations. Flying close to the earth in a helicopter can affect the stability of the aircraft under certain conditions; hence height above the ground is part of what describes the NOE mode.

Preliminary helicopter design documents provide further details about the expected operational modes and the conditions under which transitions can occur between the modes. The design documents indicate that the system always starts in automatic mode. The pilot can explicitly select manual mode or NOE mode, and can "reset to" automatic mode from both the NOE and manual modes. In addition, the design documents specify the conditions under which transitions can occur between operational modes. Some transitions are initiated by the pilot, whereas others are initiated by environmental sensors or are triggered by system failure modes. Using this information, we capture the operational modes and transitions, initially without taking failure modes into account, and express them in a state chart. This allows us to formulate claims that can be verified against the detailed logic design of the stabilizer, which is ultimately realized as code within the FM. Figure 6.4 shows pilot-initiated transitions as solid lines, and airspeed-triggered transitions as dashed lines.

Keep in mind that the state diagram in Figure 6.4 represents the pilot's view of the operational behavior of the helicopter control. Given that this view emanated from higher-level requirements and design documents, we now want to be able to verify that the underlying control logic behaves as indicated in this state diagram. We can develop several claims from this diagram that we can then verify using Simulink and the Design Verifier.

Claim Generation

By studying the state diagram, we can begin to develop claims regarding system operation that must be proved. As an example, the initialization of the system places it in auto-low-speed mode. Another example would be that manual mode can be entered only from the auto-low-speed or NOE mode. Following is a set of the claims that can potentially be checked:

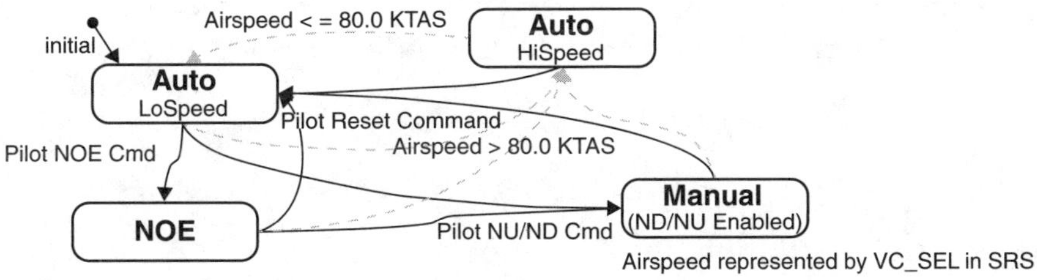

Figure 6.4: *State diagram showing stabilizer modal behavior*

- The system always begins in auto-low-speed mode. Suppose there is a proposition *AutoLowSpeed* that labels every state where the system is in auto-low-speed mode. This claim is then captured by the LTL formula $\varphi = AutoLowSpeed$. According to the semantics of LTL, φ is satisfied if the initial state of the system satisfies *AutoLowSpeed*.
- The system can never go into auto-high-speed mode until airspeed > 50 knots. (In the rest of this chapter, we use knots and KTAS synonymously.) Suppose there are two propositions: *AutoHighSpeed* and *AirSpeedGT*50. This claim is then expressed in LTL as $G(AutoHighSpeed \Rightarrow AirSpeedGT50)$.
- If the helicopter is in auto-high-speed mode and airspeed ≤ 50 knots, the system will go to auto-low-speed mode.
- If the helicopter is in auto-low-speed mode and NOE command = true, then the system will go into NOE mode.
- If the helicopter is in auto-high-speed mode and airspeed > 50 knots, then it will never be the case that the system will enter into NOE mode.
- If the helicopter is in auto-high-speed mode, airspeed ≤ 50 knots, and the pilot selects NOE mode, the system will eventually go into NOE mode. Suppose there are two propositions: *NOECommanded* and *NOEMode*. This claim is then expressed in LTL as $G(AutoHighSpeed \wedge \neg AirSpeedGT50 \wedge NOECommanded \Rightarrow F\ NOEMode)$.
- When the helicopter is in NOE mode, the stabilizer position will be commanded to become –15 degrees trailing edge down.
- When the helicopter is in stabilizer manual mode, the stabilizer will eventually reach its commanded position.

These claims represent behavior under normal operating conditions (the list is not intended to be exhaustive). The effects of various component failures could be overlaid on the modal model, and claims regarding operation given failures could be generated and verified.

We will focus on verifying the claim that when the helicopter is in auto-high-speed mode and airspeed > 50 knots, it will never be the case that the system will enter into NOE mode for the Simulink Design Verifier example.

Top-Level State Charts

To perform verification, the design of the logic for the flight manager must be captured using Simulink Stateflow state charts; within those state charts, detailed logic diagrams must be produced. In essence, we are increasing the amount of detail and verifying that the detailed implementation meets the behavioral claims of the higher-level state models. At this level of detail, we are verifying that the logic involved in the control of the stabilizer supports the claims included in the verification set. In actuality, the underlying logic is translated into software that runs on the target processor. If the translation of the logic into code is performed correctly, then the execution of the code will support the behavior of the higher-level state chart. In reality, a number of factors—for example, other application code running on the same processor and associated scheduling interactions—can impact both the behavior and the correctness of the code. Hence another round of verification will be necessary when the applications are run on the target processor. Moreover, the FM controls other entities besides the stabilizer, and those entities could be verified in a similar manner to the approach outlined here.

In our example, the stabilizer logic itself has been decomposed into 13 substates and each substate is invoked by a 13-step sequencer. Each substate contains logic in the form of flowcharts that express the logic flow of the stabilizer control. A generic depiction of the 13-substates chart designed with the Simulink state chart designer is shown in Figure 6.5.

We will now look at a smaller subset of the stabilizer control logic that includes the sequencer, Stabilizer Control Logic #1 (SCL1), and Stabilizer Control Logic #2 (SCL2). Figure 6.6 shows these two logic blocks.

The state charts in Figure 6.6 indicate the input signals (variables), which are aligned on the right side of the ovals, and the output signals (variables), which are aligned on the left side of the ovals. Signal flow begins at the left and goes to the right. The signals are user-defined variables that are either in local scope to the stabilizer itself or global variables outside the scope of the stabilizer. The signal can be either a Boolean, integer, or floating-point value and is defined within Simulink.

Essentially, the set of top-level state charts in Figure 6.6 indicates that when signal 1 from the sequencer is asserted, the logic contained within SCL B1 begins its execution. The values in the input signals are read, the internal logic is executed, and the output signals are written to as the execution proceeds. With the next tick of the sequencer, signal 2 is asserted (signal 1 is deasserted), the input signals are read, the internal logic of SCL B2 is executed, and the output variables are written as the logic is executed. This execution sequence continues for each of the 11 remaining SCL blocks, and then the cycle repeats.

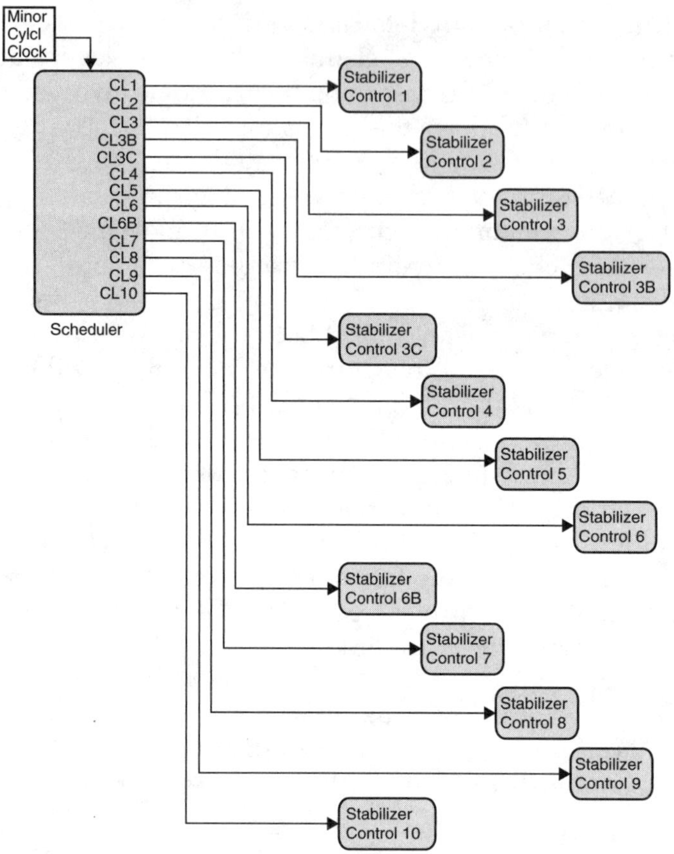

Figure 6.5: *Simulink state charts of the stabilizer control logic*

Note that output signals from one block can become input signals to another block in this system, and that input signals that do not come from the stabilizer control are either global system variables or signals from other subsystems that have a limited visibility. The importance of this observation is that the more extensive the use of global variables is, the more the system states contribute to the behavior of the stabilizer. This implies that in model checking a large system, the number of states may become so large that the memory in the computer system cannot hold all of the states necessary for the model-checking algorithms to work—a phenomenon known as the state explosion problem. Techniques have been developed to address the scalability of model-checking applications and circumvent this problem, and the reader is referred to [Clarke12] and [Clarke87] for a discussion of these techniques.

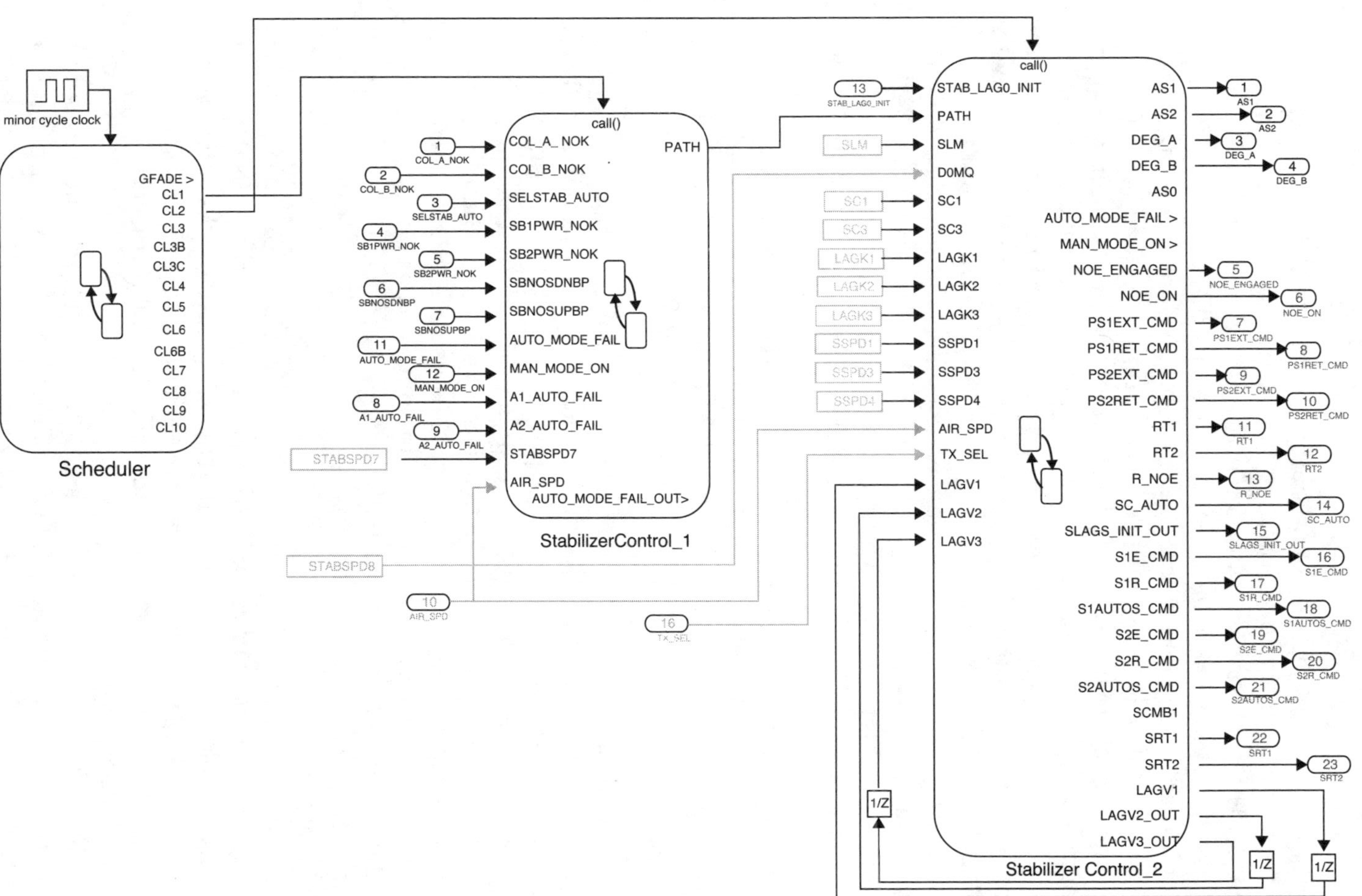

Figure 6.6: *Stabilizer control logic blocks 1 and 2*

Detailed State Charts

Figure 6.7 shows the input and output signals of SCL1, and Figure 6.8 shows the login in the form of a flowchart that is contained within SCL1. The input signals on the left of the top-level state chart are a combination of global variables (list) and local stabilizer variables (list). The logic that is contained within the state chart is shown in Figure 6.8.

The control logic is produced as part of the detailed design activity of the project. To be useful for the Simulink Design Verifier, the logic is mapped into a Simulink state chart. The mapping takes the comparison, assignment, and math blocks and turns them into states, while the operations within the blocks become the triggers on the arcs between the states. The transformation for SCL1 is shown in Figure 6.9. The underlying Simulink simulation engine can take this Stateflow diagram and, by assigning values to the variables, can use the underlying model-checking engine to verify or falsify the claims.

With this understanding of the modeling semantics of Simulink Stateflow chart, we can now look at how claims can be constructed using System Design Verifier and analyzed. Functional requirements for discrete systems are typically explicit statements about expected liveness properties (e.g., behaviors that a system exhibits) and safety properties (e.g., behaviors that it must never exhibit).

Formalizing Claims in System Design Verifier

To formally verify that the design behaves in accordance with the desired system properties (e.g., requirements), the requirements statements first need to be translated from a human language into the language understood by the formal analysis engine. Simulink Design Verifier can express formal requirements using MatLab functions and Stateflow blocks. Each requirement in Simulink has one or more verification objectives associated with it. These verification objectives are used to check whether the design matches the specified properties (claims).

Simulink Design Verifier (SDV) provides a set of building blocks and functions that one can use to define and organize verification objectives. The block library provided in SDV includes blocks and functions for test objectives, proof objectives, assertions, constraints, and a dedicated set of temporal operators for modeling of verification objectives with temporal aspects.

In general, a claim is an expression of a requirement that is to be proved. In SDV, a claim is termed a "property" and is equivalent to a requirement. Proof of a property in SDV can be expressed in two

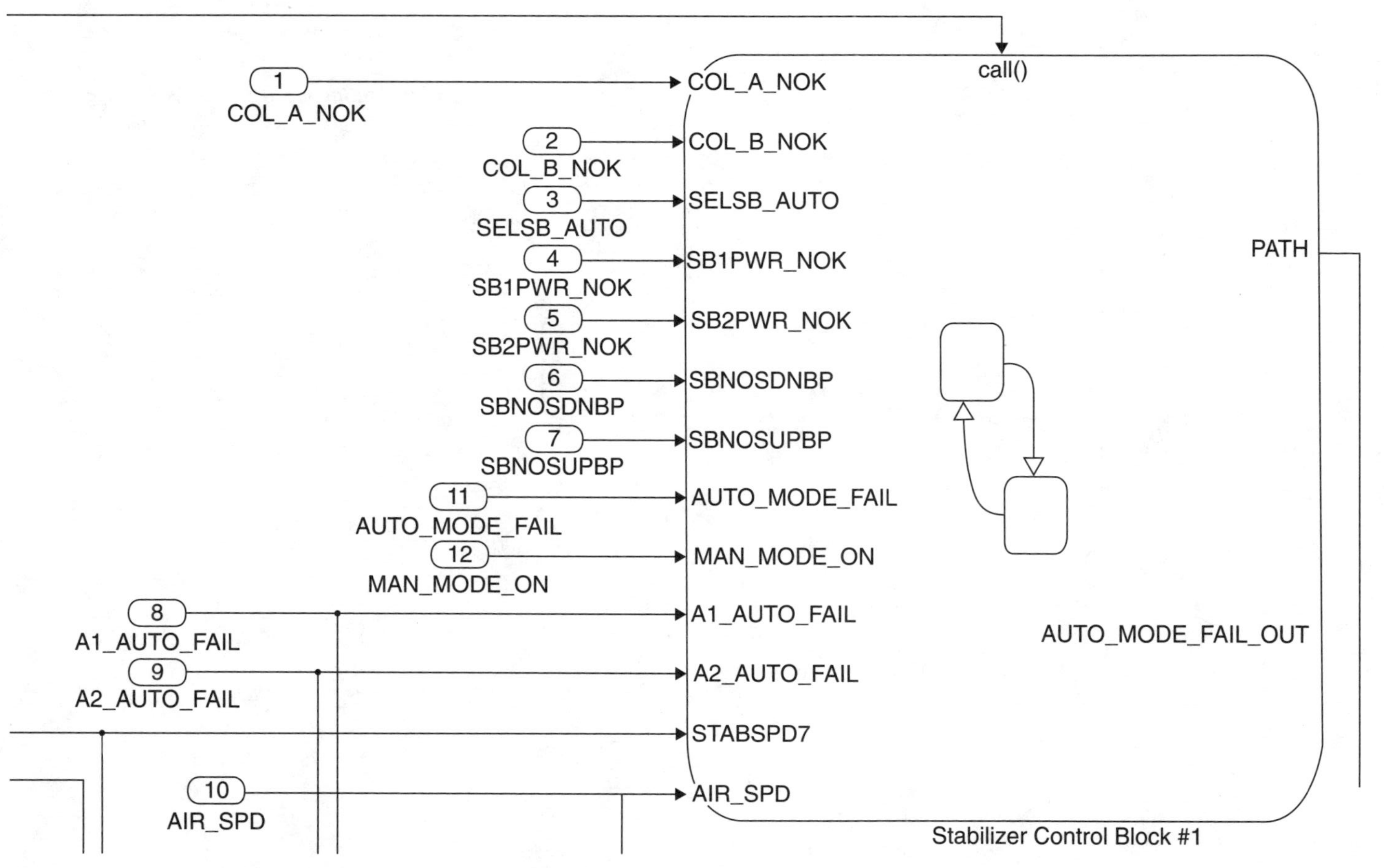

Figure 6.7: *Input and output signals of SCL B1*

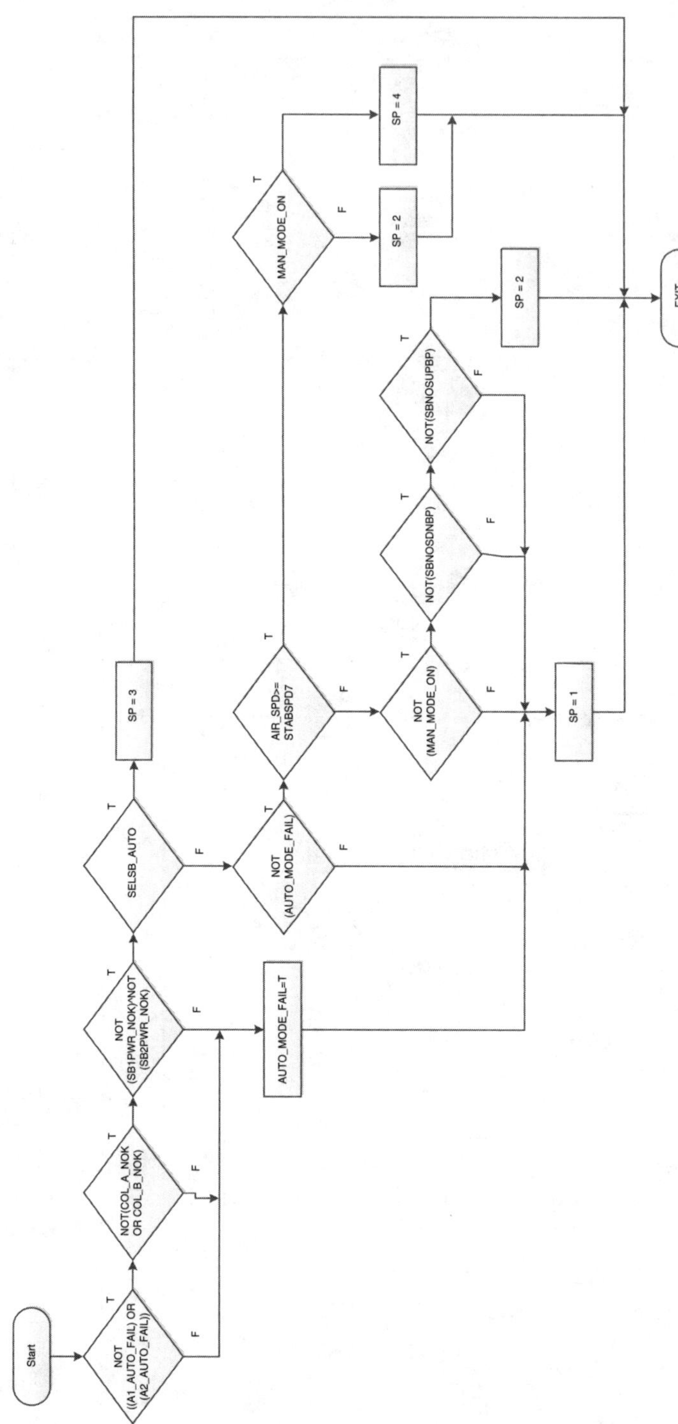

Figure 6.8: *A flowchart of the logic contained with SCL B1*

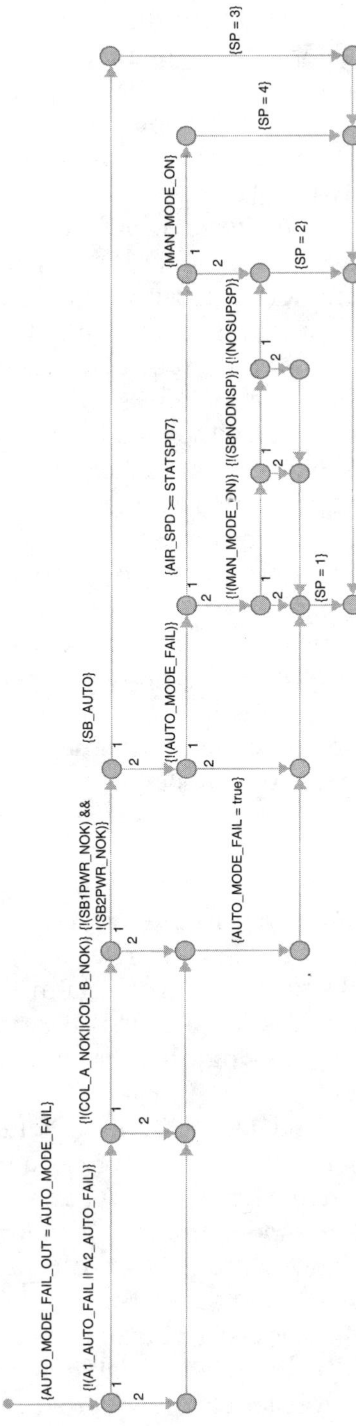

Figure 6.9: *The Simulink Stateflow chart for SCL1*

general forms: (1) a signal (variable) in the Simulink model that attains a particular value or range of values during a simulation and (2) a signal (variable) in the Simulink model that is proved to be logically equivalent to an expression over a number of input and output signals. We will focus on the latter form, as it is the more comprehensive and in general maps more closely to CTL expressions.

To construct a proof to be checked, SDV contains some primitive functions that allow construction of logical expressions that can be applied to the Simulink model. In essence, the model is simulated by providing all combinations of inputs and logging the corresponding outputs; SDV then checks whether the proof that is being checked is valid (true) for all possible combinations of inputs. SDV provides a report indicating that the proof to be verified either is satisfied (true for all combinations) or failed (false), in which case a counterexample is provided. The counterexample is the combination of inputs that make the proof false.

In Simulink, a property is a requirement that one models in Simulink, Stateflow, and MatLab function blocks. A property can be a simple requirement, such as a signal in the model that must attain a particular value or a range of values during simulation. The SDV software provides two blocks that allow one to specify property proofs in Simulink models. *Proof objective* blocks define the values of a signal to prove, and *proof assumption* blocks constrain the values of a signal during a proof.

We will use the proof objective block to prove the NOE mode claim. The claim we will review basically states that if the system is in auto-high-speed mode and the air speed is greater than 50 knots, the system should never go into NOE mode. Thus, we must insert a proof objective block on the NOE ON signal in the Simulink model. In addition, because the auto-high-speed mode comprises a number of signals, we must use some logic blocks to construct the claim.

To construct a logical expression of signals that constitute the claim, we use a combination of MatLab logic blocks, which lead to the expression shown in Figure 6.10.

To construct the logic associated with the NOE mode claim, we must recognize that auto-high-speed mode comprises a collection of signals that are global to the flight manager system. A review of the design documents reveals that auto-high-speed mode is a collection of signals that include `A1_AUTO_FAIL`, `A2_AUTO_FAIL`, `COL_A_NOK`, `COL_B_NOK`, `SB1PWR_NOK`, `SB2PWR_NOK`, `SELSB_AUTO`, `AUTO_MODE_FAIL`, and `MANUAL_MODE_ON`. These signals are inputs to the MatLab function block named Auto Mode HS. The airspeed variable is also an input, but it needs to be compared to a constant and converted to a binary value that indicates whether it is above or below a threshold (in this case, 50 knots).

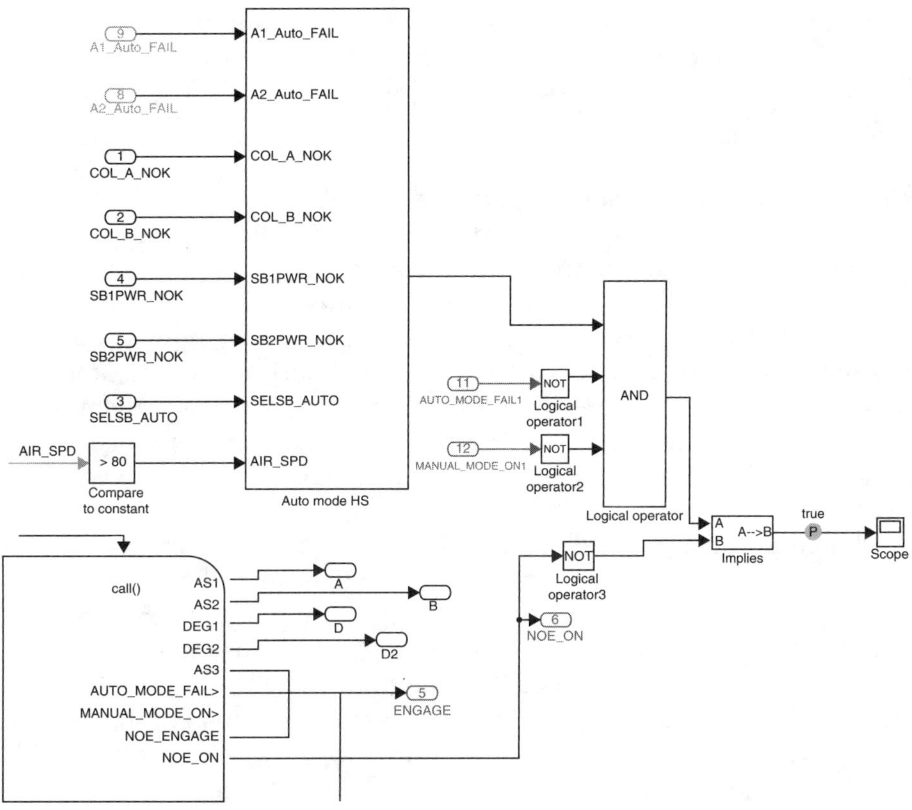

Figure 6.10: *Construction of the proof objective for NOE mode*

The following logic is contained within the Auto Mode HS function block:

```
function y = fcn(A1_AUTO_FAIL, A2_AUTO_FAIL, COL_A_NOK, COL_B_NOK,
SB1PWR_NOK, SB2PWR_NOK, SELSB_AUTO, AIR_SPD)
%#codegen
y = ~(A1_AUTO_FAIL || A2_AUTO_FAIL)…
    && ~(COL_A_NOK || COL_B_NOK)…
    && (~(SB1PWR_NOK)&& ~(SB2PWR_NOK)…
    && AIR_SPD)…
    && SELSB_AUTO;
```

This logic expression equates to the conditions that make up the auto mode and the airspeed. The output of this block is connected to a three-input `AND` logic block that `AND`s in the `AUTO_MODE_FAIL1` signal and the `MANUAL_MODE_ON1` signal. The conjunction of these signals feeds an

`IMPLIES` logic block, which tests whether the A input implies the B input. The block outputs a Boolean value of false when the A input is true and the B input is false; otherwise, it outputs true. The truth table is shown in Figure 6.11.

Running the Model-Checking Engine

At this point, the claim has been expressed in the Simulink Design Verifier notation and the model-checking analysis can be performed. The model is compiled and checked for compatibility with SDV, and then the property prover is run. (The mechanics of how to do this within the Simulink environment are omitted here.) Two outcomes are possible: the objectives are proven valid (e.g., the claim is proven true), or the objectives are shown to be invalid. The outcome of this exercise in our example is that the objective is proven valid. The output window is shown in Figure 6.12.

It is important to understand what this example encompasses and what the results are saying. The entire stabilizer model consisted of 13 state charts, with each state chart having an average of 15 variables. Over the entire stabilizer model, approximately 25 of the variables are global system variables. All of the stabilizer variables are used in the checking of the model. The underlying model-checking engine uses all of the variables and guarantees that if the property holds, then it holds for all possible inputs.

As a closing thought recall that failures of the FM (either hardware or software) were omitted from the high-level state chart in this example. To verify the fault tolerance of a system, we might potentially identify the modes where faults can occur and construct the arcs to which the system should transition. For example, a failure in the subsystem that produces the airspeed when the flight manager is in either auto-low-speed or auto-high-speed mode should cause a transition into manual mode, so that the pilot can use other means to determine airspeed and fly the helicopter accordingly.

A	B	Output
F	T	T
T	F	F
F	F	T
T	T	T

Figure 6.11: *`IMPLIES` block truth table*

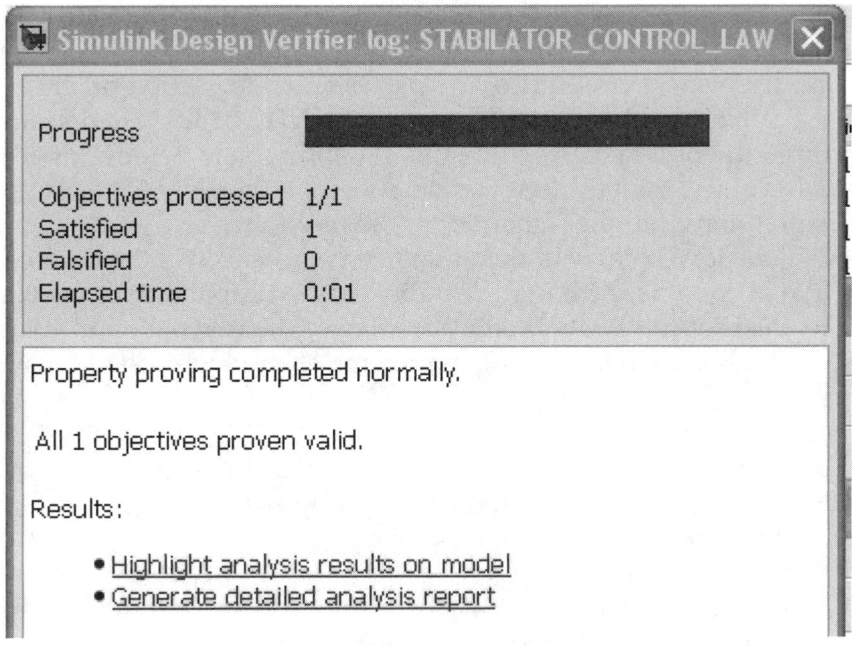

Figure 6.12: *SDV output showing the NOE mode claim to be true*

6.2.1.4 Observations

The combination of Simulink and Design Verifier provides a fairly powerful approach to incorporating a formal model-checking verification activity into the system design. The tool set can handle large systems at abstract levels and can check detailed logic designs as shown in the previous example.

As in the case with all model-checking tools, there is a limitation that is encountered known as the state explosion problem. Consider that there may be m systems and each system can have n states; then the asynchronous composition of these systems will have n^m states. The ability of modern computing platforms to keep the model representational structures in place is constrained by the amount of memory contained in a computing platform. Various computational approaches have been developed to address the state explosion problem for the underlying analysis engine in SDV, but this limitation persists. Suggested approaches to mitigate the problem within SDV are to limit the scope of variables used in the design of various subsystems and to use the technique called assume-guarantee reasoning. The reader is referred to [Flanagan03] for a discussion of this topic.

Other limitations to the tool reflect the fact that it is limited to safety properties only. Moreover, it can detect only a finite number of counterexamples. Liveness properties are also beyond the scope of the currently available tool. The tool cannot use LTL or CTL expressions directly in the proof stage. While this capability may be developed as the tool evolves, the best that can be done now is a loose mapping of LTL expressions into the semantics of the proof prover.

Also, the lowest level of detail that can be checked is the logic contained in flowcharts. Although Simulink has the capability to generate C code, the user must rely on the accuracy of the translation to ensure correctness. Research is ongoing in the area of checking code [Jhala09].

For the Practitioner

The tools within Simulink provide some advantages to the practitioner wishing to use formal methods in a design. Working within the MatLab–Simulink development environment allows many levels of design detail to evolve while maintaining traceability to the top-level design description. The capability of simulating and testing a design complements the formal model checking. In addition, through the use of hardware in the loop testing approaches, effects of timing can be observed on the design.

In dealing with systems of any size, the practitioner would benefit from developing or using automated means to help identify and manage claims, perhaps through the use of a stylized grammar for expressing functional and nonfunctional requirements. It may then be possible to develop a set of common proof block templates using the primitive blocks contained within Simulink and SDV. In addition, developing a technique to manage variable scoping during design would help reduce the potential for state explosion.

For the Researcher

The state explosion problem has been the focus of a great deal of research over the last 10–15 years. The nature of this problem is such that as larger and larger systems are developed and need to be verified, the number of states will continue to grow. It remains a fertile area for research to develop additional techniques.

Given that the flowchart representation is an abstraction above the generated code, developing techniques to verify the code and relate it back to the higher-level abstraction is an area where research remains in its infancy. Additionally, developing approaches that consider the application code in conjunction with the operating system and guaranteeing its functional and temporal behavior seem to be areas where very little research has been done.

6.3 Advanced Techniques

In this section, we present the most recent innovations in hybrid verification techniques, including timing verification with timed automata and hybrid discrete/continuous systems verification.

6.3.1 Real-Time Verification

In the previous section, we presented discrete verification problems. In this section, we turn our attention to real-time verification. Informally, this means verifying a hybrid system whose correctness depends not only on discrete state information, but also on the flow of time. In contrast to a discrete state, which changes instantaneously from one value to another, time flows continuously at a uniform rate. Formally, such hybrid systems are represented as timed automata [Alur94, Alur99]. A timed automaton is a finite-state machine augmented with a set of "clock" variables. Within each state, each clock variable records the passage of time. Transitions between states are instantaneous, and guarded by constraints on the clock variables. Finally, a subset of the clock variables is reset during each transition. The combination of these features enables timed automata to be a powerful modeling framework for real-time systems.

Several different flavors of timed automata have been proposed in the literature. They vary in their syntax but are semantically equivalent. Thus, the choice of which syntax to use is guided by some other factor, such as suitability to the application domain, familiarity, and tool support. In this chapter, we use the syntax of the model checker UPPAAL [Behrmann11], which is one of the most popular and actively maintained tools for verifying timed automata. Given our focus on practitioners, this choice is the most appropriate. In the discussion here, we do not go into the theoretical details behind the semantics of timed automata and their specification and verification, as these topics are covered comprehensively in the existing literature [Alur94, Alur99]. Instead, we describe timed automata, and its uses, via a simple example.

6.3.1.1 Example: Simple Light Control

Consider an intelligent light switch[1] that has the following specifications:

1. The switch has three states: off, medium, and high.
2. Initially, the switch is in the off state.
3. If the switch is pressed from the off state, it moves to the medium state.
4. If the switch is pressed from the medium state, there are two possible outcomes:
   a. If the latest press occurred within 10 seconds of the previous press, then the switch moves to the high state.
   b. If the latest press occurred more than 10 seconds after the previous press, the switch moves to the off state.
5. If the switch is pressed from the off state, it moves to the on state.

Clearly, time plays a crucial role in determining the behavior of this switch. Indeed, if the switch is in the medium state, given the same external input (press), the next state of the switch is determined by the amount of time that has passed since the last input. In addition, any model of the switch must have some mechanism to keep track of the amount of time elapsed and express conditions on time. The clock resetting and transition guards of timed automata provide these two functionalities.

The model of this light switch in UPPAAL is shown in Figure 6.13. As expected, there are three states of the automaton—denoted as `OFF`, `MEDIUM`, and `HIGH`—with `OFF` being the initial state. This captures specifications 1 and 2. In addition, an input action `press?` denotes the switch being pressed, and there is a clock variable `clk`. Each transition has a label of the form $\alpha\{\text{guard}\}[\text{action}]$, where α is the action triggering the transition, guard is a constraint on clock variables that must be `TRUE` for the transition to occur, and action is a set of assignments resetting certain clock variables. Note that {} means that `guard` is `TRUE`, and `[ ]` means that no clock variable is reset.

In Figure 6.13, all transitions have labels such that α is `press?`, indicating that transitions happen only if the switch is pressed. The transition from `OFF` to `MEDIUM` happens at any time but resets variable `clk`. This captures specification 3. The transition from `MEDIUM` to `HIGH` requires that `clk` be no greater than 10, which captures

1. This example, and other flavors of it, has appeared in a number of publications [Hessel03] and presentations on timed automata and UPPAAL.

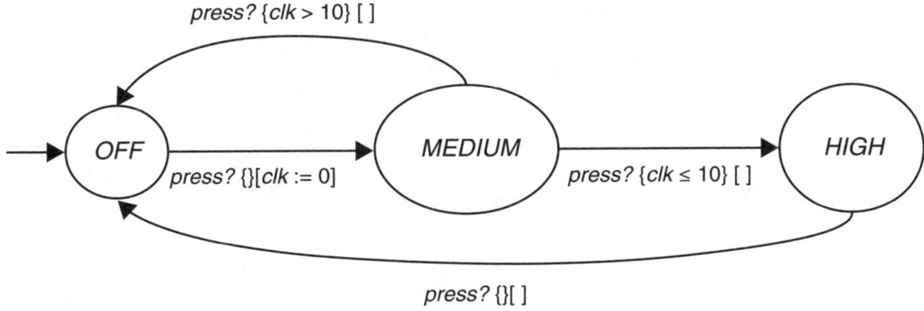

Figure 6.13: *Timed automata to model an intelligent light switch*

specification 4a. The transition from `MEDIUM` to `OFF` requires that the value of `clk` be greater than 10, which captures specification 4b. Both outgoing transitions from `MEDIUM` leave `clk` unchanged. Finally, the transition from `HIGH` to `OFF` can occur at any time, and leaves `clk` unchanged.

6.3.1.2 Composition and Synchronization

Timed automata can be composed, which in turn means that a more complex timed automaton can be described in terms of simpler ones. A number of composition semantics have been studied in the literature. In this chapter, we apply the semantics used by UPPAAL. In this semantics, in addition to the natural synchronization provided by clocks, timed automata are synchronized via matching input-output action pairs that label their transitions.

Consider again our example light switch from Figure 6.13. Recall that its transitions are labeled by the input action `press?`. Figure 6.14(a) shows the timed automata that models a "fast" user pressing the key every 3–5 seconds. Note that its transition is labeled by the output action `press!` and that it has its own clock variable `uclk`. The transition occurs only when 3 ≤ `uclk` ≤ 5 and resets `uclk`. The single state `STATE` of the automaton is also labeled with the "state invariant" `uclk` ≤ 5. This means that the automaton can remain in state `STATE` only as long as its clock is no greater than 5—in effect, modeling a user who waits no more than 5 seconds between successive switch presses. State invariants are useful to enforce "liveness" conditions; for example, the switch is pressed infinitely often, which is necessary to eliminate unrealistic behaviors from the model.

Figure 6.14(b) shows the timed automata that models a "slow" user who presses the key every 12–15 seconds. Note that its transition is also

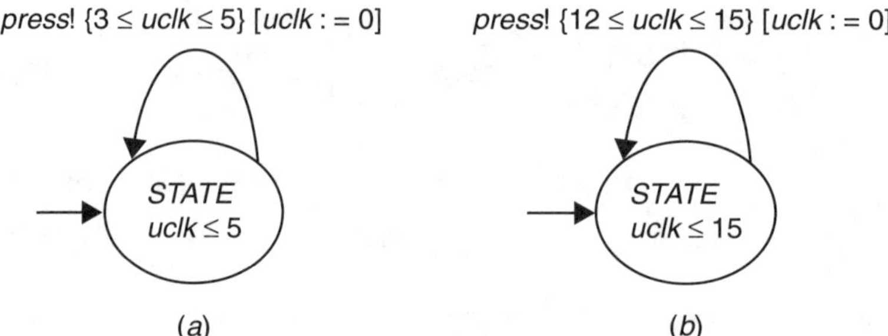

Figure 6.14: *A timed automata modeling: (a) a fast user and (b) a slow user*

labeled by the output action `press!` and that it has its own clock variable `uclk`. However, the transition occurs only when 12 ≤ `uclk` ≤ 15 and resets `uclk`. The other difference from Figure 6.14(a) is that the state invariant for `STATE` is `uclk` ≤ 15.

6.3.1.3 Functional Properties

In general, for timed automata, we can specify and verify functional properties that involve sequences of discrete state information and clock valuation. UPPAAL supports a rich specification language, a subset of CTL [Clarke83] extended with time. A full description of the syntax and semantics of this language is beyond the scope of this chapter. However, Table 6.1 shows a sampling of properties in UPPAAL syntax (column 2), and their informal meaning (column 3), for a system composed of the timed automaton shown in Figure 6.13 and one of the automata from Figure 6.14. In the properties, `Sw` refers to the timed automaton from Figure 6.13.

The last two columns in Table 6.1 show whether the property holds depending on which user automaton is selected from Figure 6.14. Specifically, column 4 shows the result for the "fast" system (using the user model from Figure 6.14a) while column 5 shows the result for the "slow" system (using the user from Figure 6.14b).[2] Let us consider these results in more detail:

- Property 1 is `FALSE` for both systems because in each case, a `press!` action by the user forces the switch to transition away from the `OFF` state. Therefore, it is impossible for the switch to remain indefinitely in the `OFF` state for any possible system execution.

---

2. These results were obtained using UPPAAL 4.0.13.

Table 6.1: *A Set of Functional Properties for the Timed Automata*

ID	UPAAL	Informal Meaning	Holds (fast)	Holds (slow)
1	*E [ ] Sw.OFF*	Is it possible that the light switch is always `OFF`?	No	No
2	*E <> (Sw.MEDIUM and Sw.clk > 6)*	Is it possible that after the switch reaches the `MEDIUM` state, it is not pressed for more than 6 seconds?	No	Yes
3	*A [ ] not deadlock*	The system does not deadlock?	Yes	Yes
4	*A <> Sw.HIGH*	From any given state, the switch will always eventually reach the `HIGH` state under all possible user behaviors.	Yes	No
5	*A <> Sw.OFF*	From any given state, the switch will always eventually reach the `OFF` state under all possible user behaviors.	Yes	Yes

- Property 2 is `TRUE` for the slow system because the slow user presses the switch only after every 12 seconds or more. However, the fast user presses the switch at least once every 5 seconds, forcing property 2 to be `FALSE` for the fast system.
- Property 3 is `TRUE` for both systems. Note that the state invariants in the user models are critical for this result. Without these invariants, the timed automaton for the user can remain in state `STATE` indefinitely without generating any `press!` actions, causing the system to deadlock.
- Property 4 is `TRUE` for the fast system because the rapid switch presses by the fast user force the switch to always visit the `HIGH` state after the `MEDIUM` state. In contrast, the property is `FALSE` for the slow system. Indeed, for the slow system, the `HIGH` state is never visited by the switch.
- Property 5 is `TRUE` for both systems because the `OFF` state is visited infinitely often. Once again, the state invariants are crucial to ensure this property.

In addition to exhaustive verification via model checking, UPPAAL supports a graphical user interface for modeling and a simulation environment. These features are beyond the scope of this chapter, but we refer the reader to the UPPAAL website for further information: http://www.uppaal.org.

6.3.1.4 Limitations and Future Work

We conclude this section with a discussion of other tools and techniques for functional verification of real-time systems, their limitations, and open problems.

Another formalism for specifying and verifying real-time systems is the process algebra called Timed CSP [Reed88]. It is a timed extension of Hoare's communicating sequential processes (CSP) [Hoare78]. Verification of timed CSP processes is supported by the FDR tool (http://www.cs.ox.ac.uk/projects/concurrency-tools). The main verification problem solved by FDR is to check whether one timed CSP process refines another. Thus, there is no separate specification language, such as a temporal logic. Instead, both the system under verification and the property being verified are expressed as timed CSP processes. Depending on the verification problem, this may be a more viable approach compared to using timed automata and UPPAAL.

A number of efforts have been made to incorporate priority-based scheduling inside model checkers for real-time systems. One approach is to model the scheduler as an explicit component of the system. This is possible when using either timed automata or timed CSP. There have also been attempts to introduce priorities as first-class concepts. A notable example is the Algebra of Communicating Shared Resources (ACSR) [Brémond-Grégoire93]. However, such efforts have not been backed up by the robust tool support necessary for their practical adoption.

Both UPPAAL and FDR can verify only models, leaving a big semantic gap between the system being verified and that being deployed. While significant advancements have been made in applying model checking directly to source code [Jhala09] in recent years, relatively less work has focused on applying model checking to real-time software. The REK tool [Chaki13] model-checks periodic programs written in C with rate-monotonic scheduling, and supports both priority-ceiling and priority-inheritance locks. REK is still in the prototype stage, however, and it is limited to verifying software deployed on a single processor only.

From a practitioner's perspective, a robust set of tools is available for exhaustively verifying the functional properties of models of real-time systems. While these tools emerged from the academic community, they are well supported and designed with usability in mind. Some of them have even been commercialized (http://www.uppaal.com). The main challenge here is continued improvement of these tools in terms of scalability and usability, and their integration with industrial tool chains, which requires sustained and active engagement and collaboration between the research and practitioner communities. A related challenge in using any such tool is the need to express problems and properties in the tool's formalism, and to find the right level of abstraction and decomposition.

For the research community, model checking real-time software is a fundamental open challenge. Extending and adapting the techniques that have been successful for verifying non-real-time software is a good start, but the tools and techniques in this area have a long way to go. Another challenge is dealing with the increasing "real concurrency" in systems, in terms of the emergence of multicore processors and distributed real-time systems. Both of these features exacerbate the state space explosion problem and hamper the scalability of model checking. The only promising paradigms to address this challenge are abstraction and compositional reasoning. Automated solutions, such as new assume-guarantee rules, are required for their widespread adoption in practice. Finally, the incorporation of faults, uncertainty, and security adds new dimensions to this problem.

6.3.2 Hybrid Verification

In this final section, we present verification examples for fully hybrid systems. Informally, a fully hybrid system is a system with discrete and continuous parts such that continuous parts of the system evolve according to some differential equations. Abusing the notation, in the rest of this section, we use the term "hybrid system" to refer to a fully hybrid system. A real-time system, described in the previous section, is an instance of a hybrid system in which the only continuous variables are clocks that evolve at a fixed uniform rate. Formally, hybrid systems are represented by hybrid automata [Alur95, Henzinger96]. A hybrid automaton is a finite-state machine with a set of continuous variables. Within each state, continuous variables record continuous changes relative to some time. The dynamics of the flow are represented by differential equations. Transitions between states are guarded by constraints on continuous variables. The ability to specify this kind of continuous change using differential equations makes hybrid automata much more powerful than the timed automata [Alur94, Alur99] used for real-time systems. In particular, the reachability problem for hybrid automata—that is, whether a given state and a valuation of continuous variables are reachable—is undecidable. Thus, existing verification techniques are best-effort measures and require the user to specify the upper bound on the exploration depth.

A number of notations have been developed for depicting hybrid automata and the tools to analyze them. In our examples, we use the SpaceEx State Space Explorer tool from Verimag (http://spaceex.imag.fr). Among the other powerful hybrid system verification tools being developed is the KeYmaera [Jeannin15] system, which allows verification of hybrid systems via a combination of automated and interactive proof methods. KeYmaera can, therefore, verify a wider class of systems and properties than is possible with fully automated systems such as SpaceEx.

Here, we focus on the practical aspects of modeling and analyzing the system, and leave the theoretical details of the semantics of hybrid automata and the underlying analysis techniques to the comprehensive exiting literature on the topic [Alur95, Henzinger96]. In the rest of this section, we give two examples of hybrid systems: a bouncing ball and a thermostat.

6.3.2.1 Example: Bouncing Ball

We want to model a ball that is dropped from a fixed height. The model in SpaceEx is shown in Figure 6.15. It consists of two continuous variables: x, which represents the current height of the ball relative to the ground, and v, which represents the velocity of the ball. The system has a single discrete state in which the continuous flow is given by the following differential equations:

$$\frac{dx}{dt} = v \text{ and } \frac{dv}{dt} = -g$$

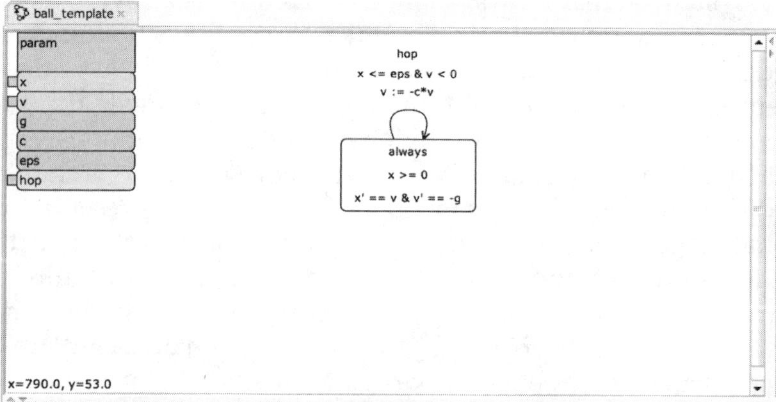

Figure 6.15: *A hybrid automaton for a bouncing ball example*

Here g stands for standard gravity. Note that in Figure 6.15, primes indicate first derivatives. Furthermore, it is required that the distance to the ground is always positive.

The system has a single transition that changes the continuous dynamics. Whenever the center of the ball is close to the ground (i.e., less than the parameter eps) and the velocity is negative, the ball hits the ground and shoots up with proportional velocity, dampened by factor c. Note that the transition has a label hop that is unused in our example. Such labels are used for synchronizing networks of components, similarly to the read and write labels used for synchronization of real-time systems.

Figure 6.16 shows the result of computing the set of reachable states using SpaceEx, where initially $10 \le x \le 10.2$ and $v = 0$. Because the reachability of hybrid automata is undecidable [Henzinger96] in general, the user must specify how many reachability steps to apply for both discrete and continuous evolution of the system. This example was constructed with 50 discrete steps and 80 continuous ones. The graph shows the height of the ball (x-axis) versus its velocity (y-axis). Initially the ball starts at a height of approximately 10 m and falls. The ground is hit at the velocity of slightly more than –4 m/s. Then, the ball bounces with the velocity of slightly less than 4 m/s because of the dampening effect of the impact. Each arc in the graph represents one complete bounce.

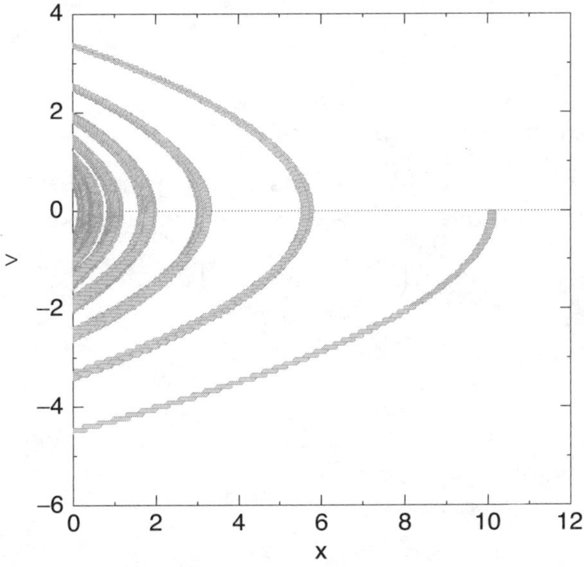

Figure 6.16: *Reachable states of the bouncing ball example*

6.3.2.2 Example: Thermostat

Our second example is a model of a thermostat from [Henzinger96] that is similar to the example hybrid system in Section 6.1. This thermostat has two states: on and off. In the off state, the temperature gradually decreases. In the on state, it gradually increases. Whenever the temperature reaches a particular limit (too cold or too hot), the thermostat switches to the corresponding model. The model in SpaceEx is shown in Figure 6.17. The continuous dynamics in state off are given by the following differential equation:

$$\frac{d}{dt} temp = -0.1 * temp$$

The continuous dynamics in the on state are given by the following:

$$\frac{d}{dx} temp = 5 - 0.1 * temp$$

The set of reachable states computed by SpaceEx for the thermostat model is shown in Figure 6.18. The initial state was temp = 20 and the system started in the off state. The graph shows the possible values for the temperature (y-axis) relative to time (x-axis). As before, because the problem of computing all reachable states is undecidable, a limit for computation was selected: 40 discrete steps and 80 continuous ones. Note that this was sufficient to explore all the states the system can reach in approximately 5 seconds.

6.3.2.3 Limitations and Future Work

Verification of fully hybrid systems is the most difficult challenge among the three types of systems we considered as targets for

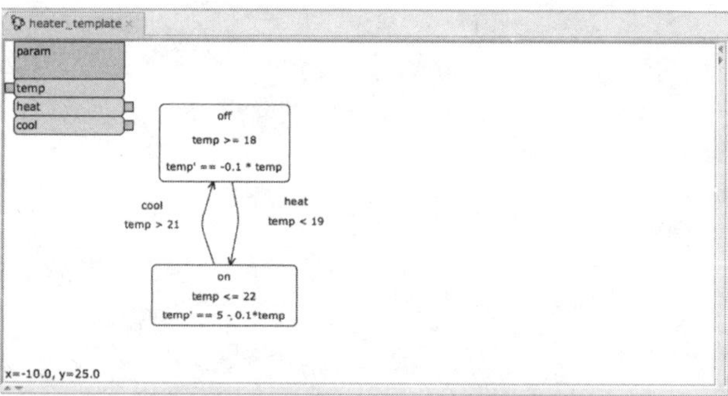

Figure 6.17: *A model of a thermostat*

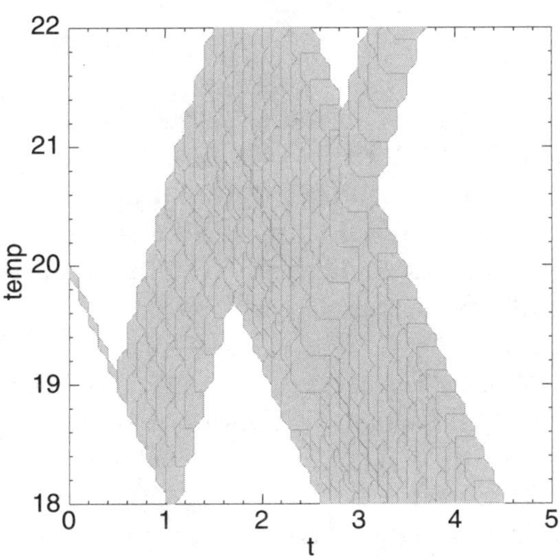

Figure 6.18: *Reachable states of the thermostat example*

verification. Not surprisingly, the tools and techniques in this area are the least mature. Practitioners have a much smaller selection of tools to choose from, and expressing a specific problem in the tool's format is correspondingly more difficult. From a theoretical perspective, identifying restricted types of hybrid systems for which efficient verification techniques exist is an important challenge.

6.4 Summary and Open Challenges

Ensuring the logical correctness of hybrid systems is a fundamental challenge in the domain of assuring cyber-physical systems. Model checking is a promising approach to address this challenge. There has been considerable progress in both the theory and the practice of applying model checkers to verify problems that are purely discrete, real time, or fully hybrid. Even so, many open problems remain to be solved in terms of fundamental theoretical advancements and technology transfer issues. One area that is ripe for research is bridging the gap between models and software, and reducing the semantic gap between what is verified and what is executed. Another area is dealing with faults. Finally, verifying hybrid systems in the presence of uncertainty is also critical. Overall, this is an important and challenging area of collaborative work for practitioners and basic and applied researchers.

References

[Alur99]. R. Alur. "Timed Automata." *Proceedings on Computer Aided Verification*, pages 8–22, 1999.

[Alur95]. R. Alur, C. Courcoubetis, N. Halbwachs, T. A. Henzinger, P. Ho, X. Nicollin, A. Olivero, J. Sifakis, and S. Yovine. "The Algorithmic Analysis of Hybrid Systems." *Theoretical Computer Science*, vol. 138, no. 1, pages 3–34, 1995.

[Alur94]. R. Alur and D. L. Dill. "A Theory of Timed Automata." *Theoretical Computer Science*, vol. 126, no. 2, pages 183–235, 1994.

[Baier08]. C. Baier and J. Katoen. *Principles of Model Checking*. MIT Press, 2008.

[Behrmann11]. G. Behrmann, A. David, K. G. Larsen, P. Pettersson, and W. Yi. "Developing UPPAAL over 15 years." *Software: Practice and Experience*, vol. 41, no. 2, pages 133–142, 2011.

[Brémond-Grégoire93]. P. Brémond-Grégoire, I. Lee, and R. Gerber. "ACSR: An Algebra of Communicating Shared Resources with Dense Time and Priorities." *CONCUR*, pages 417–431, 1993.

[Chaki13]. S. Chaki, A. Gurfinkel, S. Kong, and O. Strichman. "Compositional Sequentialization of Periodic Programs." *Verification, Model Checking and Abstract Interpretation*, pages 536–554, 2013.

[Cimatti99]. A. Cimatti, A. Biere, E. M. Clarke, and Y. Zhu. "Symbolic Model Checking Without BDDs." *Tools and Algorithms for Construction and Analysis of Systems*, March 1999.

[Clarke09]. E. M. Clarke, E. A. Emerson, and J. Sifakis. "Model Checking: Algorithmic Verification and Debugging." *Communications of the ACM*, vol. 52, no. 11, pages 74–84, 2009.

[Clarke83]. E. M. Clarke, E. A. Emerson, and A. P. Sistla. "Automatic Verification of Finite-State Concurrent Systems Using Temporal Logic Specifications: A Practical Approach." *Principles of Programming Languages*, pages 117–126, 1983.

[Clarke87]. E. M. Clarke and O. Grumberg. "Avoiding the State Explosion Problem in Temporal Logic Model Checking Algorithms." Carnegie Mellon University, CMU-CS-87-137, July 1987.

[Clarke99]. E. M. Clarke, O. Grumberg, and D. A. Peled. *Model Checking*. MIT Press, 1999.

[Clarke12]. E. M. Clarke, W. Klieber, M. Novacek, and P. Zuliani. "Model Checking and the State Explosion Problem." In *Tools for Practical Software Verification*. Springer, Berlin/Heidelberg, 2012.

[Comella_Dorda01]. S. Comella-Dorda, D. P. Gluch, J. J. Hudak, G. Lewis, and C. B. Weinstock, "Model-Based Verification: Abstraction Guidelines." SEI Technical Note CMU/SEI-2001-TN-018, Software Engineering Institute, Carnegie Mellon University, October 2001.

[FAA12]. Federal Aviation Administration. "Helicopter Flight Controls." In *Helicopter Flying Handbook*. FAA, 2012.

[Flanagan03]. C. Flanagan and S. Qadeer. *Assume-Guarantee Model Checking.* Technical Report, Microsoft Research, 2003.

[Frehse08]. G. Frehse. "PHAVer: Algorithmic Verification of Hybrid Systems Past HyTech." *Software Tools for Technology Transfer*, vol. 10, no. 3, pages 263–279, 2008.

[Gablehouse69]. C. Gablehouse. *Helicopters and Autogiros: A History of Rotating-Wing and V/STOL Aviation*. Lippincott, 1969.

[Harel87]. D. Harel. "Statecharts: A Visual Formalism for Complex Systems." *Science of Computer Programming*, vol. 8, pages 231–274, 1987.

[Harel90]. D. Harel, H. Lachover, A. Naamad, A. Pnueli, M. Politi, R. Sherman, A. Shtul-Trauring, and M. Trakhtenbrot. "STATEMATE: A Working Environment for the Development of Complex Reactive Systems." *IEEE Transactions on Software Engineering*, vol. 16, no. 4, pages 403–414, 1990.

[Harel87a]. D. Harel, A. Pnueli, J. Schmidt, and R. Sherman. "On the Formal Semantics of Statecharts." *Proceedings of the 2nd IEEE Symposium on Logic in Computer Science*, pages 54–64, Ithaca, NY, 1987.

[Henzinger96]. T. A. Henzinger. "The Theory of Hybrid Automata." *Logic in Computer Science*, pages 278–292, 1996.

[Henzinger97]. T. A. Henzinger, P. Ho, and H. Wong-Toi. "HYTECH: A Model Checker for Hybrid Systems." *Software Tools for Technology Transfer*, vol. 1, no. 1–2, pages 110-122, 1997.

[Hessel03]. A. Hessel, K. G. Larsen, B. Nielsen, P. Pettersson, and A. Skou. "Time-Optimal Real-Time Test Case Generation Using UPPAAL." *Formal Approaches to Testing Software*, pages 114–130, 2003.

[Hoare78]. C. A. R. Hoare. "Communicating Sequential Processes." *Communications of the ACM*, vol. 21, no. 8, pages 666–677, 1978.

[Holzmann04]. G. J. Holzmann. *The SPIN Model Checker: Primer and Reference Manual*. Addison-Wesley, 2004.

[Hudak02]. J. J. Hudak, S. Comella-Dorda, D. P. Gluch, G. Lewis, and C. B. Weinstock. "Model-Based Verification: Abstraction

Guidelines." SEI Technical Note CMU/SEI-2002-TN-011, Software Engineering Institute, Carnegie Mellon University, October 2002.

[Huth04]. M. Huth and M. Ryan. *Logic in Computer Science: Modelling and Reasoning About Systems*. Cambridge University Press, 2004.

[Jeannin15]. J. Jeannin, K. Ghorbal, Y. Kouskoulas, R. Gardner, A. Schmidt, E. Zawadzki, and A. Platzer. "A Formally Verified Hybrid System for the Next-Generation Airborne Collision Avoidance System." In *Tools and Algorithms for the Construction and Analysis of Systems*. Springer, 2015.

[Jhala09]. R. Jhala and R. Majumdar. "Software Model Checking." *ACM Computing Surveys*, vol. 41, no. 4, 2009.

[Jhala01]. R. Jhala and K. L. McMillan, "Microarchitecture Verification by Compositional Model Checking." Computer Aided Verification 13th International Conference, LNCS 2102, pages 396–410, Paris, France, July 2001.

[Lamport77]. L. Lamport. "Proving the Correctness of Multiprocess Programs." *IEEE Transactions on Software Engineering*, vol. 3, no. 2, pages 125–143, March 1977.

[Magee06]. J. Magee and J. Kramer. *Concurrency: State Models and Java Programs, 2nd Edition*. John Wiley and Sons, Worldwide Series in Computer Science, 2006.

[Peled01]. D. Peled. *Software Reliability Methods*. Springer-Verlag, 2001.

[Ramadge89]. P. J. G. Ramadge and W. M. Wonham. "The Control of Discrete Event Systems." *Proceedings of the IEEE*, vol. 77, no. 1, January 1989.

[Reed88]. G. M. Reed and A. W. Roscoe. "A Timed Model for Communicating Sequential Processes." *Theoretical Computer Science*, vol. 58, pages 249–261, 1988.

[Rozenburg98]. G. Rozenburg and J. Engelfriet. "Elementary Net Systems." In *Lectures on Petri Nets I: Basic Models. Advances in Petri Nets*, vol. 1491 of *Lecture Notes in Computer Science*, pages 12–121. Springer, 1998.

[Schneider85]. F. B. Schneider and B. Alpern. "Defining Liveness." *Information Processing Letters*, vol. 21, no. 4, pages 181–185, October 1985.

[Simulink13]. *Simulink Reference, R2013a*. The Mathworks, Natick, MA, 2013.

[Simulink13a]. *Simulink Design Verifier User's Guide, R2013a*. The Mathworks, Natick, MA, 2013.

[Simulink13b]. *Simulink Stateflow User's Guide, R2013a.* The Mathworks, Natick, MA, 2013.

[Usatenko10]. O. V. Usatenko, S. S. Apostolov, Z. A. Mayzelis, and S. S. Melnik. *Random Finite-Valued Dynamical Systems: Additive Markov Chain Approach*. Cambridge Scientific Publisher, 2010.

Chapter 7

Security of Cyber-Physical Systems

Bruno Sinopoli, Adrian Perrig, Tiffany Hyun-Jin Kim, and Yilin Mo

A wide variety of motivations exist for launching an attack on cyber-physical systems (CPSs), ranging from economic reasons (e.g., reducing electricity bills), to pranks, to terrorism (e.g., threatening people by controlling electricity and other life-critical resources). The first-ever CPS malware called Stuxnet was found in July 2010. This malware, which targeted vulnerable CPSs, raises new questions about CPS security [Vijayan10]. CPSs are currently isolated, preventing external access to them. Malware, however, can spread using USB drives and can be specifically crafted to sabotage CPSs. The emerging, highly interconnected, and complex CPSs, such as vehicular networks, embedded medical devices, and smart grids, will be exposed to increased external access in the future, which in turn can lead to compromise and infection of components. In this chapter we discuss how new security threats arise from this new reality and which new innovations take advantage of physics to counter these threats.

7.1 Introduction and Motivation

The tight integration between cyber and physical spaces brings new challenges for the research community. In CPSs, cyber attacks can cause disruptions that transcend the cyber realm and affect the physical world. Stuxnet [Vijayan10] is a clear example of a cyber attack that induced physical consequences. Conversely, pure physical attacks can also affect the security of the cyber system. For example, the integrity of a smart meter can be compromised if the attacker uses a shunt to bypass the meter. Likewise, a traffic signal preemption transmitter can compromise the integrity of a traffic light system. Placing a compromised sensor beside a legitimate one breaks the system secrecy while leaving the cyber system intact. Based on the discussions at the Army Research Office workshop on CPS security in 2009, we classify current attacks on cyber-physical systems into four categories. Table 7.1 provides examples to illustrate these classifications.

Defense mechanisms need to take into the account the dual nature of the CPS. As an example, a common practice for purely cyber systems is to shut down the compromised cyber components; however, such a practice may cause instability of the physical system and, therefore, is an infeasible approach for ensuring CPS security. For physical attacks, it is economically infeasible to shield all assets of large-scale physical systems, and purely cyber security-based approaches are insufficient for CPS security because they are not well equipped to detect and counter physical attacks, nor can they analyze the consequences of cyber attacks and defense mechanisms on the physical system. Therefore, system theory-based approaches, which leverage the physical model of the system, need to be used in combination with cyber security to provide solutions for detection, response, reconfiguration, and restoration of system functionalities while keeping the system operating.

The rest of this chapter reviews the basic techniques for dealing with cyber threats and countermeasures as well as advanced techniques based on system theoretic approaches for CPSs.

Table 7.1: *Taxonomy of Attacks and Consequences in Cyber and Physical Systems*

Consequence/Attack	Cyber	Physical
Cyber	Eavesdropping on private information	Stuxnet CPS malware
Physical	Sensor bypassing	Instability due to physical destructions

7.2 Basic Techniques

In this section, we analyze the cyber security requirements for cyber-physical systems, consider how the new attack models arise in the new CPS reality, and identify the basic corresponding countermeasures.

7.2.1 Cyber Security Requirements

In general, cyber security requirements for a system include three main security properties:

- *Confidentiality:* Prevents an unauthorized user from obtaining secret or private information
- *Integrity:* Prevents an unauthorized user from modifying the information
- *Availability:* Ensures that the resource can be used when requested

A variety of information is exchanged in CPS, including sensor data and control commands, which we consider as two core information types in this chapter. First, we examine the importance of protecting the core information types with respect to the main security properties. Second, we analyze the degree of cyber security's importance for software.

Confidentiality of sensor data, such as meter data, global positioning system (GPS), or accelerometer data, is clearly important. For example, meter data provide information about the usage patterns for individual appliances, which can reveal personal activities through nonintrusive appliance monitoring [Quinn09], and GPS data provide information about the location of individuals, which can be used to track individuals. Confidentiality of control commands may not be important in cases where they are public knowledge. Confidentiality of software should not be critical, because the security of the system should not rely on the secrecy of the software, but only on the secrecy of the keys, according to Kerckhoffs's principle [Kerckhoffs83].

The integrity of sensor data and commands is important as well as the integrity of software. Compromised software or malware may potentially be able to control any device or component in the CPS.

Denial-of-service (DoS) attacks are resource consumption attacks that send fake requests to a server or a network. Distributed DoS (DDoS) attacks are accomplished by utilizing distributed attacking sources such as compromised smart meters, traffic lights, or sensors in a vehicle. In CPS, availability of information is the most important aspect of the system's operation. For example, in automotive control systems, availability of the battery information in an electric vehicle is critical to prevent the vehicle from stopping in the middle of a busy road and causing an accident. Availability of control commands is also important—for example, when reducing the speed of a vehicle to maintain a safe distance from the car ahead. By comparison, availability of sensor data (e.g., gas mileage) may not be as critical because the data can usually be read at a later point.

Table 7.2 summarizes the relative importance of data, commands, and software. In this table, high risk implies that a property of certain information is critical, medium risk implies some importance, and low risk is noncritical. This classification enables prioritization of risks, to focus effort on the most critical aspects first. For example, the integrity of control commands is more important than the commands' confidentiality; consequently, we need to focus on efficient cryptographic authentication mechanisms before addressing encryption.

7.2.2 Attack Model

To launch an attack, an adversary must first exploit entry points, and upon successful entry, deliver specific cyber attacks on the CPS. In this section, we describe this attacker model in detail.

7.2.2.1 Attack Entry Points

In general, a strong perimeter defense is used to prevent external adversaries from accessing information or devices within the trusted zone. Unfortunately, the size and complexity of the networks in CPS provide numerous potential entry points:

- *Inadvertent infiltration through infected devices:* Malicious media or devices may be inadvertently infiltrated inside the trusted perimeter by personnel. For example, USB memory sticks have become a

Table 7.2: *Importance of Security Properties for Commands, Data, and Software*

	Control Commands	Sensor Data	Software
Confidentiality	Low	Medium	Low
Integrity	High	High	High
Availability	High	Low	N/A

popular tool to circumvent perimeter defenses: A few stray USB sticks left in public spaces may be picked up by employees and plugged into previously secure devices inside the trusted perimeter, enabling malware on the USB sticks to immediately infect the devices. Similarly, devices used both inside and outside the trusted perimeter can be infected with malware when outside that perimeter, allowing the malware to penetrate the system when used inside the perimeter. For example, corporate laptops that are privately used at home are vulnerable to this type of attack.

- *Network-based intrusion:* Perhaps the most common mechanism to penetrate a trusted perimeter is through a network-based attack vector. Exploiting poorly configured firewalls through both misconfigured inbound and faulty outbound rules is a common entry point, enabling an adversary to insert a malicious payload onto the control system.
- *Backdoors and holes in the network perimeter:* Backdoors and holes may be caused by components of the IT infrastructure that have vulnerabilities or misconfigurations. For example, networking devices at the perimeter (e.g., fax machines, forgotten but still connected modems) can be manipulated to bypass proper access control mechanisms. In particular, dial-up access to remote terminal units (RTUs) is used for remote management, and an adversary can directly dial into modems attached to field equipment, where many units do not require a password for authentication or have unchanged default passwords. Further, attackers can exploit vulnerabilities of the devices and install backdoors to enable their future access to the prohibited area. Exploiting trusted peer utility links is another potential network-based entry point. For example, an attacker could wait for a legitimate user to connect to the trusted control system network via a virtual private network (VPN) and then hijack that VPN connection. All of these kinds of network-based intrusions are particularly dangerous because they enable a remote adversary to enter the trusted control-system network.
- *Compromised supply chain:* An attacker can pre-install malicious codes or backdoors into a device prior to shipment to a target location—a strategy called a supply chain attack. Consequently, the need for security assurance in the development and manufacturing process for sourced software and equipment is critical for safeguarding a cyber supply chain that involves technology vendors and developers.

- *Malicious insider:* An employee or legitimate user who is authorized to access system resources can perform actions that are difficult to detect and prevent. Privileged insiders also have intimate knowledge of the deployed defense mechanisms, which they can often easily circumvent.

7.2.2.2 Adversary Actions

Once an attacker gains access to the network, that person can perform a wide range of attacks. Table 7.3 lists actions that an adversary can perform to violate the main security properties (confidentiality, integrity, availability) of the core types of information. We classify specific cyber attacks based on whether they lead to cyber or physical consequences.

Cyber Consequences

From the cyber point of view, a variety of consequences rooted in how software works may arise:

- *Spreading malware and controlling devices:* An adversary can develop malware and spread it so that it infects smart meters [Depuru11] or company servers. Malware can be used to replace or add any function to a device or a system, such as sending sensitive information or controlling devices.
- *Vulnerabilities in common protocols:* CPS use existing protocols, which means they inherit the vulnerabilities of these protocols. Commonly used protocols include TCP/IP and Remote Procedure Call (RPC).
- *Access through database links:* Control systems record their activities in a database on the control system network, and then in mirror

Table 7.3: *Types of Threats Created by Attacking Security Properties*

	Control Commands	Sensor Data	Software
Confidentiality	Exposure of control structure	Unauthorized access to sensor data	Theft of proprietary software
Integrity	Changes of control commands	Incorrect system data	Malicious software
Availability	Inability to control system	Unavailability of sensor information	N/A

logs in the business network. A skilled attacker can gain access to the database on the business network, with the business network then providing a path to the control system network. Modern database architectures may allow this type of attack if they are configured improperly.

- *Compromising communication equipment:* An attacker can potentially reconfigure or compromise some of the communication equipment, such as multiplexers.
- *Injecting false information on sensor data:* An adversary can send packets to inject false information into the system. The result of injecting false information, such as that blocking traffic flow in a certain direction at an intersection, will be indefinite waiting for vehicles in that direction.
- *Eavesdropping attacks:* An adversary can obtain sensitive information by monitoring network traffic, which may result in privacy breaches or disclosure of the controlling structure of cyber-physical systems. Such eavesdropping can be used for gathering information with which to perpetrate further crimes. For example, an attacker might gather and examine network traffic to deduce information from communication patterns. Even encrypted communication can be susceptible to these kinds of traffic analysis attacks.

A SCADA protocol of noteworthy concern is the Modbus protocol [Huitsing08], which is widely used in industrial control applications such as in water, oil, and gas infrastructures. The Modbus protocol defines the message structure and communication rules used by process control systems to exchange SCADA information for operating and controlling industrial processes. Modbus is a simple client–server protocol that was originally designed for low-speed serial communication in process control networks. Given that this protocol was not designed for highly security-critical environments, several kinds of attacks are possible:

- *Broadcast message spoofing:* This attack involves sending fake broadcast messages to slave devices.
- *Baseline response replay:* This attack involves recording genuine traffic between a master and a field device, and replaying some of the recorded messages back to the master.
- *Direct slave control:* This attack involves locking out a master and controlling one or more field devices.

- *Modbus network scanning:* This attack involves sending benign messages to all possible addresses on a Modbus network to obtain information about field devices.
- *Passive reconnaissance:* This attack involves passively reading Modbus messages or network traffic.
- *Response delay:* This attack involves delaying response messages so that the master receives out-of-date information from slave devices.
- *Rogue interloper:* This attack involves attacking a computer with the appropriate (serial or Ethernet) adapters to an unprotected communication link.

Physical Consequences

The attacks mentioned here are not exhaustive, but they serve to illustrate risks and can help developers ensure that grid systems are secure:

- *Interception of SCADA frames:* An attacker can use a protocol analysis tool for sniffing network traffic to intercept SCADA DNP3 (Distributed Network Protocol 3.0) frames and collect unencrypted plaintext frames that may contain valuable information, such as source and destination addresses. The intercepted data, which include control and setting information, could then be used at a later date on another SCADA system or intelligent equipment device (IED), thereby shutting services down (at worst) or causing service disruptions (at a minimum). Additional examples of SCADA threats are available at the US-CERT website: http://www.us-cert.gov/control_systems/csvuls.html.
- *Malware targeting of industrial control systems:* An attacker can successfully inject worms into vulnerable control systems and reprogram industrial control systems.
- *DoS/DDoS attacks on networks and servers:* An adversary can launch a DoS/DDoS attack against various components in the CPS, including networking devices, communication links, and servers. If the attack is successful, then the system cannot be controlled in the target region.
- *Sending fake commands to systems in a region:* An adversary can send fake commands to a device or a group of devices in a target region. For example, sending a fake emergency electronic brake light (EEBL) notification to a vehicle can result in a vehicle braking hard, which may lead to an accident. Thus, insecure communication in CPS may be able to threaten human life.

7.2.3 Countermeasures

A number of countermeasures has been developed for these types of attacks. In this section, we present some of the most relevant.

7.2.3.1 Key Management

Key management is a fundamental approach for information security. Shared secret keys or authentic public keys can be used to achieve secrecy and authenticity for communication. Authenticity is especially important to verify the origin of messages, which in turn is key for access control.

The key setup in a system defines the root of trust. For example, a system based on public/private keys may define the public key of a trust center as the root of trust, with the trust center's private key being used to sign certificates and delegate trust to other public keys. In a symmetric-key system, each entity and the trust center would set up shared secret keys and establish additional trust relationships among other nodes by leveraging the trust center, as in the Kerberos network authentication protocol (https://web.mit.edu/kerberos/). The challenge in this space is key management across a very broad and diverse infrastructure. Accounting for several dozens of secure communication scenarios is required, ranging from communication between the manufacturers and devices to communication between sensors and field crews. For all these communication scenarios, keys need to be set up to ensure secrecy and authenticity.

Besides the tremendous diversity of equipment, a wide variety of stakeholders must be accommodated—for example, government, corporations, and consumers. Even secure email communication among different corporations is a challenge today, while the secure communication between a sensor from one corporation and a field crew of another corporation poses numerous additional challenges. By adding a variety of key management operations to the mix (e.g., key refresh, key revocation, key backup, key recovery), the complexity of key management becomes truly formidable. Moreover, business, policy, and legal aspects need to be considered in setting up the key management system, as a message signed by a private key can hold the key owner liable for the contents. A recent publication from NIST provides a good guideline for designing cryptographic key management systems to support an organization [Barker10], but it does not consider the diverse requirements of CPS.

7.2.3.2 Secure Communication Architecture

Designing a highly resilient communication architecture for CPS is critical to mitigate attacks while achieving high-level availability. Here are the required components for such an architecture:

- *Network topology design:* A network topology represents the connectivity structure among nodes, which can have an impact on the network's robustness against attacks [Lee06]. Thus, connecting networking nodes to be highly resilient under attack can be the basis for building a secure communication architecture.
- *Secure routing protocol:* A routing protocol on a network is used to build logical connectivity among nodes; in turn, one of the simplest ways to prevent communication is by attacking the routing protocol. By compromising a single router and by injecting bogus routes, all communication in the entire network can be brought to a standstill. Thus, we need to consider the security of a routing protocol running on top of a network topology.
- *Secure forwarding:* An adversary who controls a router can alter, drop, and delay existing data packets or inject new packets. Thus, securing individual routers and detecting malicious behaviors are required steps to achieve secure forwarding.
- *End-to-end communication:* From an end-to-end perspective, secrecy and authenticity of data are the most crucial properties. Secrecy prevents an eavesdropper from learning the data content, while authenticity (sometimes referred to as integrity) enables the receiver to verify that the data indeed originated from the sender, thereby preventing an attacker from altering the data. While numerous protocols exist (e.g., SSL/TLS, IPsec, SSH), some low-power devices may need lightweight protocols to perform the associated cryptography.
- *Secure broadcasting:* Many CPS rely on broadcast communication. Especially for dissemination of sensor data, authenticity of the information is important, because an adversary could inject bogus information to cause undesired consequences.
- *DoS defense:* Even when all of the previously mentioned mechanisms are in place, an adversary may still be able to prevent communication by mounting a DoS attack. For example, if an adversary controls many endpoints after compromising them, he or she can use these endpoints to send data that floods the

network. Hence, enabling communication under these circumstances is crucial—for example, to perform network management operations that defend against the attack. Moreover, electricity itself, rather than communication networks, can be a target of DoS attacks [Seo11].

- *Jamming defense:* To prevent an external adversary from jamming the wireless network, jamming detection mechanisms can be used to detect attacks and raise alarms. A multitude of methods to counter jamming attacks has been developed [Pickholtz82], enabling operation during jamming.

7.2.3.3 System and Device Security

An important area to address to ensure CPS security is vulnerabilities that enable exploitation through software-based attacks, in which either an adversary exploits a software vulnerability to inject malicious code into a system or a malicious insider uses administrative privileges to install and execute malicious code. The challenge in such an environment is to obtain the "ground truth" when communicating with a potentially compromised system: Is the response sent by legitimate code or by malware? An illustration of this problem is the case in which we attempt to run a virus scanner on a potentially compromised system: If the virus scanner returns the result that no virus is present, does that result occur because no virus could be identified or because the virus has disabled the virus scanner? A related problem is the case in which the current virus scanners contain an incomplete list of virus signatures; failure to detect a virus could then occur because the virus scanner does not yet recognize the new virus.

A promising new approach to provide remote code verification is the technology called attestation. Code attestation enables an external entity to query the software that is executing on a system in a way that prevents malware from hiding. Since attestation reveals a signature of executing code, even unknown malware will alter that signature and, therefore, can be detected. In research conducted in this direction, hardware-based approaches have been tested for attestation [LeMay07, LeMay12]. Software-based attestation is an approach that does not rely on specialized hardware, but rather makes some assumptions that the verifier can uniquely communicate with the device under verification [Seshadri04]. [Shah08] demonstrates the feasibility of this concept on SCADA devices.

7.3 Advanced Techniques

New and currently developing security schemes need to be aware of new vulnerabilities stemming from both the cyber and physical sides of a CPS and take advantage of both sides. In this section we discuss some initial steps that the research community is taking to enhance CPS security.

7.3.1 System Theoretic Approaches

System theoretic approaches take into account the physical properties of the system as a basis for creating new protection mechanisms.

7.3.1.1 Security Requirements

The security and resiliency of cyber-physical systems are essential for continued operation of critical infrastructures, despite problematic occurrences such as faults and attacks. The resiliency of CPS in real-time security settings can be distilled down to ensuring that the system satisfies the following general properties:

- The CPS should withstand a prespecified list of contingencies.
- The CPS should be able to detect and isolate faults and attacks.
- The performance of the CPS should degrade gracefully with respect to failures or attacks.

The precedent for CPS security has been set by the U.S. Department of Energy (DoE) in the field of smart grids, which are crucial instances of cyber-physical systems. The DoE's *Smart Grid System Report* [US-DOE09] summarizes six characteristics of the smart grid, which were further developed from the seven characteristics in "Characteristics of the Modern Grid" [NETL08] published by the National Energy Technology Laboratory (NETL). Among these characteristics, the most important characteristic identified by DoE with respect to security is the ability to operate resiliently even during disturbances, attacks, and natural disasters.

These system properties can be reinforced and enhanced by using countermeasures designed specifically for this function. A thorough contingency analysis can be executed to confirm that the system will continue to operate under constricted conditions. Graceful performance degradation can be effectuated by utilizing a robust control mechanism, which will prevent the system from becoming unstable. Fault detection and isolation can be employed to pinpoint and diagnose system faults, as well as some forms of malicious attacks. Countermeasures explicitly designed for distinct attacks can be applied for critical systems.

These methodologies are further elaborated in the following sections. Before presenting their details, we outline a system and attack model as the basis for developing the countermeasures.

7.3.1.2 System and Attack Model

In this section, we describe a linear time-invariant (LTI) state space model for CPS. Although most of the practical systems are nonlinear and possibly time varying, the LTI model can serve as a good approximation of the real system around any operating point. To be specific, for continuous-time systems, the following relationship holds:

$$\frac{d}{dt}x_t = Ax_t + Bu_t + B^aU_t^a + w_t$$

Here $t \geq 0$ is the index of time, and $x_t \in \mathbb{R}^n$ and $u_t \in \mathbb{R}^p$ are the state vector and input vector at time t, respectively. The attacker's input at time t is denoted by $u_t^a \in \mathbb{R}^q$. w_t is white Gaussian noise, which models the uncertainty of the system. $A \in \mathbb{R}^{nxn}$ and $B \in \mathbb{R}^{nxp}$ are called the system matrix and the input matrix, respectively. Finally, the attacker's input matrix $B^a \in \mathbb{R}^{nxq}$ is used to model the possible direction of the attacker's influence on the system.

We further assume that the sensors follow a linear model, which is given by the following expression:

$$y_t = Cx_t + \Gamma u_t^a + v_t$$

Here $y_t \in \mathbb{R}^m$ is the sensors' measurements at time t, and v_t is white Gaussian noise, which models the uncertainty in the sensors. $C \in \mathbb{R}^{mxn}$ is called the output matrix, and $\Gamma = diag(\gamma_1, \ldots, \gamma_m)$ is a diagonal matrix, where γ_i is a binary variable that indicates whether the attacker can change the ith measurements: $\gamma_i = 1$ if the attacker can modify the ith measurements; otherwise, $\gamma_i = 0$.

Given the measurements y_t, a controller is used to generate the control input u_t to stabilize the system or improve its performance. A typical state space controller consists of two parts:

- A state estimator, which generates an estimate $\hat{x}_t$ of the current state x_t based on the current and previous measurements y_t
- A feedback controller, which generates the control input u_t based on the state estimate $\hat{x}_t$

In practice, a fixed-gain state estimator and a fixed-gain feedback controller are commonly used, which are given by the following expressions:

$$\frac{d}{dt}\hat{x}_t = A\hat{x}_t + Bu_t + K(y_t - C\hat{x}_t)$$

$$u_t = L\hat{x}_t$$

The matrices K and L are called the estimation and the control gain, respectively. The closed-loop CPS is stable (in the absence of the attacker's input) if and only if all of the eigenvalues of the matrices $A - KC$ and $A + BL$ are on the left half-plane (i.e., the real parts of all the eigenvalues are strictly negative).

Figure 7.1 illustrates a simple control system view of the CPS.

Similarly, a discrete-time LTI system model can be described by the following equations:

$$x_{k+1} = Ax_k + Bu_k + B^a u_k^a + w_k$$

$$y_k = Cx_k + \Gamma u_k^a + v_k$$

Here $k \in \mathbb{N}$ is the index of discrete time. The fixed-gain estimator and controller take the following form:

$$\hat{x}_{k+1} = A\hat{x}_k + Bu_k + K(y_k - C\hat{x}_k)$$

$$u_k = L\hat{x}_k$$

The closed-loop CPS is stable (in the absence of the attacker's input) if and only if all of the eigenvalues of the matrices $A - KCA$ and $A + BL$ are located inside the unit circle (i.e., the absolute value of all the eigenvalues is strictly less than 1).

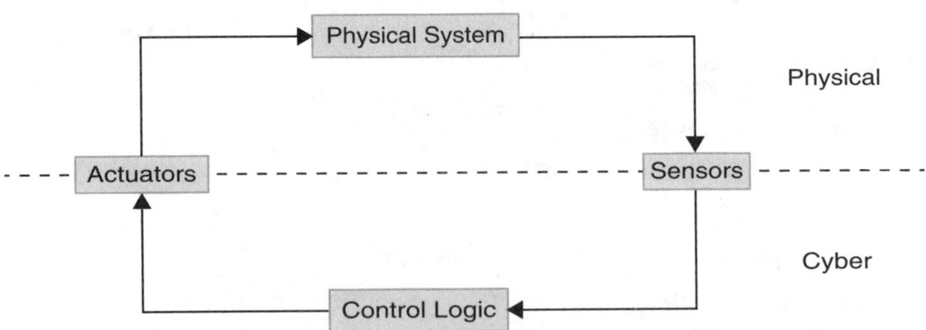

Figure 7.1: *Control system view of a CPS*

The continuous-time and discrete-time models are dynamical models in the sense that the current state x_t (i.e., x_k) affects the future state. For some large-scale systems (e.g., power grids), a static model is commonly adopted, which is given by the following expression:

$$y_k = Cx_k + \Gamma u_k^a + v_k$$

Comparing this equation to the dynamic model, the state x_t at time k is assumed to be unknown and independent from the previous state. A fixed-gain estimator is thus given by the following equation:

$$\hat{x}_k = Ky_k$$

Characterization of Adversary Models

[Teixeira12] proposes to characterize the capability of the adversary by its a priori system knowledge, disclosure power, and disruption power. The a priori knowledge that the attacker has about the system comprises the adversary's knowledge of the system's static parameters, such as the matrices A, B, C, the control and estimation gain L, K, and the statistics of the noise. By comparison, the disclosure power enables the attacker to obtain real-time system information, such as the current state x_k, the sensor measurements y_k, the estimated state $\hat{x}_k$, and the control input u_k. Lastly, the disruption power consists of the adversary's capability to disturb the system by injecting a malicious control input or by compromising the data integrity of the sensory information or control commands.

Figure 7.2 illustrates four different attacks based on system knowledge, disclosure power, and disruption power. For a pure eavesdropping

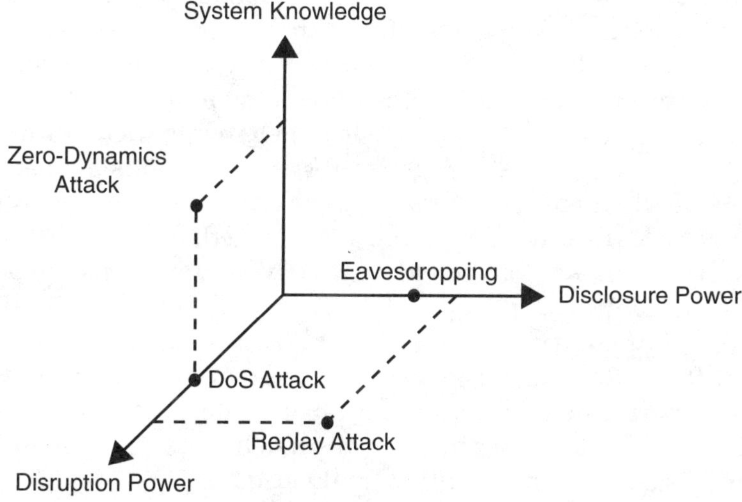

Figure 7.2: *A characterization of cyber-physical attacks*

attack, only disclosure power is needed; for a DoS attack, only disruption power is required. Later in the chapter we will introduce more complicated attacks, such as zero-dynamics attacks and replay attacks, and the corresponding defense mechanisms.

7.3.1.3 Countermeasures

Advanced countermeasures that involve both cyber and physical elements are key to address the new forms of attacks in CPS. In the following discussion, we present a number of the most important innovations in this area.

Contingency Analysis

Contingency analysis checks whether the steady-state system is outside the operating region for each contingency [Shahidehpour05]. The most used security criterion is generally referred to as the $N - 1$ criterion. The stipulations of this criterion, among others, are as follows:

- The system must adhere to constraints on the operating conditions, during normal operation when all (N) elements of the system are in operation.
- For any incident leading to the disconnection of one and only one element of the system, the operating point stays within the required parameters.

The first stipulation is tested for different normal states—for example, for peak and off-peak loads in the case of power grids. The second stipulation means that a system component may fail without overloading other components or without violating operating parameters—that is, the intention is that the system will remain in a state where any element may fail but the other elements will remain below their operating limits [Hug08].

In the example of power grids, a list of elements such as transformers, transmission lines, cables, generators, and so on is available. In a simulation, each of these elements is taken out of service, one by one, at which point a load-flow calculation is performed. If the load-flow computation shows an overloaded element, preventive actions can be taken in the real system.

This kind of analysis, which is based on a large number of contingent load-flow calculations, is called contingency analysis. Contingency analysis is performed in utilities control centers on a regular basis [Schavemaker09]. A comprehensive approach to power system contingency analysis has been described by [Deuse03].

While this security benchmark can be easily generalized considering the loss of more than one element, changing the criterion from $N - 1$ to $N - 2$ and further to $N - k$, several factors can rule out such analyses [Zima06]:

- Due to the combinatorial nature of the elements, each successive element failure increases the possible number of contingencies exponentially. Even considering a smaller number of elements that are subject to successive failures prohibitively increases the computational burden.
- If the system need to comply with an $N - k$ criterion, the utilization of the system resources in normal (N) operation would be very low, which would result in a bigger operating investment for the same conditions.

In practical systems such as power grids, the number of potential contingencies is very large. Due to real-time constraints, it is nearly impossible to evaluate each contingency, even considering an $N - 1$ constraint. Thus, the list of possible contingencies is usually screened and ranked, and only a select number of contingencies are evaluated.

Fault Detection and Isolation

Fault detection and isolation (FDI) is used in CPS to detect the presence of a fault and pinpoint its type and location. In principle, if a dynamic system is under normal operation—that is, no failure or attacker's input occurs—then the sensor measurements $\{y_t\}$ converges to a stationary Gaussian process. As a result, if the distribution of $\{y_t\}$ differs from the nominal distribution, then some faults (or possibly attacks) are presented in the CPS. A survey by [Willsky76] provides details on the process of detecting the deviation of $\{y_t\}$ from the normal distribution.

For some systems, however, it is possible for an adversary to carefully construct a sequence of inputs $\{u_a\}$, such that the resulting measurements $\{y_t\}$ follow the same distribution as the measurements $\{y_t\}$ under normal operation. Such an attack strategy is called a zero-dynamics attack. Since the compromised and normal $\{y_t\}$ are statistically identical, no detector can distinguish the compromised system from the normal system. As a result, the zero-dynamics attacks effectively render all FDI schemes useless.

To launch a zero-dynamics attack, the adversary needs accurate system knowledge; otherwise, the measurements $\{y_t\}$ of the compromised system cannot follow the exact same distribution as the normal $\{y_t\}$. The attacker also needs disruption power to inject malicious control input and/or modify the sensory information and control commands.

In contrast, zero-dynamics attacks do not require disclosure power, due to the linear nature of the system model.

Since zero-dynamics attacks cannot be detected, it is crucial for the system designer to carefully choose the parameter of the system such that no zero-dynamics attacks are possible. The existence of a zero-dynamics attack can be checked by the following algebraic conditions [Pasqualetti13]: There exists a zero-dynamics attack on the continuous-time system if and only if there exists $s \in C$, $x \in C^n$, and $u_a \in C^p$, such that we get the following:

$$(sI - A)x - B^a u^a = 0,\ Cx + \Gamma u^a = 0$$

Furthermore, the existence of zero-dynamics attacks can also be characterized using topological conditions, which [Pasqualetti13] discusses in detail. For the static system model, which has been widely adopted in power systems, a bad data detector such as χ^2 or the largest normalized residue detector [Abur04] detects the corruption in measurements y_k by checking the residue vector $r_k \triangleq y_k - C\hat{x}_k$. For uncorrupted measurements, the residue vector r demonstrates a Gaussian distribution.

However, similar to the case with zero-dynamics attacks, it is possible that a corrupted measurement vector y_k might generate the same residue r_k as a normal y_k. By exploiting this vulnerability, an adversary can inject a stealthy input u_k^a into the measurements to change the state estimate $\hat{x}_k$ and fool the bad data detector at the same time [Liu11]. As a result, it is important to design the system to rule out the possibility of a "stealthy" attack.

Robust Control

Robust control deals with uncertainty and disturbance in the control systems. In the context of CPS security, the attacker's input can be modeled as a disturbance. Suppose the attacker is energy constrained—that is, $\int \left(u_t^a\right)^2 dt < \infty$. Then techniques such as $\mathcal{H}_\infty$ loop-shaping can be used to reduce the sensitivity of the state $\{x_t\}$ with respect to the attacker's input $\{u_t^a\}$, and hence guarantee that the system will not deviate greatly from its normal trajectory. [Zhou98] provides a more detailed treatment of this topic.

Physical Watermarking and Authentication

A replay attack, like that employed by the Stuxnet malware, requires a significant amount of disclosure power as well as disruption power. The attacker, in this scenario, is assumed to be able to inject an external control input into the system and read all sensor readings and modify them arbitrarily. Given these capabilities, the attacker can implement an attack strategy similar to the replay attacks encountered in computer security, whereby he or she gives the system a sequence of desired control input while replaying previously recorded measurements.

As an illustration throughout this section, consider a discrete-time system that implements a Kalman filter (with gain K) and a linear quadratic Gaussian controller (with gain L). [Mo13] proves that unless the matrix $\mathcal{A} \triangleq (A+BL)(I-KC)$ is strictly unstable, the widely used χ^2 failure detector [Greenwood96, Mehra71] will be ineffective in exposing such replay attacks. This result, although derived for a specific case of estimator, controller, and detector, is generally applicable to a larger class of systems, albeit with a stronger condition.

Possible countermeasures to distinguish such an attack are to redesign the control system to make $\mathcal{A}$ unstable, or to include some form of physical watermarking in the control inputs, which acts as an authentication signal. The key idea behind physical watermarking is to embed a small random signal in the control signal, which is kept secret from the attacker. As the signal passes through the communication channels, actuators, physical systems, and sensors, it is transformed according to the models of these components. As long as the transformed watermark is detected, all the components on the control loop can be considered to be operating with desired functionalities. The problem of computing the optimal watermarking signal in the class of independent and identically distributed (IID) Gaussian processes can be cast as a semi-definite programming problem and hence solved efficiently [Mo13]. A diagram of the watermarking scheme is illustrated in Figure 7.3, where u_k is the

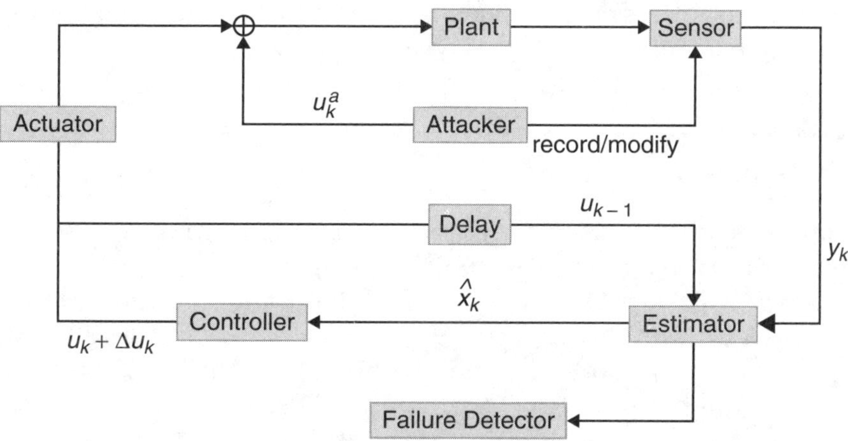

Figure 7.3: *System diagram of a watermarking scheme*

control signal generated by the feedback controller and Δu_k denotes the random watermark.

7.4 Summary and Open Challenges

With the proliferation of remote management and control of CPS, security plays a critically important role in these systems. Unfortunately, the convenience of remote management can be exploited by adversaries for nefarious purposes from the comfort of their homes.

Compared to current cyber infrastructures, the physical component of cyber-physical infrastructures adds significant complexity that greatly complicates security. On the one hand, the increased complexity will require more effort from the adversary to understand the system. On the other hand, this increased complexity introduces numerous opportunities for exploitation. From the perspective of the defender, more complex systems require dramatically more effort to analyze and defend because of the state space explosion when considering combinations of events.

Current approaches to secure cyber infrastructures are certainly applicable to securing cyber-physical systems: techniques for key management, secure communication (offering secrecy, authenticity, and availability), secure code execution, intrusion detection systems, and so on. Unfortunately, these approaches largely ignore the physical aspects of cyber-physical systems. In contrast, system theoretical approaches, although useful, tend to neglect the cyber aspect of the system. Therefore, a new science of cyber-physical security, which combines the efforts from both cyber and system theoretical security, needs to be developed to address the challenges raised by CPS, and will offer exciting research challenges for many years to come.

References

[Abur04]. A. Abur and A. G. Expósito. *Power System State Estimation: Theory and Implementation*. CRC Press, 2004.

[Barker10]. E. Barker, D. Branstad, S. Chokhani, and M. Smid. *Framework for Designing Cryptographic Key Management Systems*. National Institute of Standards and Technology (NIST) Draft Special Publication, June 2010.

[Depuru11]. S. S. S. R. Depuru, L. Wang, and V. Devabhaktuni, "Smart Meters for Power Grid: Challenges, Issues, Advantages and Status." *Renewable and Sustainable Energy Reviews*, vol. 15, no. 6, pages 2736–2742, August 2011.

[Deuse03]. J. Deuse, K. Karoui, A. Bihain, and J. Dubois. "Comprehensive Approach of Power System Contingency Analysis." *Power Tech Conference Proceedings*, IEEE Bologna, vol. 3, 2003.

[Greenwood96]. P. E. Greenwood and M. S. Nikulin. *A Guide to Chi-Squared Testing*. John Wiley & Sons, 1996.

[Hug08]. G. Hug. "Coordinated Power Flow Control to Enhance Steady-State Security in Power Systems." Ph.D. dissertation, Swiss Federal Institute of Technology, Zurich, 2008.

[Huitsing08]. P. Huitsing, R. Chandia, M. Papa, and S. Shenoi. "Attack Taxonomies for the Modbus Protocols." *International Journal of Critical Infrastructure Protection VL*, vol. 1, no. 0, pages 37–44, December 2008.

[Kerckhoffs83]. A. Kerckhoffs. "La Cryptographie Militaire." *Journal des Sciences Militaires*, vol. IX, pages 5–38, January 1883.

[Lee06]. H. Lee, J. Kim, and W. Y. Lee. "Resiliency of Network Topologies Under Path-Based Attacks." *IEICE Transactions on Communications*, vol. E89-B, no. 10, pages 2878–2884, October 2006.

[LeMay07]. M. LeMay, G. Gross, C. A. Gunter, and S. Garg. "Unified Architecture for Large-Scale Attested Metering." 40th Annual Hawaii International Conference on System Sciences, pages 115–125, 2007.

[LeMay12]. M. LeMay and C. A. Gunter. "Cumulative Attestation Kernels for Embedded Systems." *IEEE Transactions on Smartgrid*, vol. 3, no. 2, pages 744–760, 2012.

[Liu11]. Y. Liu, P. Ning, and M. K. Reiter. "False Data Injection Attacks Against State Estimation in Electric Power Grids." *ACM Transactions on Information and System Security*, vol. 14, no. 1, pages 1–33, 2011.

[Mehra71]. R. K. Mehra and J. Peschon. "An Innovations Approach to Fault Detection and Diagnosis in Dynamic Systems." *Automatica*, vol. 7, no. 5, pages 637–640, September 1971.

[Mo13]. Y. Mo, R. Chabukswar, and B. Sinopoli. "Detecting Integrity Attacks on SCADA Systems." *IEEE Transactions on Control Systems Technology*, no. 99, page 1, 2013.

[NETL08]. National Energy Technology Laboratory (NETL). *Characteristics of the Modern Grid*. Technology Report, July 2008.

[Pasqualetti13]. F. Pasqualetti, F. Dorfler, and F. Bullo. "Attack Detection and Identification in Cyber-Physical Systems." *IEEE Transactions on Automatic Control*, vol. 58, no. 11, pages 2715–2729, 2013.

[Pickholtz82] R. L. Pickholtz, D. L. Schilling, and L. B. Milstein. "Theory of Spread-Spectrum Communications: A Tutorial." *IEEE Transactions on Communications*, vol. 30, pages 855–884, May 1982.

[Quinn09]. E. L. Quinn. "Smart Metering and Privacy: Existing Laws and Competing Policies." *Social Science Research Network (SSRN) Journal*, 2009.

[Schavemaker09]. P. Schavemaker and L. van der Sluis. *Electrical Power System Essentials*. John Wiley & Sons, 2009.

[Seo11]. D. Seo, H. Lee, and A. Perrig. "Secure and Efficient Capability-Based Power Management in the Smart Grid." Ninth IEEE International Symposium on Parallel and Distributed Processing with Applications Workshops (ISPAW), pages 119–126, 2011.

[Seshadri04]. A. Seshadri, A. Perrig, L. van Doorn, and P. K. Khosla. "SWATT: Software-Based Attestation for Embedded Devices." *Proceedings of IEEE Symposium on Security and Privacy*, pages 272–282, 2004.

[Shah08]. A. Shah, A. Perrig, and B. Sinopoli. "Mechanisms to Provide Integrity in SCADA and PCS Devices." *Proceedings of the International Workshop on Cyber-Physical Systems: Challenges and Applications (CPS-CA)*, 2008.

[Shahidehpour05]. M. Shahidehpour, W. F. Tinney, and Y. Fu. "Impact of Security on Power Systems Operation." *Proceedings of the IEEE*, vol. 93, no. 11, pages 2013–2025, 2005.

[Teixeira12]. A. Teixeira, D. Perez, H. Sandberg, and K. H. Johansson. "Attack Models and Scenarios for Networked Control Systems." *Proceedings of the 1st International Conference on High Confidence Networked Systems*, pages 55–64, New York, NY, USA, 2012.

[US-DOE09]. U.S. Department of Energy. *Smart Grid System Report: Characteristics of the Smart Grid*. Technology Report, July 2009.

[Vijayan10]. J. Vijayan. "Stuxnet Renews Power Grid Security Concerns." *Computerworld*, 2010.

[Willsky76]. A. S. Willsky. "A Survey of Design Methods for Failure Detection in Dynamic Systems." *Automatica*, vol. 12, no. 6, pages 601–611, November 1976.

[Zhou98]. K. Zhou and J. C. Doyle. *Essentials of Robust Control*. Prentice Hall, Upper Saddle River, NJ, 1998.

[Zima06]. M. Zima. "Contributions to Security of Electric Power Systems." Ph.D. dissertation, Swiss Federal Institute of Technology, Zurich, 2006.

Chapter 8

Synchronization in Distributed Cyber-Physical Systems

Abdullah Al-Nayeem, Lui Sha, and Cheolgi Kim

Cyber-physical systems (CPS) are commonly implemented as networked real-time systems consisting of a network of sensors, actuators, and their controllers. Many of the information processing functions in these systems require consistent views and actions in real time. Guaranteeing consistency in these distributed computations in hard real time is challenging. In this chapter, we discuss the challenges and solutions for the synchronization of these distributed computations.

8.1 Introduction and Motivation

Compared with networked components on a chip, the main source of difficulties in distributed systems is the asynchronous interactions between different nodes. In distributed systems, many of the computations are periodic in nature and are triggered by periodic timers based on the local clocks. The local clocks' relative skew can be bounded but cannot be entirely eliminated. Embedded system designers model these distributed computations using the "globally asynchronous, locally synchronous" (GALS) design philosophy, which was originally proposed in the hardware community [Chapiro84, Muttersbach00]. In this design, the computations within a node execute synchronously based on the local clock, but execute asynchronously with respect to the computations of other nodes.

Because of the clock skews in a GALS system, a small timing difference in the execution and communication delay can lead to distributed race conditions. To illustrate this problem, consider an example of a triplicated redundant control system that receives a new reference position or setpoint command from a supervisory controller, as shown in Figure 8.1. Because of the non-zero clock skews, one controller (say, controller A) could be in its local clock–based period $j + 1$, while the other two (controllers B and C) are still in their local clock–based period j at the time when they receive the setpoint. Since they receive the setpoint at different periods, controller A's control command might diverge from the other two control commands and, therefore, would be voted out. This race condition leads to an invalid failure detection of controller A (even though it is not actually faulty). As a result, the system is no longer capable of handling a genuine fault when a controller produces incorrect data.

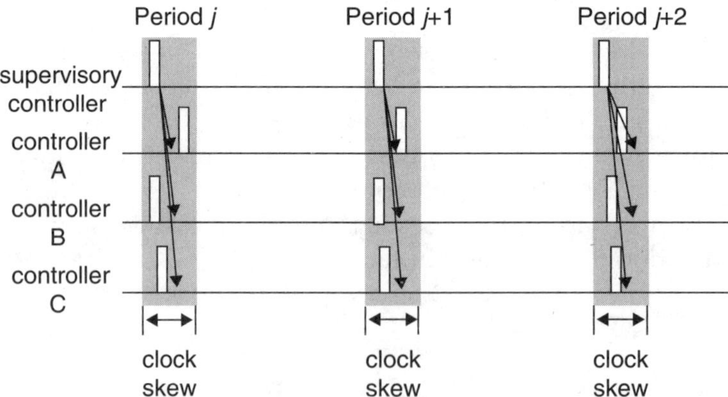

Figure 8.1: *Triplicated control in a GALS system*

8.1.1 Challenges in Cyber-Physical Systems

In safety-critical CPS, such as avionics, potential distributed race conditions create significant overhead in the design, verification, and certification processes. They are the source of many serious bugs that are very difficult to reproduce. In these cases, a system may operate correctly for a long time, even for years, but suddenly fail after a change in some logically unrelated hardware, software, or workloads. Tracking down the root causes of these problems is like finding a needle in a haystack. Formal analysis tools, such as model checker, may fail to produce any counterexample within a prolonged period. In particular, a model checker must explore the application states to verify distributed protocols under all possible event interleaving. This readily leads to the state space explosion problem in a GALS system [Meseguer12, Miller09].

This problem becomes more challenging when an application requires distributed computations interacting at different rates. For example, in a fly-by-wire aircraft, the control surfaces are each locally controlled by higher-level supervisory controllers operating at different rates. These controllers are deployed redundantly for fault tolerance. Consistent views and actions are mandatory when a component interacts not only with its redundant components for replica management, but also with other components of the upper layers to exchange discrete commands such as set-points and mode change commands. The race conditions in these multirate distributed computations significantly complicate the problem.

8.1.2 A Complexity-Reducing Technique for Synchronization

In this chapter, we discuss some existing techniques that are used for synchronization in cyber-physical and other distributed systems, in general. Cyber-physical systems have traditionally used custom networking systems, such as Boeing 777's SAFEbus [Hoyme93] and time-triggered architecture (TTA) [Kopetz03], to synchronize distributed computations. Synchronization in these systems happens at the networking layer at the expense of specialized hardware.

Researchers have also implemented many protocols for distributed synchronization [Abdelzaher96, Awerbuch85, Birman87, Cristian95, Renesse96, Tel94, Tripakis08]. Many of these protocols were initially designed for traditional computing environments, so they do not take advantage of the real-time performance guarantees of CPS. Instead, they rely on complex handshaking protocols for synchronization of distributed computations.

These techniques, whether implemented in hardware or software, also have a limitation in that they often rely on architecture-dependent semantics to guarantee consistent views and actions. Thus, the cost of verification and certification increases when the configuration changes with advanced hardware and software.

In the engineering community, we need to invent complexity-reducing technologies for synchronization that will reduce the cost and can be applied in a wider range of applications. In this chapter, we discuss one such advanced technology, called a "physically asynchronous, logically synchronous" (PALS) system, which we have developed recently in collaboration with Rockwell Collins Inc. and Lockheed Martin Inc. The materials presented here are based on prior works [Al-Nayeem09, Al-Nayeem12, Al-Nayeem13, Meseguer12, Miller09, Sha09].

The PALS system is a formal architectural pattern that eliminates race conditions arising from the asynchronous interactions. This pattern makes use of the basic features of the networked cyber-physical systems—for example, bounded clock skew and bounded end-to-end delay—to simplify the interactions of the distributed computations.

In this approach, engineers design and verify distributed applications as though the computations will execute on a globally synchronous architecture. The pattern distributes this synchronous design over a physically asynchronous architecture without any modification of application logic and properties. Any temporal logic formula that holds under a globally synchronous system also holds for an asynchronous system regulated by this pattern [Meseguer12]. Hence the cost of the verification is significantly reduced in the PALS system, as we need to verify only the application logic for the globally synchronous architecture. Furthermore, the PALS system can work with commercial off-the-shelf (COTS) components and without specialized hardware (as long as the pattern's assumptions are satisfied).

8.2 Basic Techniques

In this section, we consider the basic techniques that have a reasonable acceptance in the practitioner's community. We discuss their effectiveness and limitations in the context of synchronization in cyber-physical systems.

8.2.1 Formal Software Engineering

Since the publication of the famous book *Design Patterns: Elements of Reusable Object-Oriented Software* [Gamma95], many design patterns have been proposed for different domains. There also exist many architecture patterns based on fault tolerance, real-time computing, and networking of the cyber-physical systems [Douglass03, Hanmer07, Schmidt00]. A pattern can be viewed as a design template of the solution to a generic problem. Although useful in enabling software reuse, the standard practice of software patterns is not sufficient for cyber-physical systems. These patterns are usually documented in informal languages, so their correct instantiation often depends on users' expertise and interpretation of the application's context. There have also been many efforts to formally model software patterns [Alencar96, Allen97, de Niz09, Dietrich05, Taibi03, Wahba10]. These attempts intend to avoid implementation ambiguities by using domain-specific languages and structural analyses, but they do not directly address how to reduce the design and verification complexities of cyber-physical systems.

Synchronous design languages and tools, such as Simulink, SCADE, and Lustre, are also widely used in CPS development, though they are applied only in modeling, simulation, and analysis of software components. Furthermore, the software components in a synchronous design language were originally intended to be only centralized and driven by a global clock. As a result, these techniques, by default, lack support for architectural-level analysis of distributed software components. Several solutions have been proposed to simulate the asynchronous behavior of the distributed software in a synchronous design language [Halbwachs06, Jahier07]. These solutions simulate the nondeterministic asynchronous behavior through sporadic execution of processes and controlled delivery of messages. While they are useful in modeling asynchronous software components, we still need to deal with combinatorial event interleaving and complex interactions in a distributed application.

In this respect, the PALS architecture pattern complements these languages and tools. The languages and tools may be used to design a system in logically synchronous models of the PALS system just as we did under the Architecture Analysis and Design Language (AADL) standard [Bae11]. We would then need a correctness-preserving transformation of the synchronous models to execute them on the physically asynchronous architecture.

8.2.2 Distributed Consensus Algorithms

Distributed consensus is a fundamental concept in distributed systems and theory, and virtual synchronization was one of the early solutions for achieving distributed consensus. [Birman87] first introduced the process group abstraction to achieve virtual synchronization for event-triggered computations. This virtual synchrony model guarantees that the behavior of the replicated processes is indistinguishable from the behavior of a single reference process on a nonfaulty node.

ISIS and its subsequent version ISIS2 are two middleware platforms that achieve virtual synchrony with a group communication service [Birman12]. The group communication service maintains a list of active processes and notifies the processes of join or crash events, known as view change events. These platforms synchronize the view change events and the application messages in such a way that distributed processes remain consistent.

Horus is another system that supports virtual synchrony [Renesse96]. [Guo96] gives a lightweight version of this implementation. [Pereira02] uses application-level semantics to relax some strong consistency requirements of virtual synchrony.

Unfortunately, these techniques do not provide hard real-time guarantees or timing bounds for when a synchronization will be complete. Real-time versions of these communication services have been proposed by [Abdelzaher96] and [Harrison97]. For example, [Abdelzaher96] provides multicast and membership services for real-time process groups organized in a logical ring. When an application needs to send real-time messages, it presents the message with timing constraints to an admission controller, which then performs online schedulability analysis. Real-time messages that can be scheduled are admitted; otherwise, they are rejected. These techniques were designed for soft real-time systems where deadlines do not need to be ensured.

We can make several salient observations about the virtual synchronization model implemented in these group communication services. First, these services are primarily used as the transport-layer services for reliable and consistent message communications in a group of computations. One can synchronize the computations that occur upon individual events such as application messages, membership, or view change events with these services. Thus, for an application, it makes more sense to use these services with event-triggered or aperiodic computations. Otherwise, the application must provide the necessary mechanisms for the timed processing of the events. As a result, there is a need to coordinate both computations and message communications in an application such that the messages generated during two consecutive clock events are processed consistently at the receiving tasks. Thus, these computations must be synchronized at clock events within a global time interval across the nodes.

Second, the group communication services, by default, guarantee reliable multicast of individual application messages, which have more complexity than is needed in many real-time applications. For example, real-time distributed systems may not require the ordered delivery of individual messages within a period as long as they are received before the start of next period. The tasks in the system may care only whether the messages are delivered reliably and in time.

Third, the group communication services bundle various fault-tolerance mechanisms such as group membership and state transfer upon initialization. These mechanisms are useful for many applications, but it is generally better to separate the logical synchronization mechanism from these fault-tolerance mechanisms. That way, software engineers can apply appropriate fault-tolerance mechanisms to meet the different reliability requirements of the applications.

Other well-known consensus algorithms include Lamport's Paxos algorithm [Lamport01] and Chandra-Toueg's algorithm [Chandra96]. These consensus algorithms are widely used in distributed transactions, distributed locking protocols, and leader election. Nevertheless, they do not provide hard real-time guarantees. Instead, they assume a globally asynchronous system, which does not provide any bound on message transmission delay, clock drift rate, and execution speed.

A famous theory on the impossibility of distributed consensus appears in [Fischer85]. This theory suggests that no algorithm can always reach consensus in a bounded time in this model of asynchronous system, even with a single process crash/failure. The main reason is that processes cannot correctly differentiate a slow process from a crashed one without a bound on the end-to-end delay. Some consensus algorithms circumvent this impossibility of consensus through application of concepts such as failure detectors and quorum consistency [Gifford79].

Real-time systems require a bound on message transmission delay, clock drift rate, clock skew, and response time—the theory on the impossibility of distributed consensus does not apply in this type of system. Furthermore, in networked control systems such as modern avionics, the fault-tolerant and real-time requirements are guaranteed at the networking layer. The network is the "wire" in "fly-by-wire," for example. Ideally, we will take advantage of existing fault-tolerant real-time networks in networked control systems to simplify the design of real-time virtual synchrony.

8.2.3 Synchronous Lockstep Executions

Other researchers have implemented a synchronous model onto different asynchronous architectures, such as a Loosely Time-Triggered Architecture (LTTA) [Benveniste10] and an asynchronous bounded delay (ABD) network [Chou90]. [Tripakis08] deals with the problem of mapping a synchronous computation on an asynchronous architecture. In their approach, these authors consider an LTTA. The mapping is achieved through an intermediate finite FIFO platform (FFP) layer. Although correctness is achieved in spite of unpredictable network delays and clock skews, these approaches do not provide the hard real-time guarantee required for synchronization and consistent views in cyber-physical systems. Furthermore, this work does not handle any failure and multirate computations.

An ABD network primarily assumes that the message transmission delay is bounded. [Chou90] and [Tel94] give similar protocols to simulate a globally synchronous design on an ABD network with a bounded clock drift rate. These protocols define the logical synchronization period in terms of the round intervals for different network topologies, where each round interval gives an upper bound for the message transmission delay. However, they do not assume that a fault-tolerant clock synchronizer is applied to the local clocks that are required in networked control systems such as avionics. As a result, they require a complex reinitialization procedure to correct the clock drift errors. For example, after a certain number of rounds, the protocols reset the clocks based on the multicast of special "start" messages. In these approaches, the real-time periodic computations of a cyber-physical system may be discontinuous during this reinitialization procedure. Furthermore, none of them discusses node failure, reliable message communication, and multirate computations.

[Awerbuch85] gives three protocols for achieving logical synchronization: α synchronizer, β synchronizer, and γ synchronizer. These synchronizers generate local tick events to execute the synchronous logic, but they depend on either the acknowledgment messages or a leader node to prevent the arrival of past messages after a tick event. As a result, these solutions require longer synchronization periods and have high overhead to maintain a verifiable leader-election logic with respect to failures and other asynchronous events.

[Rushby99] also gives a round-based synchronous execution in a time-triggered architecture. In this model, each synchronous round or period has two phases: communication and computation. The computation phase begins only after the communication phase has finished. This work, however, does not support multirate executions and requires the computation phases to be complete prior to sending messages.

8.2.4 Time-Triggered Architecture

Time-triggered architecture (TTA) was one of the earliest system architectures that introduced distributed real-time clock sources for maintaining consistency [Kopetz03]. The core functions of TTA are implemented in a custom network architecture, such as TTP/A and TTP/C, for reliable message communications. In both of these protocols, the nodes communicate according to a prespecified time division multiple access (TDMA) schedule. Hence, every node knows the exact message transmission time and has a time window for receiving each incoming message. The hardware in these architectures, such as the network guardian and the network switch, also maintains the message schedule and detects a faulty node when a message is not received in the allowable time window [Kopetz03]. The correctness of these solutions depends on a tight clock synchronization of all nodes, including the network switches. Thus, these network architectures also implement a fault-tolerant clock synchronization algorithm in the hardware [Pfeifer99, Steiner11]. TTA-Group recently introduced TTEthernet, a real-time switched Ethernet network system that employs TDMA-based message deliveries and hardware-based time synchronization, which was inspired by TTA [Kopetz10].

TTA has its own distinctive characteristics that make its implementation of distributed consistency and logical synchronization different from that found in other approaches. In TTA, the distributed consistency is based on the concept of a sparse time base [Kopetz92]. In the sparse time base, TTA controls the send instants of the message transmissions. Messages are transmitted with sufficient time differences so that other nodes can agree on the order of the messages based on the local timestamps of message transmissions. To define the timestamps, TTA defines a logical clock whose granularity (i.e., the tick duration) is a function of the maximum clock skew. Based on this clock, TTA can ensure that the temporal order of events can be recovered from their timestamps, provided that the difference between their timestamps is equal to or greater than 2 ticks [Kopetz11].

While this model provides a simple approach to define the temporal order of messages in TTA, additional effort is required to coordinate the distributed interactions. In particular, TTA does not consider the task response time and the message transmission delay in its logical clock. Therefore, despite the control exerted over message transmissions, variations of these system parameters can increase the verification complexity of the distributed algorithms in TTA.

[Steiner11] recently proposed an extension of the sparse time base to implement a globally synchronous model in TTA. This approach requires an additional timing layer on top of the original logical clock of TTA, which ensures that the synchronization period is sufficiently long to allow for the task response time and the message transmission delay.

Other approaches, such as the PALS system [Sha09], do not require knowledge of the global message schedule of the time-triggered network architectures to perform logical synchronization. The PALS pattern uses only the system's performance parameters to abstract away the underlying network architecture, even though the concept of timestamps may make the implementation of this approach more practical in PALSware, a middleware version of PALS systems [AlNayeem13].

8.2.5 Related Technology

A number of related technologies may be used in conjunction with distributed synchronization. In this section, we discuss some of the most relevant ones.

8.2.5.1 Real-Time Networking Middleware

Distributed middleware, such as real-time CORBA [Schmidt00], web services, and publish–subscribe middleware [Eugster03], provides a virtualized platform through which distributed tasks can collaborate. The abstractions provided by these middleware platforms are quite different, though. Real-time CORBA, web services, and publish–subscribe middleware require developers to be explicitly aware of the asynchronous nature of the distributed nodes. Therefore, the applications running on top of these middleware layers should be carefully designed and verified to ensure they will provide consistency under such asynchronous environments.

8.2.5.2 Fault-Tolerant System Design

Fault tolerance is a major design criterion for safety-critical systems. Various techniques have been proposed in different application domains. For example, triple modular redundancy and pair–pair redundancy are widely used to mask single-point failure. [Sha01] has proposed a simplex architecture to separate the concerns of effectiveness and reliability for command and control systems. Applications in a process group may also use membership algorithms to obtain a consistent picture of the members' state in the presence of various faults such as crash failure, message omission, and Byzantine faults [Cristian86, Kopetz93].

8.2.5.3 Formal Verification of Distributed Algorithms

Formal verifications of distributed consensus algorithms have been investigated in the past [Hendriks05, Lamport05]. They show that model checking of distributed consensus algorithms in physically asynchronous architectures that guarantee only bounded message transmission delay may be feasible to some extent. These architectures are commonly known as partially synchronous distributed systems. Unfortunately, these algorithms do not provide any generic solution for achieving scalable verification in distributed systems.

Researchers have also verified other distributed algorithms, such as distributed convergence, in this architecture. For example, [Chandy08] transforms a shared-memory architecture to verify the distributed convergence problem in a partially synchronous distributed system. The architectural assumptions of the shared memory architecture and the partially synchronous architecture, however, are different from those required in real-time systems. In contrast to these architectures, a real-time system places time bounds on the various system parameters. Thus, it is possible to reduce the possible nondeterministic interaction scenarios by enabling the use of an equivalent globally synchronous design.

In recent years, researchers have also explored model checking of distributed software [Killian07, Lauterburg09, Lin09, Sen06]. Their efforts consider various heuristics for efficient state space explorations, such as random walk, bounded search, and dynamic partial-order reduction. Despite these optimizations, model checking of distributed algorithms remains extremely difficult beyond a certain model size and complexity.

8.3 Advanced Techniques

Traditional techniques for synchronization in distributed CPS leave important challenges unresolved. In this section we present a new technique to address these issues that, at the same time, simplifies the verification of these systems.

8.3.1 Physically Asynchronous, Logically Synchronous Systems

In this section, we discuss an advanced technique—the physically asynchronous, logically synchronous (PALS) system—that is still under development in the scientific community.

The main concept of logical synchronization in the PALS system is quite simple, as shown in Figure 8.2. In a globally synchronous system, the distributed computations are dispatched in lockstep by the global clock for that period. For each dispatch, a task reads the messages from its input ports, processes the messages, and sends its output messages

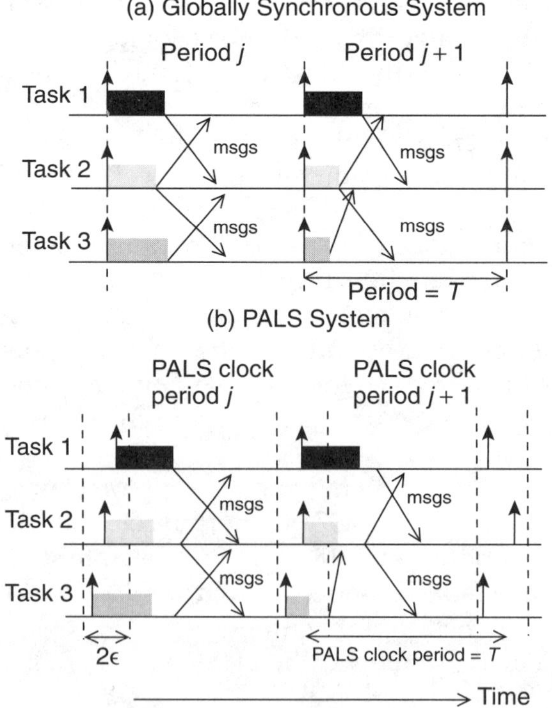

Figure 8.2: *Logically equivalent globally synchronous system and PALS system*

to other nodes. In this lockstep execution, messages generated during period j are always consumed by their destination tasks in period $j + 1$.

The PALS system guarantees an equivalent lockstep (synchronous) execution in the physically asynchronous architecture. In this system, each node defines a logical clock, called the PALS clock of period T, for each distributed computation. Each node triggers the computation logic at the start of a PALS clock period. The jth PALS clock period starts at the local clock time of a node, which is defined as $c(t) = jT$.

The local clocks are asynchronous. Thus, the PALS clock periods do not start at the same global time. Instead, the PALS clock periods at different nodes begin within well-defined intervals in global time.

Despite the physical asynchrony of the PALS clocks, this pattern ensures that messages generated during period j are consumed by their destination tasks in period $j + 1$. That is, the views of inputs are identical to those in the globally synchronous model running at the PALS clock period.

8.3.1.1 PALS System Assumptions

For a cyber-physical system to be structured as a PALS system, its architecture must satisfy a set of assumptions. The PALS system assumptions can be organized into three categories: system context, timing, and external interface constraints.

System Context

The PALS pattern is applicable in hard real-time networked systems with the following features and bounded performance parameters:

- Each node has a monotonically nondecreasing local clock, c: Time $\rightarrow$ Time (Time $= R \geq 0$). Here, $c(t) = x$ is the local clock time x at the "ideal" global time t.
- The local clocks are synchronized to the extent that corresponding local clock times happen within a 2ε interval in global time. Let $c(t) = x$. Then t happens in the global time interval $[x - \varepsilon, x + \varepsilon]$, where ε is defined as the worst-case clock skew with respect to the global time.
- Response time of a computation task α is bounded: $0 < \alpha^{min} \leq \alpha \leq \alpha^{max}$.
- Message transmission delay μ is bounded: $0 < \mu^{min} \leq \mu \leq \mu^{max}$.
- Nodes demonstrate fail-stop behavior and may recover later. A failed node must not be able to send extra messages during the current period.

Existing cyber-physical systems, such as those found in avionics, meet these requirements. For example, the nodes in an avionic system communicate through real-time network architectures that guarantee the message delivery in bounded time. Furthermore, the nodes synchronize the local clocks to minimize the jitter of sampling and control operations. These architectures also support fail-stop execution with redundant processor pairs. Thus, the pattern can be instantiated in these systems with minimal overhead.

Timing Constraints

The following constraints on the system parameters must be also satisfied in the physically asynchronous architecture:

- *PALS causality or output hold constraint:* Since the PALS clock events are not perfectly synchronized, delivering a task's outputs too early may result in the violation of the lockstep synchronization. It is possible that a message could be consumed in the same period at a destination task. To prevent this erroneous condition, a task is not allowed to send a message earlier than $t = (jT + H)$ in period j, where $H = \max(2\varepsilon - \mu^{min}, 0)$.
- *PALS clock period constraint:* The distributed computation of a node requires at least the delay of end-to-end computation and message transmission to know the state of the computations in other nodes. Thus, the distributed computations in a PALS system must run at a period longer than this delay. The PALS system defines an optimal lower bound for the PALS clock period, as follows: $T > 2\varepsilon + \max(\alpha^{max}, 2\varepsilon - \mu^{min}) + \mu^{max}$.

External Interface Constraints

The PALS pattern defines additional constraints for interaction with external components. Such a system uses a special component, called an environment input synchronizer, to coordinate the logically synchronous delivery of external messages. In its simplest form, an environment input synchronizer is a periodic task satisfying both the PALS clock period constraint and the PALS causality constraint. Suppose the environment input synchronizer receives external messages during period j. The environment input synchronizer then forwards these messages to the final destination tasks during period $j + 1$ so that the receiving tasks receive the external messages consistently in the same PALS clock period.

Similarly, one may use an environment output synchronizer to maintain a consistent view of the PALS computations at the external observers. The environment output synchronizer also executes in a similar way as other PALS computations and satisfies the previously mentioned timing constraints.

8.3.1.2 Pattern Extension for Multirate Computations

The PALS system has been extended to support the logical synchronization of multirate distributed computations [Al-Nayeem12], in a pattern called multirate PALS system. In this pattern, application tasks execute at different rates and more than one message transfer is possible per PALS clock period.

This extension also defines a sequence of periodic PALS clock events, when the local clock C equals jT_{hp}, $\forall\, j \in N$. Here T_{hp} is the PALS clock period. In multirate PALS system, the PALS clock period is chosen as the hyper-period or the least-common-multiple (LCM) of the periods of the participating tasks. A smaller synchronization period is not possible, because discrete state updates in a globally synchronous system with multirate tasks happen synchronously only at the hyper-period boundaries. A smaller synchronization period may result in asynchronous actions and affect the system's safety adversely. To preserve the same synchronous semantics, the PALS clock period is set to the LCM of the periods of the participating tasks.

Suppose that M_1 and M_2 are two periodic tasks of period T_1 and T_2, respectively. M_1 transmits its output messages to M_2, and vice versa. Let the PALS clock period equal T_{hp}, where $T_{hp} = \text{LCM}(T_1, T_2)$. The multirate PALS system pattern guarantees that M_1 receives a total of $n_2 = T_{hp}/T_2$ messages from M_2 generated during the PALS clock period j. Similarly, M_2 receives a total of $n_1 = T_{hp}/T_1$ messages from M_1 in the same PALS clock period. The pattern filters these received messages and delivers a selected message identically during the next PALS clock period. Figure 8.3 illustrates this example.

To ensure this behavior is realized, the end-to-end delay of a message from a source task must be smaller than the task's period. The pattern must also satisfy a similar set of assumptions as the PALS system. [Bae12-b] provides a related mathematical analysis of the multirate PALS system.

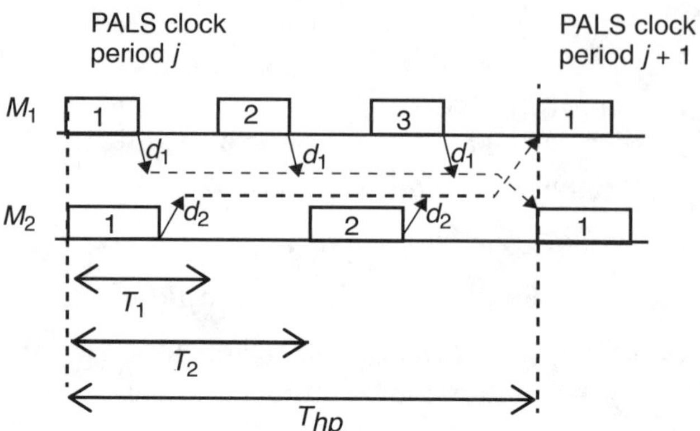

Figure 8.3: *Multirate PALS system-based interaction*

8.3.1.3 PALS Architectural Specifications

The PALS system pattern is particularly well suited for formal specification and analysis in architectural modeling languages. One can validate the architectural models of a PALS system to manage the complexity of logical synchronization in cyber-physical systems. For example, in the past, we defined architectural specifications and requirements for the PALS system design in SAE Architecture Analysis and Design Language (AADL), an industry-standard modeling language for the cyber-physical systems [Aerospace04]. These requirements identify a set of simple constraints that are readily checked in the AADL models. By ensuring that the models satisfy these constraints, we can easily guarantee the correctness of the logical synchronization in the physically asynchronous models.

In AADL, there are two specifications for the PALS system design: Synchronous AADL and PALS AADL.

Synchronous AADL Specification

The first step of the PALS system design is the design and verification of the synchronous model. With our colleagues at UIUC and University of Oslo, we have proposed a specification of the synchronous design in AADL [Bae11]. This Synchronous AADL specification models the lock-step execution of the globally synchronous computations.

In the Synchronous AADL specification, we use only a subset of the AADL modeling constructs to define the components and their connections. This specification includes just two types of components: (1) the components that participate in the distributed computation in a lockstep manner and (2) the external environment, which sends external inputs to them. Both types of components are modeled using the AADL thread construct that dispatches messages periodically at the same rate. The states and state transitions are modeled using AADL Behavior Annex [Frana07].

The Synchronous AADL specification models the message communications between the threads using the delayed data port connection semantic. This ensures that in the lockstep execution, the receiving threads consume the output of a source thread at the next period. The connections between the environment thread and other threads are modeled using the immediate data port connection semantic. An immediate connection gives a scheduling order that the source thread executes prior to the receiving threads when they dispatch at the same time.

AADL models are not executable, however, and they cannot be automatically verified. To support formal verification, the Synchronous AADL models are automatically translated to Real-Time Maude for model checking and simulation. The formal semantics of the Synchronous AADL specification and the model checking results are available in [Bae11].

PALS AADL Specification

The next step in the PALS system design process is to map the Synchronous AADL model to a physically asynchronous system model. One of the main challenges in model-based engineering is to preserve the verification properties as engineers refine and extend the models during the development process. With our colleagues from Rockwell Collins Inc., we have proposed an easily checkable AADL specification of the PALS pattern that meets this challenge [Cofer11].

The concept of this specification is based on architectural transformation. In AADL, this transformation involves component inheritance—that is, annotation of the components and connections with the pattern-specific properties. We have defined a set of relevant AADL properties that identify the scope and timing properties of the logically synchronous interactions in a pattern instance. For example, in the asynchronous model, we annotate the threads and connections of the corresponding Synchronous AADL model with two properties: `PALS_Properties::PALS_Id` and `PALS_Properties::PALS_Connection_Id`. These properties accept a string literal as the value. With these two properties, we define a logical synchronization group that consists of a set of components and connections with the same value for these properties.

The PALS system rules are applied to the components and connections of a logical synchronization group. An AADL model checker can load the system model and validate the PALS system assumptions. For example, for each thread M_i in a logical synchronization group, we validate the PALS clock period mentioned earlier and PALS causality constraints by simply evaluating the following conditions:

$$M_i.Period > 2\varepsilon + \max(M_i.Deadline, 2\varepsilon - \mu^{min}) + \mu^{max}$$
$$\min(M_i.PALS\ Properties::PALS\_Output\_Time) > \max(2\varepsilon - \mu^{min}, 0)$$

Here, $M_i.Period$ is the extracted value of the AADL property period defined by the thread M_i. Similarly, $M_i.Deadline$ is the value of the AADL property deadline of the thread M_i. In this specification, we use a thread's deadline as the parameter for the maximum response time α^{max}. A thread also defines an AADL property, called PALS `Properties::PALS_Output_Time`. It gives the earliest and the latest time at which a thread transmits its outputs since the dispatch. The remaining performance parameters are computed based on other standard AADL properties, such as latency for message transmission delays (μ^{min}, μ^{max}) and `Clock_Jitter` for the local clock skew ε.

8.3.1.4 PALS Middleware Architecture

We have developed a distributed middleware, called PALSware, which implements the task execution and communication services for logically synchronous interactions. Figure 8.4 shows the software architecture of the PALS middleware. It consists of three layers: infrastructure, middleware, and application. This layered architecture makes the software portable and extendable in different platforms.

At the *infrastructure layer*, the middleware defines generic interfaces to integrate a fault-tolerant clock synchronizer and a real-time network architecture. As long as the pattern's assumptions are satisfied, the application logic is not affected when these infrastructure services change or they are developed with COTS components.

At the *middleware layer*, PALSware addresses several implementation challenges. In particular, task failures increase the system's complexity. For example, a task may send the same message sequentially to different destination tasks. If this task suddenly stops, only a subset of

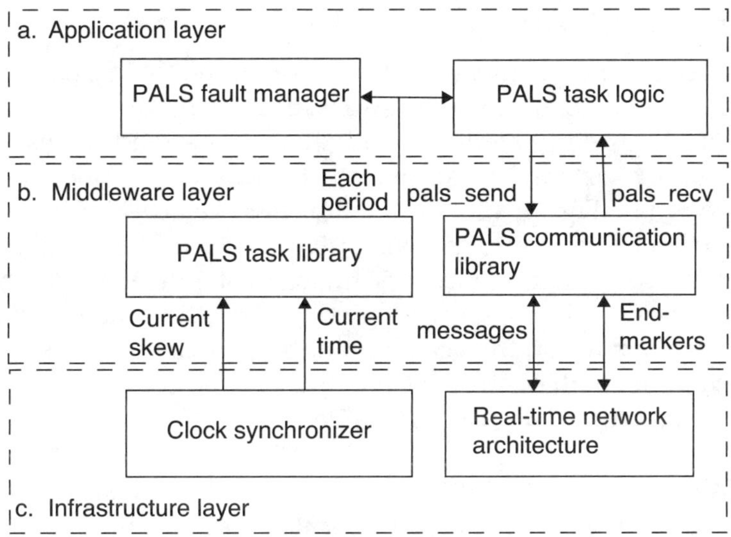

Figure 8.4: *PALS middleware architecture layers*

the destination tasks may receive this message, which subsequently leads to inconsistent states.

PALSware implements a real-time reliable multicast protocol to prevent this problem. For the fail-stop nodes that are communicating through a fault-tolerant real-time network architecture, the protocol guarantees that if a nonfaulty destination task receives a message from a source task, then the other nonfaulty destination tasks also will receive the message in the same period. PALSware also uses this communication protocol to guarantee atomicity in the distributed interactions, so that the destination tasks have consistent information about the working or failed state of a source task.

At the *application layer*, application developers implement the periodic application logic for the distributed tasks. This layer also supports runtime monitoring for any faults that may affect the logical synchronization. For example, if the clock skew assumption is violated for some reason, the distributed tasks will no longer be logically synchronized within the 2ε interval in global time. Users can customize PALSware with application-specific fault managers to detect these problems and define any fail-safe actions. For example, the fault manager may receive the current clock skew estimate from the clock synchronizer and then compare this value with ε to detect a timing fault.

8.3.1.5 Example: Dual-Redundant Control System

In this section, we illustrate the application of the PALS AADL specification and the PALS middleware API with an example of a dual-redundant control system, which we call the active–standby system. The active–standby system consists of two physically separated controllers: Side1 and Side2 (Figure 8.5). Each controller periodically receives sensor data from a sensor subsystem. Only one controller should operate as the active controller and deliver the control output, while the other controller operates in the standby mode. Users can send commands to these controllers to flip the active/standby modes of these controllers. In this example, both controllers synchronize the computations so that they agree on which controller is active.

Application of PALS AADL Specification and Analysis

In this example, the two sides communicate via a real-time network architecture. To illustrate the motivation of the PALS AADL specification and corresponding analysis, consider a hypothetical requirement change that upgrades the network architecture. The linear shared-bus architecture is replaced by a switched networked architecture. Figure 8.6 gives a subset of the system model during this configuration change. In the old configuration, the active/standby status of Side1 (`side1_status`) flows through the shared bus to Side2. In the new configuration, this message flows through a number of networked switches to the destination node.

In this example system, the status message is critical for the correctness of the coordination of these controllers. For example, if the standby controller does not receive the status message for the active controller, it assumes that the active controller has failed and, therefore, becomes active itself. If any inconsistency arises within the implementation,

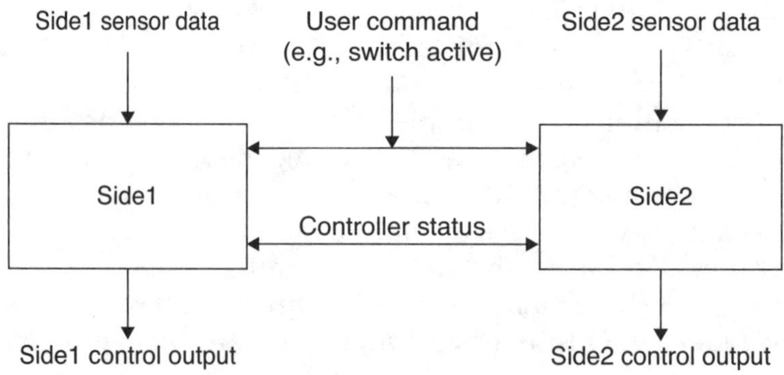

Figure 8.5: *Active–standby system*

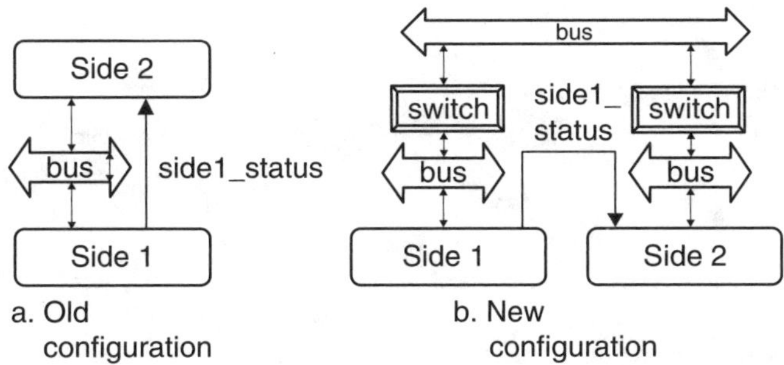

Figure 8.6: *System model during configuration changes*

Table 8.1: *Sample Values for the AADL Properties*

AADL Property	Old Configuration	New Configuration
Latency (from Side1 to Side2)	0.1–0.5 ms	0.05–0.9 ms
Clock_Jitter	0.5 ms	1 ms
PALS_properties::PALS_Output_Time	1–2 ms	1–2 ms

both controllers may incorrectly act as the active or standby controller during the same period.

The PALS AADL specification requires the annotation of a number of AADL properties, such as `Latency`, `Clock_Jitter`, and `PALS_properties::PALS_Output_Time`. Table 8.1 gives some sample values for these properties in these two configurations.

A static analyzer based on the PALS AADL specification can easily detect that the new configuration is not a valid PALS system. In particular, the PALS causality constraint may be violated. For example, Side1's status might be incorrectly delivered in the same period in which the message is transmitted (Figure 8.7).

C++ API

PALSware provides an easy-to-use and extendable C++ API for the PALS applications. It defines the task execution and communication classes based on the operating system–independent interfaces. The main classes of this middleware are as follows:

- `PALS_task`: This class is used for the execution of a task participating in the PALS system. During each period, the PALS event generator logic of `PALS_task` invokes a placeholder function called `each_pals_period`. An application developer extends this class and defines the periodic logic in this function.

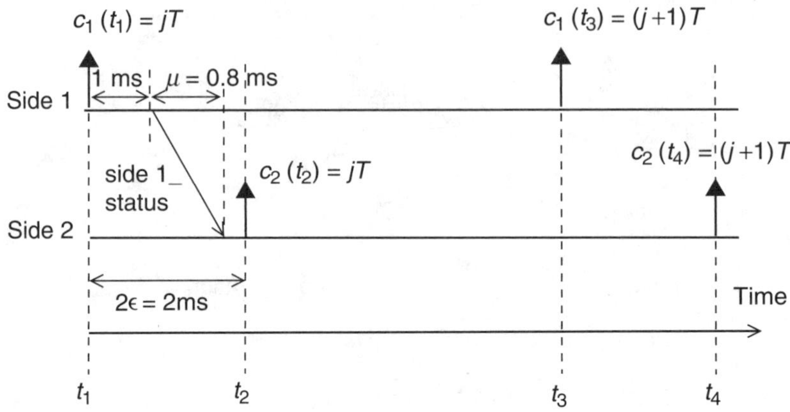

Figure 8.7: *Violation of the PALS causality constraint*

- `PALS_comm_client`: This class provides simple wrapper interfaces for logically synchronous message communications. Application tasks use two functions, `pals_send(connection name, payload)` and `pals_recv(connection name, payload)`, to send and receive messages, respectively. In both functions, `connection_name` is the input identifier for a connection from a source task to a group of destination tasks.

The following code snippet uses these classes:

```
Bool Side1_task::each_pals_period() {
    …
    // 1. Define task-specific default values for messages.
    int8_t side1 = NO_MSG;
    int8_t side2 = NO_MSG;
    bool user_cmd = false;
    // 2. Receive previous period's data, given that the source has not
    // crashed.
    comm_client->pals_recv("side1_status",&side1,1);
    comm_client->pals_recv("side2_status",&side2,1);
    comm_client->pals_recv("user_cmd",&user_cmd,1);
    // 3. Decide which side is active, based on information received
    // from Side1, Side2, and the user.
    next_side1_state = active_standy_logic(side1, side2,
user_cmd);
    // 4. Send current period's state.
    comm_client->pals_send("side1_status",&next_side1_state,1);
    …
    return true;
}
```

Here, the class `Side1_task` extends `PALS_task` and implements the periodic control computations of Side1 in the function `each_pals_period`. In each period, `Side1_task` reads the controllers' statuses and the user command by using the `pals_recv` function and sends its own active/standby status by using the `pals_send` function. In the `pals_recv` function, if the source task fails, the default input value is used.

These simple interfaces abstract away the details of the interactions with reliable communication services and management of logical synchronization groups.

8.3.1.6 Practical Considerations

In this section, we give a brief summary of the existing technologies available for the PALS system development.

AADL Tools for PALS System Modeling

Several tools are available for the PALS system modeling and analysis in AADL and Real-Time Maude:

- *SynchAADL2Maude:* Developed by [Bae12], this tool automatically translates the Synchronous AADL model of the PALS system to a Real-Time Maude model. In Real-Time Maude, the synchronous design can be verified by model checking.
- *Static checker:* We have developed a static checker in our lab to validate the correctness of the pattern implementation with multirate computations. This tool also translates the single-rate PALS model to the Synchronous AADL model. This allows users to refactor the legacy components with a logically synchronous design directly in the physically asynchronous model. Users can generate the Synchronous AADL model and verify it.
- *Rockwell Collins META Toolset:* This AADL tool also supports the PALS system analysis [Cofer11]. In general, this toolset supports design transformations, compositional verification, and static analysis for various architectural patterns. With respect to the PALS system, designers can instantiate the PALS design specification and validate the assumptions in a static checker, called Lute.
- *ASIIST:* This tool is used to perform schedulability and bus analysis of AADL models [Nam09]. ASIIST complements the static checkers for the PALS system. Designers can use ASIIST to compute

end-to-end delays, worst-case response, and output time in a system model with hardware components. The timing constraints of the PALS system can be validated once these timing properties are known.

PALS Middleware

The distributed middleware called PALSware is available for PALS system development in C++. It allows the use of COTS computers and networks, thereby lowering the synchronization overhead by potentially one to two orders of magnitude relative to custom or semi-custom legacy systems. The middleware has the following features:

- A simple C++ library that is extendable in different system architectures
- A fault-tolerant real-time multicast protocol
- Support for atomic distributed interactions
- Integration library for clock synchronization algorithms

Recently, a C version of PALSware has been developed to support formal verification at the source code level using CBMC [CBMC]. In addition, several commonly used fault-tolerant applications that can take advantage of PALS are checked at the source code level [Nam14].

8.4 Summary and Open Challenges

In this chapter, we presented a complexity-reducing architectural pattern for achieving logical synchronization in cyber-physical systems. Through this approach, developers can reduce the amount of effort required for distributed system design and verification. Engineers need to design and verify only the simple globally synchronous model as long as the system architecture satisfies the pattern's assumptions.

There are several open challenges in the research of the PALS system:

- The recent avionics certification standard DO-178C (with its formal methods supplement DO-333) provides guidance about using formal methods to satisfy certification activities. In the past, software model checking has been either very expensive or impossible due to the complexities of the distributed interactions. The PALS system should now enable more efficient testing of the PALS applications. More research is needed on software model checking that can apply this complexity-reducing pattern and support a scalable verification of distributed applications.

- The PALS system supports a small class of distributed interactions in a physically asynchronous architecture. For example, the message communication pattern is restrictive, as messages are received only at the end of the current PALS clock period. In many information-processing applications, the distributed computations operate in many phases over multiple nodes. Integration of multirate computations with intermediate message communications must also be investigated.
- Wireless communication is used in cyber-physical systems such as medical systems. The PALS system is not currently compatible with wireless communication, as it makes only probabilistic guarantees about message delivery. The pattern must be adapted to guarantee consistency in this network.

References

[Abdelzaher96]. T. Abdelzaher, A. Shaikh, F. Jahanian, and K. Shin. "RTCAST: Lightweight Multicast for Real-Time Process Groups." *Proceedings of IEEE Real-Time Technology and Applications Symposium*, pages 250–259, 1996.

[Aerospace04]. Aerospace. *The Architecture Analysis and Design Language (AADL).* AS-5506, SAE International, 2004.

[Alencar96]. P. S. Alencar, D. D. Cowan, and C. J. P. Lucena. "A Formal Approach to Architectural Design Patterns." In *FME '96: Industrial Benefit and Advances in Formal Methods*, pages 576–594. Springer, 1996.

[Allen97]. R. Allen and D. Garlan. "A Formal Basis for Architectural Connection." *ACM Transactions on Software Engineering and Methodology*, vol. 6, no. 3, pages 213–249, 1997.

[Al-Nayeem13]. A. Al-Nayeem. "Physically-Asynchronous Logically-Synchronous (PALS) System Design and Development." Ph.D. Thesis, University of Illinois at Urbana–Champaign, 2013.

[Al-Nayeem12]. A. Al-Nayeem, L. Sha, D. D. Cofer, and S. M. Miller. "Pattern-Based Composition and Analysis of Virtually Synchronized Real-Time Distributed Systems." IEEE/ACM Third International Conference on Cyber-Physical Systems (ICCPS), pages 65–74, 2012.

[Al-Nayeem09]. A. Al-Nayeem, M. Sun, X. Qiu, L. Sha, S. P. Miller, and D. D. Cofer. "A Formal Architecture Pattern for Real-Time Distributed Systems." 30th IEEE Real-Time Systems Symposium, pages 161–170, 2009.

[Awerbuch85]. B. Awerbuch. "Complexity of Network Synchronization." *Journal of the ACM*, vol. 32, no. 4, pages 804–823, 1985.

[Bae12-b]. K. Bae, J. Meseguer, and P. Olveczky. "Formal Patterns for Multi-Rate Distributed Real-Time Systems." *Proceedings of International Symposium of Formal Aspects of Component Software (FACS)*, 2012.

[Bae11]. K. Bae, P. C. Ölveczky, A. Al-Nayeem, and J. Meseguer. "Synchronous AADL and its Formal Analysis in Real-Time Maude." In *Formal Methods and Software Engineering*, pages 651–667. Springer, 2011.

[Bae12]. K. Bae, P. C. Ölveczky, J. Meseguer, and A. Al-Nayeem. "The SynchAADL2Maude Tool." In *Fundamental Approaches to Software Engineering*, pages 59–62. Springer, 2012.

[Benveniste10]. A. Benveniste, A. Bouillard, and P. Caspi. "A Unifying View of Loosely Time-Triggered Architectures." *Proceedings of the 10th ACM International Conference on Embedded Software*, pages 189–198, 2010.

[Birman12]. K. P. Birman. *Guide to Reliable Distributed Systems: Building High-Assurance Applications and Cloud-Hosted Services*. Springer, 2012.

[Birman87]. K. Birman and T. Joseph. "Exploiting Virtual Synchrony in Distributed Systems." *Proceedings of the 11th ACM Symposium on Operating Systems Principles*, vol. 21, no. 5, pages 123–138, 1987.

[CBMC]. CBMC Homepage, http://www.cprover.org/cbmc/.

[Chandra96]. T. D. Chandra and S. Toueg. "Unreliable Failure Detectors for Reliable Distributed Systems." *Journal of the ACM*, vol. 43, no. 2, pages 225–267, 1996.

[Chandy08]. K. M. Chandy, S. Mitra, and C. Pilotto. "Convergence Verification: From Shared Memory to Partially Synchronous Systems." In *Formal Modeling and Analysis of Timed Systems*, pages 218–232. Springer, 2008.

[Chapiro84]. D. M. Chapiro. "Globally-Asynchronous Locally-Synchronous Systems." Ph.D. Thesis, Stanford University, 1984.

[Chou90]. C. Chou, I. Cidon, I. S. Gopal, and S. Zaks. "Synchronizing Asynchronous Bounded Delay Networks." *IEEE Transactions on Communications*, vol. 38, no. 2, pages 144–147, 1990.

[Cofer11]. D. Cofer, S. Miller, A. Gacek, M. Whalen, B. LaValley, L. Sha, and A. Al-Nayeem. *Complexity-Reducing Design Patterns for*

Cyber-Physical Systems. Air Force Research Laboratory Technical Report AFRL-RZ-WP-TR-2011-2098, 2011.

[Cristian86]. F. Cristian, H. Aghili, R. Strong, and D. Dolev. *Atomic Broadcast: From Simple Message Diffusion to Byzantine Agreement.* CiteSeer, 1986.

[Cristian95]. F. Cristian, H. Aghili, R. Strong, and D. Dolev. "Atomic Broadcast: From Simple Message Diffusion to Byzantine Agreement." *Information and Computation*, vol. 118, no. 1, pages 158–179, 1995.

[de Niz09]. D. de Niz and P. H. Feiler. "Verification of Replication Architectures in AADL." 14th IEEE International Conference on Engineering of Complex Computer Systems, pages 365–370, 2009.

[Dietrich05]. J. Dietrich and C. Elgar. "A Formal Description of Design Patterns Using OWL." *Proceedings of IEEE Australian Software Engineering Conference*, pages 243–250, 2005.

[Douglass03]. B. P. Douglass. *Real-Time Design Patterns: Robust Scalable Architecture for Real-Time Systems.* Addison-Wesley Professional, 2003.

[Eugster03]. P. T. Eugster, P. A. Felber, R. Guerraoui, and A. Kermarrec. "The Many Faces of Publish/Subscribe." *ACM Computing Surveys*, vol. 35, no. 2, pages 114–131, 2003.

[Fischer85]. M. J. Fischer, N. A. Lynch, and M. S. Paterson. "Impossibility of Distributed Consensus with One Faulty Process." *Journal of the ACM*, vol. 32, no. 2, pages 374–382, 1985.

[Frana07]. R. Frana, J. Bodeveix, M. Filali, and J. Rolland. "The AADL Behaviour Annex: Experiments And Roadmap." 12th IEEE International Conference on Engineering Complex Computer Systems, pages 377–382, 2007.

[Gamma95]. E. Gamma, R. Helm, R. Johnson, J. Vlissides, and G. Booch. *Design Patterns: Elements of Reusable Object-Oriented Software.* Addison-Wesley Professional, 1995.

[Gifford79]. D. K. Gifford. "Weighted Voting for Replicated Data." *Proceedings of the 7th ACM Symposium on Operating Systems Principles*, pages 150–162, 1979.

[Guo96]. K. Guo, W. Vogels, and R. van Renesse. "Structured Virtual Synchrony: Exploring the Bounds of Virtual Synchronous Group Communication." *Proceedings of the 7th ACM SIGOPS European Workshop: Systems Support for Worldwide Applications*, pages 213–217, 1996.

[Halbwachs06]. N. Halbwachs and L. Mandel. "Simulation and Verification of Asynchronous Systems by Means of a Synchronous

Model." 6th International Conference on Application of Concurrency to System Design (ACSD), pages 3–14, 2006.

[Hanmer07]. R. Hanmer. *Patterns for Fault Tolerant Software*. John Wiley and Sons, 2007.

[Harrison97]. T. H. Harrison, D. L. Levine, and D. C. Schmidt. "The Design and Performance of a Real-Time CORBA Event Service." *ACM SIGPLAN Notices*, vol. 32, no. 10, pages 184–200, 1997.

[Hendriks05]. M. Hendriks. "Model Checking the Time to Reach Agreement." In *Formal Modeling and Analysis of Timed Systems*, pages 98–111. Springer, 2005.

[Hoyme93]. K. Hoyme and K. Driscoll. "SAFEbus (for Avionics)." *IEEE Aerospace and Electronic Systems Magazine*, vol. 8, no. 3, pages 34–39, 1993.

[Jahier07]. E. Jahier, N. Halbwachs, P. Raymond, X. Nicollin, and D. Lesens. "Virtual Execution of AADL Models via a Translation into Synchronous Programs." *Proceedings of the 7th ACM and IEEE International Conference on Embedded Software*, pages 134–143, 2007.

[Killian07]. C. Killian, J. W. Anderson, R. Jhala, and A. Vahdat. "Life, Death, and the Critical Transition: Finding Liveness Bugs in Systems Code." *Networked Systems Design and Implementation*, pages 243–256, 2007.

[Kopetz03]. H. Kopetz. "Fault Containment and Error Detection in the Time-Triggered Architecture." IEEE 6th International Symposium on Autonomous Decentralized Systems (ISADS), pages 139–146, 2003.

[Kopetz11]. H. Kopetz. *Real-Time Systems: Design Principles for Distributed Embedded Applications*. Springer, 2011.

[Kopetz92]. H. Kopetz. "Sparse Time Versus Dense Time in Distributed Real-Time Systems." *Proceedings of the 12th International Conference on Distributed Computing Systems*, pages 460–467, IEEE, 1992.

[Kopetz10]. H. Kopetz, A. Ademaj, P. Grillinger, and K. Steinhammer. "The Time-Triggered Ethernet (TTE) Design." 8th IEEE International Symposium on Object-Oriented Real-Time Distributed Computing, December 2010.

[Kopetz03]. H. Kopetz and G. Bauer. "The Time-Triggered Architecture." *Proceedings of the IEEE*, vol. 91, no. 1, pages 112–126, 2003.

[Kopetz93]. H. Kopetz and G. Grunsteidl. "TTP: A Time-Triggered Protocol for Fault-Tolerant Real-Time Systems." 23rd International Symposium on Fault-Tolerant Computing, pages 524–533, IEEE, 1993.

[Lamport01]. L. Lamport. "Paxos Made Simple." *ACM SIGACT News*, vol. 32, no. 4, pages 18–25, 2001.

[Lamport05]. L. Lamport. "Real-Time Model Checking Is Really Simple." In *Correct Hardware Design and Verification Methods*, pages 162–175. Springer, 2005.

[Lauterburg09]. S. Lauterburg, M. Dotta, D. Marinov, and G. Agha. "A Framework for State-Space Exploration of Java-Based Actor Programs." *Proceedings of the IEEE/ACM International Conference on Automated Software Engineering*, pages 468–479, 2009.

[Lin09]. H. Lin, M. Yang, F. Long, L. Zhang, and L. Zhou. "MODIST: Transparent Model Checking of Unmodified Distributed Systems." *Proceedings of the 6th USENIX Symposium on Networked Systems Design and Implementation*, 2009.

[Meseguer12]. J. Meseguer and P. C. Ölveczky. "Formalization and Correctness of the PALS Architectural Pattern for Distributed Real-Time Systems." *Theoretical Computer Science*, 2012.

[Miller09]. S. P. Miller, D. D. Cofer, L. Sha, J. Meseguer, and A. Al-Nayeem. "Implementing Logical Synchrony in Integrated Modular Avionics." IEEE/AIAA 28th Digital Avionics Systems Conference (DASC'09), pages 1. A. 3-1-1. A. 3-12, 2009.

[Muttersbach00]. J. Muttersbach, T. Villiger, and W. Fichtner. "Practical Design of Globally-Asynchronous Locally-Synchronous Systems." Proceedings of the Sixth International Symposium on Advanced Research in Asynchronous Circuits and Systems, pages 52–59, 2000.

[Nam09]. M. Nam, R. Pellizzoni, L. Sha, and R. M. Bradford. "ASIIST: Application Specific I/O Integration Support Tool for Real-Time Bus Architecture Designs." 14th IEEE International Conference on Engineering of Complex Computer Systems, pages 11–22, 2009.

[Nam14]. M. Y. Nam, L. Sha, S. Chaki, and C. Kim. "Applying Software Model Checking to PALS Systems." Digital Avionics Systems Conference, 2014.

[Pereira02]. J. Pereira, L. Rodrigues, and R. Oliveira. "Reducing the Cost of Group Communication with Semantic View Synchrony." *Proceedings of International Conference on Dependable Systems and Networks*, pages 293–302, 2002.

[Pfeifer99]. H. Pfeifer, D. Schwier, and F. W. von Henke. "Formal Verification for Time-Triggered Clock Synchronization." *Dependable Computing for Critical Applications*, vol. 7, pages 207–226, IEEE, 1999.

[Rushby99]. J. Rushby. "Systematic Formal Verification for Fault-Tolerant Time-Triggered Algorithms." *IEEE Transactions on Software Engineering*, vol. 25, no. 5, pages 651–660, 1999.

[Schmidt00]. D. Schmidt, M. Stal, H. Rohnert, and F. Buschmann. Pattern-Oriented Software Architecture Volume 2: Patterns for Concurrent and Networked Objects. John Wiley and Sons, 2000.

[Schmidt00]. D. Schmidt and F. Kuhns. "An Overview of the Real-Time CORBA Specification." *Computer*, vol. 33, no. 6, pages 56–63, 2000.

[Sen06]. K. Sen and G. Agha. "Automated Systematic Testing of Open Distributed Programs." In *Fundamental Approaches to Software Engineering*, pages 339–356. Springer, 2006.

[Sha01]. L. Sha. "Using Simplicity to Control Complexity." *IEEE Software*, vol. 18, no. 4, pages 20–28, 2001.

[Sha09]. L. Sha, A. Al-Nayeem, M. Sun, J. Meseguer, and P. Ölveczky. *PALS: Physically Asynchronous Logically Synchronous Systems.* Tech Report, University of Illinois at Urbana–Champaign, 2009.

[Steiner11]. W. Steiner and B. Dutertre. "Automated Formal Verification of the TTEthernet Synchronization Quality." In *NASA Formal Methods*, pages 375–390. Springer, 2011.

[Steiner11]. W. Steiner and J. Rushby. "TTA and PALS: Formally Verified Design Patterns for Distributed Cyber-Physical Systems." IEEE/AIAA 30th Digital Avionics Systems Conference (DASC), pages 7B5-1–7B5-15, 2011.

[Taibi03]. T. Taibi and D. C. L. Ngo. "Formal Specification of Design Patterns: A Balanced Approach." *Journal of Object Technology*, vol. 2, no. 4, pages 127–140, 2003.

[Tel94]. G. Tel, E. Korach, and S. Zaks. "Synchronizing ADB Networks." *IEEE/ACM Transactions on Networking*, vol. 2, no. 1, pages 66–69, 1994.

[Tripakis08]. S. Tripakis, C. Pinello, A. Benveniste, A. Sangiovanni-Vincentelli, P. Caspi, and M. Di Natale. "Implementing Synchronous Models on Loosely Time Triggered Architectures." *IEEE Transactions on Computers*, vol. 57, no. 10, pages 1300–1314, 2008.

[Renesse96]. R. Van Renesse, K. P. Birman, and S. Maffeis. "Horus: A Flexible Group Communication System." *Communications of the ACM*, vol. 39, no. 4, pages 76–83, 1996.

[Wahba10]. S. K. Wahba, J. O. Hallstrom, and N. Soundarajan. "Initiating a Design Pattern Catalog for Embedded Network Systems." *Proceedings of the 10th ACM International Conference on Embedded Software*, pages 249–258, 2010.

Chapter 9

Real-Time Scheduling for Cyber-Physical Systems

Bjorn Andersson, Dionisio de Niz, Mark Klein, John Lehoczky, and Ragunathan (Raj) Rajkumar

Cyber-physical systems (CPS) are composed of software (cyber) and physical parts. Implicit in this composition is the fact that they must act together in a synchronous manner. Verifying the timing of software has been the objective in studying real-time scheduling theory and, therefore, is key to the verification of its synchronization with the physical part of a CPS. The new challenges in CPS, however, call for new perspectives and solutions, which are the topics of discussion in this chapter.

9.1 Introduction and Motivation

CPS are composed of software and physical processes that must be synchronized. For instance, the airbag of a car must begin to inflate when a crash is detected, and it must be completely inflated before the driver gets too close to the steering wheel. In this example, the physical process is the driver moving toward the steering wheel during a crash as well as the airbag inflating. The software in the system is in charge of detecting the crash and triggering and synchronizing the inflation of the airbag so as to intercept the driver's movement at the right time. Traditional real-time scheduling theory verifies that this synchronization happens through a timing abstraction that captures the evolution of the physical process. Specifically, this theory uses periodic sampling (i.e., with a fixed period), assuming that the physical process cannot evolve by more than a fixed amount that the application software can handle. For instance, in the case of the airbag, the real-time system will periodically sense if a crash has happened. This period is designed to be small enough that, in the worst case, the delay between the actual instant of the crash and the time of the detection does not affect the guaranteed inflation time.

As CPS increase in complexity and need to interact with more uncertain physical processes, the periodic sampling abstraction breaks. For instance, if a control algorithm in a car's engine needs to execute at each revolution of the engine, then it is easy to see that this periodicity would be constantly evolving; that is, as the revolutions per minute (rpm) of the engine increases, the time between one revolution and the next is reduced [Kim12]. Another aspect of CPS that needs to be considered for the timing verification is the uncertainty in the environment. For instance, the execution time of a collision avoidance algorithm in an autonomous vehicle depends on the number of objects in its field of vision. This uncertainty demands new approaches that can adapt accordingly. We discuss all of these timing concerns in this chapter, so as to predictably allocate not only processing time, but also other types of resources, such as networks and power.

In the rest of this chapter, we discuss basic, traditional real-time scheduling techniques for single-core processors that have been widely accepted in the practitioner community. We then discuss advanced techniques currently under development in the research community, including techniques to schedule multiprocessor and multicore processors, accommodate variability in the timing parameters of tasks and uncertainty from the environment, allocate and schedule other related types of resources (e.g., network bandwidth and power), and schedule tasks with the timing parameters that vary with the physical process.

9.2 Basic Techniques

In this section, we focus on the basic techniques developed for single-core processors with fixed timing parameters. These techniques gave birth to the real-time scheduling domain. First we focus on single-core scheduling with fixed timing parameters, which forms the basis for the real-time scheduling platform from which the other types of scheduling evolved.

9.2.1 Scheduling with Fixed Timing Parameters

Real-time scheduling is concerned with the timing verification of the interactions between the software and the physical processes of a CPS. This domain stems from the seminal paper [Liu73] that formalized NASA's engineering principles in the Apollo missions. At that time, the software that interacted with the physical processes ran on simple hardware, and it was possible to fully characterize it with fixed timing parameters such as periods, deadlines, and worst-case execution time (WCET). Such systems are called *real-time systems*.

Real-time systems are designed to guarantee the timing of their interactions with the physical world, as described in the airbag example presented earlier. Such guarantees include rare corner cases. Due to this rarity, it is almost impossible for system testing to be able to uncover them. Nevertheless, such corner cases do occur—and when they do, the consequences can be fatal (e.g., the driver may hit the steering wheel and suffer a devastating crush injury).

The study of real-time systems includes the operating system scheduling algorithms that decide the execution order of threads (e.g., based on their priorities), the timing analysis algorithms that verify the execution of these threads is completed at the required time (e.g., 20 ms for the front airbag of a car), and techniques to obtain the worst-case execution time of the programs run by these threads, which is assumed to be known by the timing analysis algorithms. In this section (and in this chapter in general), we focus on the first two types of algorithms and provide a brief discussion of the WCET techniques.

9.2.1.1 Determining the Worst-Case Execution Time

Two main approaches have been taken to obtain the WCET of a program—measurements and static analysis. Over the years, directly measuring the WCET has been the preferred approach for most practitioners. This approach relies on experiments designed to generate the maximum execution time in the programs. These experiments are run a large number of times to measure the WCET, and then the WCET is inflated to account for uncertainty of the measurements. The WCET experiments can be supported by techniques from extreme-value theory [Hansen09] to improve the confidence in the measurements.

As the uncertainty about execution time increases, as tends to occur with the speculative execution features of modern processors such as cache and out-of-order execution, it becomes more difficult to design the experiments to obtain the WCET. This challenge has spawned new research efforts focused on WCET based on analysis [Wilhelm08] of the code of the program, its access to memory, and the characteristics of the hardware architecture on which such programs run. One program can also influence the memory behavior of another program. A well-known example is the so-called cache-related preemption delay (CRPD); in this case, a program suffers from longer execution because some of its cache blocks are evicted because the program was preempted by another, higher-priority program. [Chattopadhyay11] has analyzed CRPD by exploring all possible execution interleavings to discover when these preemptions can happen and calculate the corresponding delay.

9.2.1.2 Representation and Formalisms

To describe the model of a system (also known as formalism), let us first look at an example. Consider the inverted pendulum presented in Figure 9.1. The control software of the inverted pendulum is designed to move a cart back and forth as needed to keep the pendulum as vertical as possible. To achieve this goal, the system needs to first sense the inclination of the pendulum, and then move the cart in the same direction as this inclination to correct it. This sensing and moving sequence must be executed periodically in a loop, known as a control loop. The time between one execution of the loop and the next one—known as the period of the control loop—must be short enough to prevent the pendulum from inclining beyond the point where this inclination cannot be corrected.

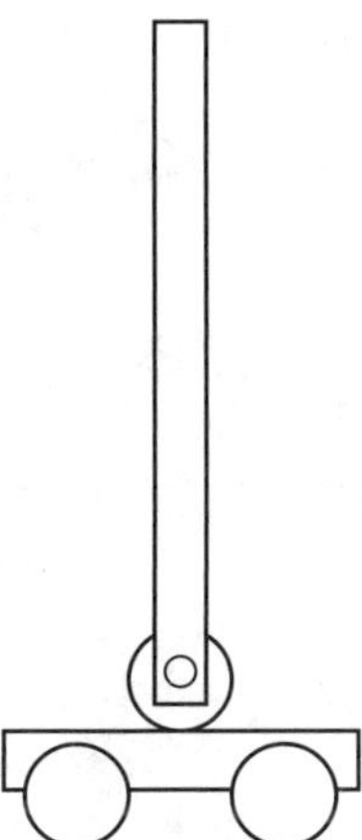

Figure 9.1: *Inverted pendulum in a control loop*

[Liu73] used this period to develop a periodic task model in which each task has a fixed period and a worst-case execution time. A task is considered to execute periodically; each of these periodic executions is known as a *job*. Hence, a task can be seen as an infinite number of jobs that arrive periodically. A job of a task is then considered to execute correctly if it finished before the arrival of the next job. This next-job arrival point is known as the *deadline* of a job. Based on this definition, we say that a task is schedulable if all its (infinite number of) jobs execute correctly. A task is then characterized as follows:

$$\tau_i = (C_i, T_i, D_i)$$

Here C_i represents the WCET of the task represented by τ_i, T_i represents its period, and D_i represents the deadline of a job of task τ_i relative to the arrival of this job. In this case, the deadline is equal to the period; thus, it is frequently omitted from the model, and the model is said to have an implicit deadline. Other options are a constrained deadline when $D \leq T$ and an arbitrary deadline when $D \neq T$ that we discuss later. The proportion of the processor that a task uses over time is known as its utilization and is calculated as follows:

$$U_i = \frac{C_i}{T_i}$$

[Liu73] defined a real-time system as a set of periodic tasks that are executed in a processor under a specific scheduling policy. The set of tasks is deemed schedulable if all the tasks are schedulable, as expected. This is denoted as $\tau = \{\tau_1, \tau_2, \ldots, \tau_n\}$. The total utilization of the set of tasks (also known as the taskset) is then calculated as follows:

$$U = \sum_{\forall \tau_i \in \tau} \frac{C_i}{T_i}$$

[Liu73] studied the scheduling of this type of task under the fixed-priority scheduling discipline. In fixed-priority scheduling, tasks are given a priority and the scheduler always runs the task that is ready to run; this task has the highest priority until it releases the processor or a higher-priority task becomes available to run. For the period task model, this means that the job of a task is executed to completion if no other job from a higher-priority task arrives.

The periodic task model is nowadays successfully used in practice to verify the schedulability of tasksets. For instance, [Locke90] describes a generic avionics system designed after the periodic task model. Table 9.1 presents a subset of the tasks in this system.

[Liu73] studied two types of priority assignments: fixed-priority assignments and dynamic-priority assignments. In fixed-priority assignment, priorities are assigned to task at design time and all their jobs inherit the same priority. In dynamic-priority assignment, the priorities are assigned to the jobs when they arrive (at runtime). Different jobs from the same task are allowed to have different priorities.

9.2.1.3 Fixed-Priority Assignment

[Liu73] proved that the rate-monotonic (RM) priority assignment over the fixed-priority scheduler is an optimal fixed-priority assignment for tasks with implicit deadlines. In RM-based assignment, tasks with higher rates (or smaller periods) are given higher priority than those with lower rates. The combination of RM priority assignment with fixed-priority scheduling is known as rate-monotonic scheduling (RMS).

The execution of three periodic tasks scheduled under RMS when they arrive together at time 0 is depicted in Figure 9.2 (arrivals are marked with an upward-pointing arrow and the deadline with a downward-pointing arrow). In the figure, note that no two tasks can be executing (indicated by a rectangle) at the same time. However, we can observe that τ_1 starts executing at time 0 and both τ_2 and τ_3 delay their starts because τ_1 is still excuting. Similarly, once τ_2 starts executing at time 5, the start of τ_3 is delayed further while this task waits for τ_2 to

Table 9.1: *Sample Tasks from a Generic Avionics System*

Task	Description	Period (T)	WCET (C)
Aircraft flight data	Determine estimates of position, velocity, altitude, and so on	55 ms	8 ms
Steering	Compute steering cues for cockpit display	80 ms	6 ms
Radar control (ground search mode)	Detect and identify ground objects	80 ms	2 ms
Target designation	Designate a tracked object as a target	40 ms	1 ms
Target tracking	Track the target	40 ms	4 ms
Heads-up display (HUD)	Display navigation and target data	52 ms	6 ms
Multipurpose display (MPD) HUD display	Back up the HUD display	52 ms	8 ms
MPD tactical display	Display radar contacts and target	52 ms	8 ms

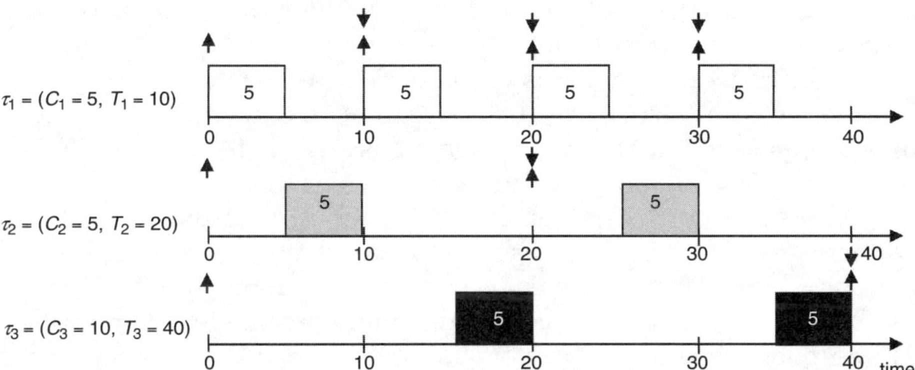

Figure 9.2: *Sample execution of three periodic tasks under RMS*

finish. Moreover, once τ_2 finishes at time 10, a second job from τ_1 arrives at time 10 and starts executing, which keeps τ_3 waiting even longer. When τ_3 is finally able to run at time 15, it is preempted by one job from τ_1 at time 20, another job from τ_2 at time 25 and yet another job from τ_1 again at time 30. After the last interruption, τ_3 is able to finish its remaining five units of execution from time 35 to 40.

With fixed-priority assignment, there are three main forms to test for schedulability: an absolute bound on processor utilization, a parameterized bound that depends on the number of scheduled tasks, and a response-time test. The absolute bound proved in [Liu73] states that a taskset τ scheduled under RMS is schedulable if $U \leq \ln 2$. The parameterized bound, in contrast, states that a taskset of n tasks is schedulable if the following is true:

$$U \leq n(2^{\frac{1}{n}} - 1)$$

Finally, [Joseph86] developed a response-time test to determine the worst-case finish time of a task. This test is a recurrent equation defined as follows:

$$R_i^0 = C_i$$

$$R_i^k = C_i + \sum_{j<i} \left\lceil \frac{R_i^{k-1}}{T_j} \right\rceil C_j$$

In this equation, the indices of the tasks are assumed to represent the priority precedence—that is, the smaller the index, the higher the priority. This equation adds up the execution time of the task we are evaluating (τ_i) and the preemptions that it may suffer from higher-priority tasks (τ_j). The equation is evaluated recurrently until $R_i^k = R_i^{k-1}$ (it converges) or $R_i^k > D_i$. Then τ_i is considered schedulable if $R_i^k \leq D_i$. For instance, we can calculate the response time of τ_3 in Figure 9.2 by adding its execution time (10) plus 4 executions of τ_1 (4 * 5) and 2 executions from τ_2 (2 * 5) to obtain a total time of 40.

The RMS priority assignment was later generalized [Audsley91, Leung82] to allow tasks that have deadlines that are shorter than their periods. This type of assignment is known as deadline-monotonic priority. In this case, tasks with shorter deadlines receive higher priorities. The response-time test, however, can still be used by considering the corresponding priorities. Similarly, the strictly periodic nature of a task τ_i with a period T_i, which requires that the interval between two jobs' arrivals be exactly T_i, was relaxed to allow this interval to be equal to or larger than T_i. This type of task is known as sporadic with a minimum inter-arrival T_i, and the results from RMS and DMS still apply. Extensions to many other engineering situations are provided by [Klein93].

9.2.1.4 Dynamic-Priority Assignment

Figure 9.2 presents the scheduling of a taskset whose total utilization is 100% and that is schedulable according to the response-time test but is not schedulable according to either the absolute bound or the parameterized bound. Such a taskset is harmonic, meaning that the period of every task is a multiple of the immediately smaller period. In this case, the utilization bound is 100%. However, when these periods are not harmonic, some jobs with shorter deadlines may have lower priority. This scenario is depicted in Figure 9.3, where the second job of τ_1 preempts the first job of τ_2 even though the former has a deadline of 200 and the latter has a deadline of 141. We identify this as a suboptimal priority assignment.

The suboptimal job priority assignment led [Liu73] to develop a job-based priority assignment known as earliest-deadline first (EDF) assignment. In this case, the priority of each job is assigned individually at runtime based on its deadline. Specifically, priorities are assigned at the arrival of a job, thereby ensuring that jobs with the earliest deadlines are assigned higher priorities than jobs with later deadlines. For the case of the taskset of the jobs in Figure 9.3, the priority of the second job of task τ_1 is assigned a lower priority than the first job of τ_2, thereby avoiding a deadline miss. This scenario is depicted in Figure 9.4.

Given that EDF is able to utilize 100% of the processor, its schedulability test is much simpler. Specifically, it suffices to ensure that the utilization of the taskset does not exceed 100% to ensure that it is schedulable under

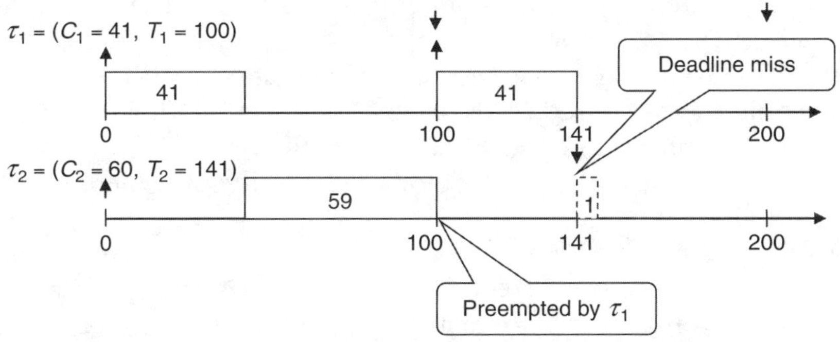

Figure 9.3: *Suboptimal job priority assignment in RMS*

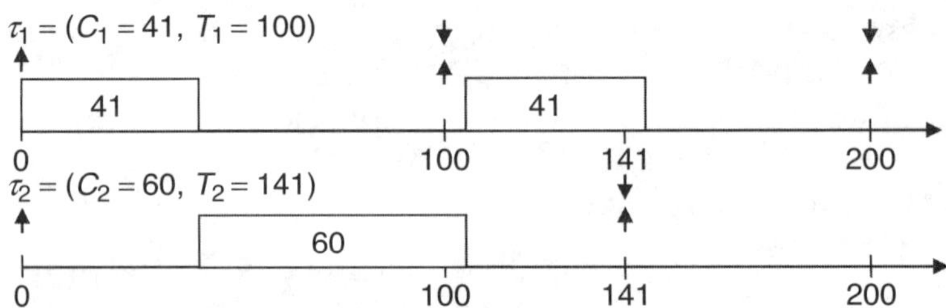

Figure 9.4: *Optimal job priority assignment with EDF*

EDF (for tasksets with implicit deadlines). Unfortunately, few commercial operating systems support this job-priority assignment algorithm.

[Liu73] proved that RM is an optimal task fixed-priority assignment algorithm and that EDF is an optimal dynamic priority assignment algorithm. Thus, if a priority assignment for a taskset exists to make it schedulable, then the optimal priority assignment algorithm (e.g., RM) will also be able to produce a priority assignment to make it schedulable.

9.2.1.5 Synchronization

The task model in [Liu73] assumes that tasks do not delay the execution of other tasks. In real applications, tasks do often need to share data or synchronize the access of other shared resources to solve a common problem. When this synchronization takes the form of a mutually exclusive access, it is reflected as a delay that a task τ_j, which is currently accessing the resource (or holding the resource), imposes on another task τ_i, which is waiting for the resource. This situation is known as a priority inversion when τ_j has a lower priority than τ_i, and the waiting time is known as blocking. This blocking can increase τ_i's response time, to the point that it misses its deadline. This situation gets worse when medium-priority tasks from a subset $\{\tau_k, \ldots, \tau_{k+r}\}$ that do not access the resource and have priorities higher than τ_j's priority and lower than τ_i's priority preempt τ_j. This situation is depicted in Figure 9.5 with a new parameter CS_i, which captures the time a task spends holding the shared resource in a code section known as the critical section (depicted as a black segment in the figure).

In [Sha90], the authors developed a family of synchronization protocols that prevents medium-priority tasks from increasing the priority-inversion blocking time with a technique known as priority inheritance. In these protocols, when a task τ_i tries to access a shared resource that is being used by task τ_j, then if the priority of τ_i is higher than the priority of τ_j, τ_j inherits τ_i's priority. This approach prevents medium-priority tasks from preempting τ_i. Figure 9.6 depicts the same case shown in Figure 9.5, but with priority inheritance being applied. The priority-inheritance protocols are typically implemented in semaphores or mutexes in real-time operating systems (RTOS) to allow tasks to access shared resources in a mutually exclusive fashion.

Two priority-inheritance protocols were developed in [Sha90]: the basic priority-inheritance protocol (PIP) and the priority-ceiling protocol (PCP). In PIP, the inherited priority is determined dynamically when a task is blocked while waiting for a resource. In PCP, in contrast, a priority ceiling is defined for a resource beforehand, as the highest priority among the potential users of the resource. Under this scheme, when a task locks a resource, it inherits the priority ceiling of the resource.

Beyond its control over priority inversion, PCP has the added benefit of preventing deadlocks. This is possible because when a task τ_i locks a resource S_u, the inherited priority prevents any other task τ_j that can access S_u from preempting it. As a consequence, τ_j cannot lock

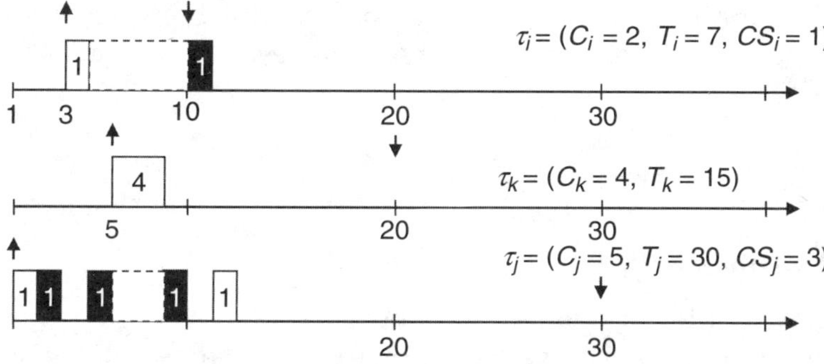

Figure 9.5: *Unbounded priority inversion*

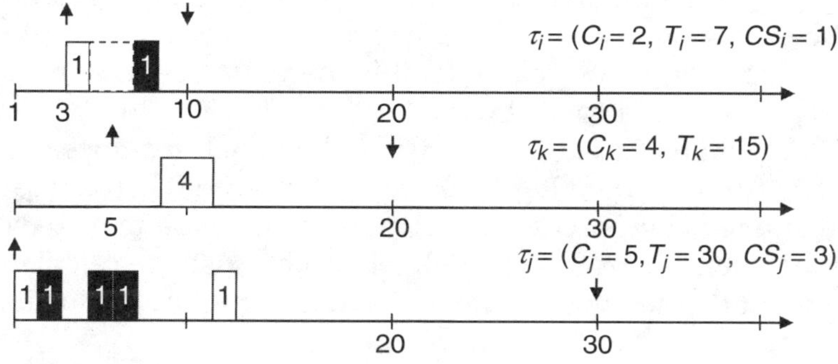

Figure 9.6: *Bounded priority inversion with priority inheritance*

another resource S_v that might potentially be requested by τ_i before acquiring S_u, which would create a circular waiting situation.

A bounded blocking term B_i for each task τ_i is calculated according to the inheritance protocol used. For PCP, for each set of resources that are locked in a nested fashion, B_i is calculated as the maximum blocking among all the lower-priority tasks that can block τ_i due to a resource in the set (i.e., only one task can block τ_i). For PIP, it is necessary to add the blockings of all tasks. In addition, in PIP, it is necessary to consider tasks τ_k that may block τ_j, which in turn may be blocking τ_i. This scheme is known as transitive blocking. Similarly, when a task τ_i inherits a priority, it blocks medium-priority tasks, which experience this effect as a blocking from lower-priority tasks—a scheme known as push-through blocking. All of these effects must be considered when calculating the blocking term of a task. Not surprisingly, such calculations can be quite complicated. Refer to [Rajkumar91] for a branch-and-bound algorithm that performs this calculation.

Once the blocking term is calculated, it can be used as an additional execution-time term for the task in the schedulability equations. For instance, it can be used in the response-time test for a task τ_i as follows:

$$R_i^k = C_i + B_i + \sum_{j<i} \left\lceil \frac{R_i^{k-1}}{T_j} \right\rceil C_j$$

9.2.2 Memory Effects

Today's computer systems—even the ones with a single processor—have complex memory systems. For example, they often use cache memories, which store frequently accessed data items close to the processor so that these items can be accessed with higher speed, and virtual memory systems, which cache frequently used translations for high-speed access. As a consequence of these effects, the execution time of a task depends on the schedule. It is therefore useful to extend the theory in Section 9.2.1 to deal with these effects.

The data items stored in the cache when a job resumes execution may differ from the data items stored in the cache when the same job was preempted, because some of these data items may have been evicted by the preempting task(s). Methods exist for calculating an upper bound on the number of evicted data items and can be incorporated in the schedulability analysis [Altmeyer11].

Even when jobs are not preempted, they may share data. This situation can make the execution time of one job dependent on the job that executed before it. Methods exist for analyzing this effect as well [Andersson12].

Also, even for a single processor, there can be multiple requestors to the memory system. For example, a job *J* might instruct an I/O device to perform direct memory access (DMA), so that when job *J* finishes and another job *J′* starts to execute, the DMA operation continues during the execution of job *J′*. Such a situation can cause the execution time of a job to increase and a DMA I/O operation to take longer to complete. Methods exist for analyzing these effects. The method in [Huang97] can be used to find the execution time of a task considering DMA I/O devices operating in cycle-stealing mode. In addition, the execution of a task can affect the amount of time it takes for a DMA operation to complete. If the arbitration for the memory bus is done with fixed priorities, and if the processor has higher priority than the I/O devices, then the method in [Hahn02] can be used to compute the response times for I/O operations to complete.

9.3 Advanced Techniques

Beyond the basic techniques that have been well accepted in the practitioner community, the real-time scientific community is currently creating solutions to a number of issues stemming from advanced features in CPS application and hardware platforms.

9.3.1 Multiprocessor/Multicore Scheduling

When a taskset τ is run on a computer with multiple processors, we consider this as a multiprocessor scheduling problem. Two basic forms of scheduling exist for multiprocessor scheduling: partitioned and global scheduling (Figure 9.7). In partition scheduling, the taskset is divided into subsets, with each subset being assigned to one processor and a uniprocessor scheduling scheme being used to schedule it. In global scheduling, all the tasks are scheduled on the set of m processors by selecting which subset of m tasks should run on the m processors at any given time. For instance, global EDF selects the m jobs with the shortest deadlines. Note that multicore processors are a special type of multiprocessor where cores share common hardware resources that will also influence scheduling. In Figure 9.7, the right side of the figure illustrates partitioned scheduling: Tasks τ_1 and task τ_2 are assigned to processor 1, and task τ_3 is assigned to processor 2.

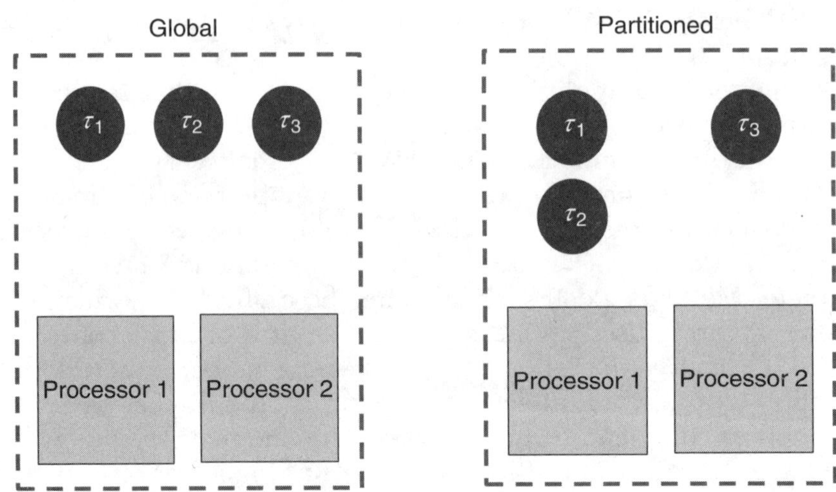

Figure 9.7: *Tasksets in partitioned and global scheduling*

Many of the properties that were true for a single processor (and incorporated in the basic techniques discussed earlier) do *not* hold for a multiprocessor. This section cites examples of where these properties do not hold in multiprocessors, which in turn motivates the design of a new scheduling algorithm. Thus, we also present a new scheduling algorithm for multiprocessors.

The literature on real-time scheduling on multiprocessors is vast. To keep things manageable, we focus on only implicit-deadline tasks because this discussion helps us understand the problem. For readers who are interested in tasksets with constrained deadlines or arbitrary deadlines, see [Davis11].

In our discussions, we use the tasksets presented in Table 9.2.

9.3.1.1 Global Scheduling

Global scheduling is a class of scheduling algorithms in which processors share a single queue storing jobs that are eligible for execution; on each processor, a dispatcher selects the highest-priority job from this queue. Let *eligjobs*(t) denote the set of jobs that are eligible at time t. Since there are m processors, at each instant, the $min(|\,eligjobs(t)\,|, m)$ highest-priority eligible jobs are selected to execute on the processors.

Task-Static Priority Scheduling

In a task-static priority scheduling algorithm, each task is assigned a priority and each job of this task has the priority of the task.

Table 9.2: *Example Implicit-Deadline Sporadic Tasksets*

Taskset 1			**Taskset 2**			**Taskset 3**	
$n = m + 1$			$n = m + 1$			$n = m + 1$	
$T_1 = 1$	$C_1 = 2\varepsilon$		$T_1 = 1$	$C_1 = ½ + \varepsilon$		$T_1 = 1$	$C_1 = ½ + \varepsilon$
$T_2 = 1$	$C_2 = 2\varepsilon$		$T_2 = 1$	$C_2 = ½ + \varepsilon$		$T_2 = 1$	$C_2 = ½ + \varepsilon$
$T_3 = 1$	$C_3 = 2\varepsilon$		$T_3 = 1$	$C_3 = ½ + \varepsilon$		$T_3 = 1$	$C_3 = 2\varepsilon$
...	...		...	...		...	...
$T_m = 1$	$C_m = 2\varepsilon$		$T_m = 1$	$C_m = ½ + \varepsilon$		$T_m = 1$	$C_m = 2\varepsilon$
$T_{m+1} = 1 + \varepsilon$	$C_{m+1} = 1$		$T_{m+1} = 1 + \varepsilon$	$C_{m+1} = ½ + \varepsilon$		$T_{m+1} = 1$	$C_{m+1} = 2\varepsilon$

As we saw previously, RM is the optimal priority assignment scheme for a single processor. Unfortunately, for multiprocessors, it is not, and its utilization bound is zero. An example follows.

Consider taskset 1 in Table 9.2 to be scheduled on m processors using global scheduling with RM and assume that $2\varepsilon * m \leq 1$. The priority assignment RM assigns the highest priority to the tasks $\tau_1, \tau_2, \ldots, \tau_m$ and the lowest priority to task τ_{m+1}.

Let us consider the situation where all tasks generate a job simultaneously, and let time 0 denote this instant (Figure 9.8). During the time interval $[0, 2\varepsilon)$, the jobs of tasks $\tau_1, \tau_2, \ldots, \tau_m$ are the highest-priority eligible jobs, so they execute, one job on each processor, and finish at time 2ε. During the time interval $[2\varepsilon, 1)$, a single job of task τ_{m+1} executes on one of the processors (the other $m-1$ processors are idle). Note that at time 1, the job of task τ_{m+1} has executed $1-2\varepsilon$ time units and its execution time is 1, so the job of task τ_{m+1} has not yet finished execution at time 1. Then, at time 1, each of the tasks $\tau_1, \tau_2, \ldots, \tau_m$ generates a new job. Hence, during the time interval $[1, 1+2\varepsilon)$, the jobs of tasks $\tau_1, \tau_2, \ldots, \tau_m$ are the highest-priority eligible jobs, so they execute, one job on each processor, and finish at time $1+2\varepsilon$. Note that the job of τ_{m+1} had its deadline at time $1+\varepsilon$. From the arrival time until its deadline, however, this job executed only $1-2\varepsilon$ time units, even though its execution time is 1 time unit. Thus, this job missed its deadline.

We can repeat this argument for any ε and for any m. Letting $\varepsilon \to 0$ and $m \to \infty$ gives us the case in which there exists a taskset for which:

$$\frac{1}{m} * \sum_{\tau_i \in \tau} u_i \to 0$$

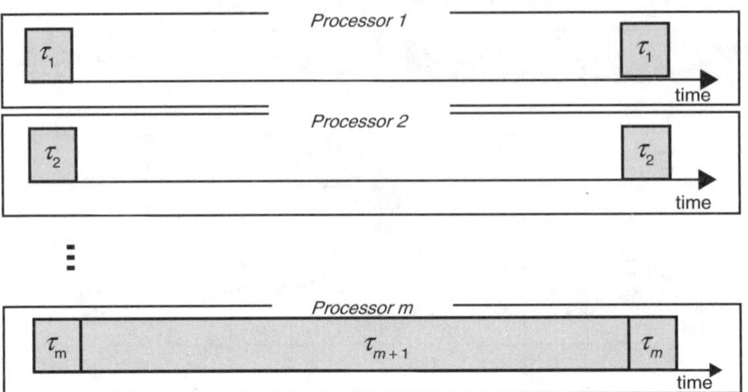

Figure 9.8: *The utilization bound of global RM*

This expression indicates that a deadline is missed when the taskset is scheduled by global RM. In turn, the utilization bound of global RM is zero. This outcome is referred to as Dhall's effect [Dhall78], named after the person who observed that RM can exhibit poor performance in global scheduling.

We can also see that changing the priority assignment makes the taskset schedulable. For example, assigning the highest priority to task τ_{m+1} and assigning a lower priority to all the tasks $\tau_1, \tau_2, \ldots, \tau_m$ makes the taskset schedulable. Hence, RM is not the optimal priority assignment scheme for global scheduling.

In this example, the utilization bound of global RM is zero, and this algorithm is not optimal for task-static priority global scheduling. It therefore behooves us to develop better priority assignment schemes. We can observe that RM performs poorly because the system did not do enough work in a time interval, and that the cause of the relatively small amount of work accomplished was that the parallel processing capacity was not used during a large time interval. Recognizing these problems, the scientific community has developed priority assignment schemes that avoid this issue and, therefore, perform better than RM.

We can categorize a task τ_i as "heavy" or "light," where a task is considered heavy if the following is true:

$$u_i > \frac{m}{3m-2}$$

Otherwise, it is light. We then assign the highest priority to the heavy tasks, assign a lower priority to the light tasks, and assign the relative priority order between the light tasks according to RM. Indeed, the priority assignment scheme RM-US($m/(3m-2)$) [Andersson01] does exactly that; for $m \geq 2$, its utilization bound has been proved to be $m/(3m-2)$. This utilization bound is greater than ⅓, so the assignment scheme circumvents Dhall's effect.

Further improvements have been made. The algorithm RM-US(y) [Lundberg02], where y is a solution to a nonlinear equation where $y \approx 0.374$, operates similarly; its utilization bound is $y \approx 0.374$.

Currently, the best algorithm (in terms of utilization bound) is from [Andersson08]:

$$\text{SM-US}\left(\frac{2}{3+\sqrt{5}}\right)$$

Its utilization bound follows:

$$\frac{2}{3+\sqrt{5}} \approx 0.382$$

One can show (using taskset 2 in Table 9.2) that for each job-static priority scheduling algorithm, it holds that the utilization bound of the algorithm is at most 0.5 [Andersson01].

It is possible to use the POSIX call `sched_setscheduler` to assign a priority to a task. In addition, one can set a processor affinity mask to allow a task to execute on any of the processors.

Job-Static Priority Scheduling

In a job-static priority scheduling algorithm, each job is assigned a priority, and this priority does not change with time. Clearly, every task-static priority scheduling algorithm is also a job-static priority scheduling algorithm.

As we saw in Section 9.3.1, EDF is the optimal priority assignment scheme for a single processor. Unfortunately, for multiprocessors, the same is not true. Also, the utilization bound for the multiprocessor case is zero—this can be seen by applying the same reasoning as in the previous example. The scientific community has developed a priority assignment scheme that avoids this problem and, therefore, performs better than EDF.

We can categorize a task τ_i as "heavy" or "light" where a task is heavy if the following is true:

$$u_i > \frac{m}{2m-1}$$

Otherwise, it is light. (Note that our categorization of heavy and light in this section is different from the one used in the previous section.) We can then assign the highest priority to the heavy tasks, a lower priority to the light tasks, and the relative priority order between the light tasks according to EDF. Indeed, the priority assignment scheme from [Srinivasan02] does exactly that:

$$\text{EDF-US}\left(\frac{m}{2m-1}\right)$$

Its utilization bound is proven to be the following:

$$\frac{m}{2m-1}$$

This utilization bound is greater than ½ and, therefore, this scheme circumvents Dhall's effect, too. One can show (using taskset 2 in Table 9.2) that for each task-static priority scheduling algorithm, it holds that the utilization bound of this algorithm is at most 0.5.

With some extra effort, one can use the Linux kernel and schedule the tasks to behave according to the scheme as follows: Designate a special process that runs at the highest priority (use the POSIX `call sched_setscheduler` to assign the priority). Whenever a task has finished the execution of its job, it should notify this special process that the task is waiting for a given event (e.g., a timer to expire). When the special process observes this event, it should unblock the task (e.g., a call signal on a semaphore where the task previously called is waiting on that semaphore) and assign the correct (according to the scheme) priority to the task that woke up; this can be done with `sched_setscheduler`.

Because global scheduling allows task migration, a job may potentially experience additional cache misses when it resumes execution on another processor than the one it was preempted on. This may prolong the execution time of a job—a factor that might need to be taken into account in the schedulability analysis. In either job-static or task-static global scheduling, a context switch can occur only when a job arrives or when a job finishes execution. Thus, for a given time interval, we can compute an upper bound on the number of possible context switches, and we can compute an upper bound on the number of migrations. If an upper bound on the overhead of a single migration is known, we can then compute an upper bound on the overhead for migrations—another factor that can be incorporated into the schedulability analysis.

9.3.1.2 Partitioned Scheduling

In partitioned scheduling, the taskset is partitioned and each partition is assigned to a processor; hence, each task is assigned to a processor. At runtime, a task is allowed to execute only on the processor to which it is assigned.

Task-Static Priority Scheduling

In partitioned, task-static priority scheduling, before runtime a task is assigned to a processor and cannot migrate; at runtime each processor is scheduled using a uniprocessor scheduling algorithm based on task-static priorities. For a single processor, RM is the optimal task-static priority schemes; thus, we can, with no performance loss, use RM on each processor. We then need to discuss only the assignment of tasks to processors.

An intuitive way to assign tasks to processors would be to use load-balancing offline. That is, we could ensure that tasks are considered one by one, and assign the currently considered task to the processor that currently has the lowest utilization. Unfortunately, the utilization bound of such a load-balancing algorithm is zero, as we will see in the following example.

Consider the taskset 1 in Table 9.2 to be scheduled on m processors using load-balancing (i.e., tasks are considered one by one), and assign the currently considered task to the processor that currently has the lowest utilization. Also, assume that $2\varepsilon * m \leq 1$. Initially, no tasks have been assigned, so the utilization of each processor is zero. Now we consider tasks one at a time, beginning with task τ_1. It should be assigned to the processor that currently has the lowest utilization. Since all processors have utilization zero, this task can be assigned to any processor. Next, consider τ_2. It can be assigned to any processor except the processor to which τ_1 is assigned. Continuing with tasks $\tau_3, \tau_4, \ldots, \tau_m$, we end up with an assignment where each task is assigned to its own processor. Now, only one task, τ_{m+1}, remains. If τ_{m+1} is assigned to processor 1, then the utilization of processor 1 would be as follows:

$$\frac{2\varepsilon}{1} + \frac{1}{1+\varepsilon}$$

This value is strictly greater than 1; if τ_{m+1} is assigned to processor 1, then, a deadline miss would occur on processor 1. It turns out that the same reasoning applies to any of the other processors: If τ_{m+1} is assigned there, a deadline miss would occur. Regardless of where τ_{m+1} is assigned, a deadline miss will result. We can repeat this argument for any ε and for any m. Letting $\varepsilon \to 0$ and $m \to \infty$ tells us that there exists a taskset for which we get the following expression:

$$\frac{1}{m} * \sum_{\tau_i \in \tau} u_i \to 0$$

A deadline is missed when the taskset is scheduled by this type of load-balancing. Hence the utilization bound of this type of load-balancing is zero.

The zero utilization bound of the load-balancing scheme presented in this example behooves us to develop better algorithms for task assignment. We can observe that this type of load-balancing performs poorly due to three conditions:

- There are more tasks than processors.
- At least one task (we will call it the *large task* in this discussion) requires so much capacity that it cannot be assigned to the same processor with any other task (and still be schedulable).
- This large task is assigned to a processor once each processor was already assigned at least one task.

Since the large task needs almost the capacity of an entire processor to meet its deadline, a deadline miss occurs. To avoid this scenario, the scientific community has developed task assignment schemes that perform better than this type of load-balancing.

One idea for designing a better task assignment scheme is to assign a task to a processor where tasks have already been assigned, so as to try to fill up the half-full processors first and avoid adding tasks to empty processors if possible. This scheme has the potential to avoid the performance problem seen in this example. Care must be taken, however. Consider taskset 3 in Table 9.2. The task τ_1 can be assigned to any processor. Then we consider task τ_2. If we follow the idea of assigning a task to a processor where other tasks have already been assigned, then we should assign task τ_2 to the same processor to which τ_1 has been assigned. The utilization of that processor would then be as follows:

$$\frac{\frac{1}{2}+\varepsilon}{1} + \frac{\frac{1}{2}+\varepsilon}{1}$$

This value is strictly greater than 1, so a deadline miss would occur on this processor. Therefore, whenever we attempt to assign a task, we must perform a uniprocessor schedulability analysis. The scientific community has developed task assignment schemes based on these two ideas: Assign a task to a processor on which other tasks have been assigned, and use a uniprocessor schedulability test to ensure that after a task has been assigned, the taskset on each processor is schedulable.

Consider the task assignment algorithm known as first-fit for task assignment for a taskset (taskset-to-be-assigned) and a set of processors (processor set). Its pseudocode follows:

1. If taskset-to-be-assigned is not empty then
   1.1. Fetch **next** task τ_i from taskset-to-be-assigned
2. Else
   2.1. Return success
3. Fetch first processor p_j in the processor set
4. If τ_i is schedulable when assigned to processor p_j then
   4.1. Assign τ_i to processor p_j
   4.2. Remove τ_i from taskset-to-be-assigned
   4.3. Jump to step 1
5. Else if there is a next processor p_j in the processor set then
   5.1. Fetch the next p_j processor from the processor set and jump to step 4
6. Else
   6.1. Return failure

If the uniprocessor in the [Liu73] utilization-based schedulability test is used in step 4, then the overall algorithm has utilization bound $\sqrt{2}-1 \approx 0.41$ [Oh98]. A more advanced algorithm is also known that has the utilization bound 0.5 [Andersson03]. One can show (using taskset 2 in Table 9.2) that for each partitioned task-static priority scheduling algorithm, it holds that the utilization bound of the algorithm is at most 0.5.

Note that the Linux kernel supports partitioned fixed-priority scheduling. It is possible to use the POSIX call `sched_setscheduler` to assign a priority to a task, and one can also set a processor affinity mask to mandate that a task is allowed to execute only on a certain processor.

Job-Static Priority Scheduling

For job-static priority scheduling on a single processor, we know that EDF is optimal. Thus, we can use it with no cost in terms of performance, and we need to discuss only task assignment in this section.

For designing a good task assignment scheme for job-static priority partitioned scheduling, we can follow the reasoning used for task-static priority partitioned scheduling but instead of using schedulability tests for uniprocessor RM, we use schedulability tests for the uniprocessor scheduling algorithm (in our case, uniprocessor EDF). Fortunately, an exact schedulability test for the uniprocessor case is (as we saw in Section 9.2) very simple: If the sum of utilization of all tasks on a processor does not exceed 1, then the taskset on this processor is schedulable. With this reasoning, a first-fit algorithm can be designed with utilization bound 0.5. One can also show (using taskset 2 in Table 9.2) that for each partitioned job-static priority scheduling algorithm, it holds that the utilization bound of the algorithm is at most 0.5.

9.3.1.3 Algorithms Using Fairness

In some scheduling algorithms, jobs can migrate to any processor at any time; many of these algorithms have the utilization bound 100%. Unfortunately, they incur a larger number of preemptions than the task-static and job-static partitioned scheduling algorithms and the task-static and job-static global scheduling algorithms. For details, see the algorithms SA [Khemka92], PF [Baruah96], PD [Srinivasan02a], BF [Zhu03], and DP-WRAP [Levin10].

A recent algorithm known as RUN [Regnier11] generates many fewer preemptions than the other algorithms that rely on fairness. With RUN, the number of preemptions per job is logarithmic as a function of the number of processors. RUN has a drawback, however: It often generates scheduling segments that are very small and cannot be realized in practice. Also, RUN cannot schedule sporadic tasks—only periodic tasks. Future research may make RUN useful for practitioners.

9.3.1.4 Algorithms Using Task-Splitting

So far, we have seen two classes of algorithms: algorithms with few preemptions and utilization bounds at most 0.5, and algorithms with a larger number of preemptions and utilization 1. It would be desirable to use a scheduling algorithm with a utilization bound greater than 0.5 but still have few preemptions. Task-splitting is a category of algorithms that offers this capability.

Consider taskset 2 in Table 9.2. With a partitioned scheduling algorithm, there is no way to assign tasks to processors so that after each task has been assigned to a processor, it holds that for each processor, the utilization is at most 1. We can achieve this goal, however, if we allow a task to be "split" into two or more "pieces." Consider taskset 2 in Table 9.2. We can assign τ_1 to processor 1, assign τ_2 to processor 2, and assign τ_m to processor m. Then we split task τ_{m+1} into two pieces as follows:

$$T_{m+1}' = 1, C_{m+1}' = 1/4 + \varepsilon/2$$

$$T_{m+1}'' = 1, C_{m+1}'' = 1/4 + \varepsilon/2$$

After this splitting, task τ_{m+1}' can be assigned to processor 1 and task τ_{m+1}'' can be assigned to processor 2. After this assignment, for each processor, it holds that the utilization is at most $1/2 + \varepsilon + 1/4 + \varepsilon/2$. For $\varepsilon \leq 1/6$, we obtain that after this assignment, for each processor, it holds that the utilization is at most 1. Hence, we can apply uniprocessor EDF on each processor and then all deadlines will be met. This scheme will work if we can schedule the two pieces of τ_{m+1} independently and we are allowed to execute the two pieces simultaneously. Recall, however, that the two pieces belong to an original task τ_{m+1}, which is a program with internal data dependencies. Thus, unless we know the internal structure of the program that represents task τ_{m+1}, we must schedule task τ_{m+1} while assuming that the two pieces of τ_{m+1} must not execute simultaneously.

For scheduling periodic tasks, the research literature provides an algorithm using task-splitting: EKG [Andersson06]. For scheduling sporadic tasks, the research literature offers different approaches to dispatch the pieces of a split task:

- *Slot-based split-task dispatching:* With this approach, time is subdivided into time slots of equal size. In each time slot, a reserve is used for the execution of the split tasks. Hence, a task τ_i that is split between processor p and processor $p + 1$ is assigned to the reserve on processor p and the reserve on processor $p + 1$. The phasing and duration of these reserves are selected such that for two consecutive processors, for those reserves that are used for a task that is split between these two processors, there is no overlap in time between these reserves. The research literature provides an algorithm using slot-based split-task dispatching, sometimes called sporadic-EKG [Andersson08a].

- *Suspension-based split-task dispatching:* In suspension-based split-task dispatching, the double-prime piece becomes noneligible when the single-prime piece of a task executes. The research literature provides an algorithm using suspension-based split-task dispatching, called Ehd2-SIP [Kato07].
- *Window-based split-task dispatching:* In window-based split-task dispatching, the double-prime piece becomes eligible only when the single-prime piece has finished execution. This ensures no overlap. The deadline of the respective pieces must be assigned so that the sum of their deadlines does not exceed the deadline of the task before it got split. The research literature provides an algorithm using window-based split-task dispatching and dynamic-priority scheduling, called EDF-WM [Kato09]. The research literature also provides an algorithm using window-based split-task dispatching and fixed-priority scheduling, called PDMS_HPTS_DS [Lakshmanan09].

9.3.1.5 Memory Effects

In multiprocessors, additional types of memory contention impact timing and some of the effects mentioned in Section 9.2.2 become more severe. Contention on the bus from the memory controller to the memory modules increases with the number of processors. Methods exist for analyzing the timing effect caused by this [Andersson10, Nan10, Pellizzoni10, Schliecker10].

Many multiprocessors are implemented on a single chip (multicore processors), in which case the processors (processor cores) typically share cache memories. Then, a cache block belonging to a task can obviously be evicted when tasks executing on the same processor preempt it (as mentioned in Section 9.2.2). In a multicore processor, however, cache blocks belonging to a task can also be evicted by other tasks executing in parallel on other processors. Methods exist for analyzing the timing effect caused by this scheme. The method known as *cache coloring* eliminates this effect by setting up the virtual-to-physical address translation so that cache sets of different tasks do not intersect.

The main memory of computers today uses dynamic random access memory (DRAM) that is organized as a set of banks, where a bank comprises multiple rows. In each bank, at each instant, there can be at most one row open. When a task performs a memory operation, it selects one memory bank (determined by the physical address) and then opens a row (determined by the physical address) in that memory bank and then reads data from this row. If the row corresponding to a given memory operation is already open, then the row does not need to be opened and the memory operation can complete faster. Often, programs have locality of reference such that, for many memory operations, there is no need to open a new row (because a previous memory operation has opened the row that a memory operation accesses). When multiple tasks execute on different processors and access different rows in the same bank, however, a task τ_i that opens a new row closes the row opened by another task τ_j and, therefore, increases the execution time of τ_j. The method called *bank coloring* (which works similarly to cache coloring) eliminates this effect. Since both cache coloring and bank coloring configure the virtual memory system for different purposes, it is necessary to coordinate this configuration if both cache coloring and bank coloring are to be achieved [Suzuki13].

Cache and bank coloring schemes have limitations due to the small number of partitions that may be created if full isolation is required. [Kim14] developed new algorithms for bounding the memory interferences that may arise when tasks share bank partitions. Specifically, these authors developed a double-bounding scheme in which the per-access bound is combined with a per-job bound to reduce the pessimism of the over-approximation. In addition, their paper explores the impact of the per-bank queue length in the memory controllers on the worst-case memory interference. These results are incorporated into a scheduling algorithm and a task allocation algorithm that reduce these effects. A recent survey [Abel13] discusses these topics.

9.3.2 Accommodating Variability and Uncertainty

In this section, we discuss techniques to accommodate variations in the timing parameters of the taskset and uncertainty from the environment.

9.3.2.1 Resource Allocation Tradeoff Using Q-RAM

In the previous sections of this chapter, we discussed tasks that have fixed resource requirements (required CPU time to complete executing). In this section, we consider tasks that can be configured to different quality-of-service levels, which in turn modifies the required resources. This is the case, for instance, in a video playback application where, depending on the frames per second (FPS) required, the application consumes different amounts of CPU cycles. In this scenario, the problem is shifted to the optimal selection of the quality-of-service (QoS) level.

To solve this problem, [Rajkumar97] developed the Quality-of-Service Resource Allocation Model (Q-RAM). Q-RAM maps QoS levels of the different applications to system utility (as perceived by the user) and allocates resources to these applications, ensuring that the total system utility is maximized (optimal). Q-RAM primarily takes advantage of the fact that as applications (e.g., video streaming) increase their QoS level, the incremental utility to the user keeps decreasing, yielding diminishing returns. In other words, the utility gain obtained by moving from a QoS level i to $i + 1$ is larger than that obtained by moving from level $i + 1$ to $i + 2$. For instance, in a video streaming application, increasing the FPS from 15 to 20 gives the user more additional utility (i.e., perceived quality) than increasing from 20 to 25 frames per second.

Q-RAM uses the utility functions to perform an optimal allocation of resources to different tasks exploiting the concave property of diminishing returns, which manifests as a monotonically decreasing utility-to-resource ratio. This ratio, which is known as marginal utility, is the derivative in a continuous utility function. Q-RAM uses the marginal utility to perform the allocation one increment at a time, starting with the increment that derives the largest utility for the smallest allocation of resources (largest marginal utility). In each of the subsequent steps, it selects the next largest marginal utility increment, continuing in this fashion until the entire resource (e.g., CPU utilization) has been allocated. When Q-RAM reaches an exact (optimal) solution, the marginal utility (or derivative) of the last allocation (QoS level) of all the tasks is the same.

Figure 9.9 depicts two sample application utility curves. In this case, if the total number of resource units is 6 (e.g., 6 Mbps), the step-by-step allocation would be as follows: Give 1 unit of resource to Application A, give 1 unit of resource to Application B, raise the allocation to Application A to 2 units, raise the allocation to Application B to 3 units, and raise the allocation to Application A to 3 units.

Applications can have multiple QoS dimensions that provide different utility to the user. For instance, the utility of a video streaming application depends not only on the frames per second, but also on the resolution of each frame—for example, VGA, half HD (720), full HD (1024), and so on. As a result, each application τ_i has a dimensional

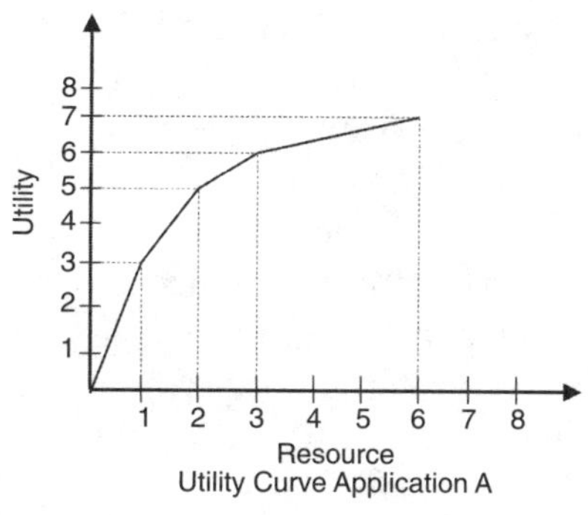

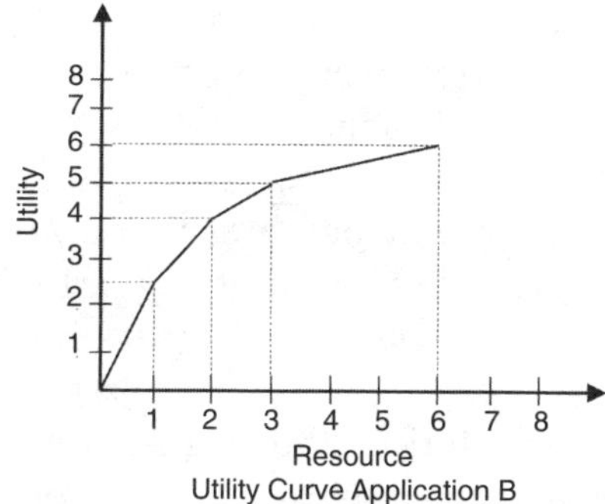

Figure 9.9: *Sample utility curves*

resource utility $U_{i,k} = U_{i,k}(R)$ for an allocation of R resources for each of the different QoS dimensions Q_k $1 \leq k \leq d$ of the application. Moreover, to reach a specific QoS level, the application may need more than one resource. For instance, in a video streaming application, it may need CPU cycles as well as network bandwidth if the streaming occurs across a network. In that case, the dimensional utility requires multiple resources: $U_{i,k} = U_{i,k}(R_{i,1}, R_{i,2}, \ldots, R_{i,m})$. The resource allocation is then considered an allocation vector of the form $R^i = \left(R_{i,1}, R_{i,2}, \ldots, R_{i,m}\right)$.

When the QoS dimensions of an application are independent, they generate their own utility, and the utility of the application as a whole is the sum of the dimensions: $U_i = \sum_{k=1}^{d} U_{i,k}$. In contrast, when the dimensions are dependent, their allocation needs to be considered as a combined allocation to reach a single utility level, as previously presented. The utility curve then becomes a surface, for the two-dimensional case, and the allocation steps are selected from the steepest trajectory on the surface with combined increments in both dimensions.

Q-RAM allows the assignment of weights to the applications of a system to encode their relative importance. The utility of the system is then calculated as $U = \sum_{i=1}^{n} w_i U_i(R^i)$.

9.3.2.2 Mixed-Criticality Scheduling

Traditional real-time scheduling theory considers all tasks as being equally critical and focuses on ensuring that the deadlines of all tasks are met. In reality, it is common to have real-time systems with different levels of criticality. For instance, an aircraft has some features that are critical to keep the plane flying—for example, automatic wing trimmings—and others that are less critical—such as object recognition in a surveillance plane. These variations can come from the uncertainty in the environment (e.g., larger number of objects to avoid) or from different degrees of pessimism required from certification practices for the worst-case execution time. For these cases, a common practice is to separate low-criticality tasks from high-criticality tasks, and to deploy these categories of tasks in different processors to prevent a low-criticality task from interfering with high-criticality tasks, while anticipating variations in the worst-case execution time.

In recent years, this separation has been implemented through the operating system to support temporal protection. This temporal protection is typically implemented as a symmetric temporal protection. That is, it prevents a low-criticality task from overrunning its CPU budget (C) and blocking high-criticality tasks from finishing before its deadline. At the same time, it prevents high-criticality tasks from overrunning their CPU budget so as to allow low-criticality tasks to complete by their deadlines. Unfortunately, in this latter case, a high-criticality task is delayed to allow a low-criticality task to run. When this situation, which is known as a criticality inversion, makes a higher-criticality task miss its deadline, we say that a criticality violation has occurred.

In [de Niz09], the authors propose a new scheduler called Zero-Slack Rate Monotonic (ZSRM) that provides temporal protection for mixed-criticality systems that do not suffer from criticality violations. In particular, ZSRM prevents low-criticality tasks from interfering with higher-criticality tasks, but allows the latter to steal CPU cycles from the former. This scheme is known as asymmetric temporal protection. ZSRM allows jobs to start executing with their rate-monotonic priorities even in the presence of criticality inversion, in what is called a rate-monotonic execution mode. A job then switches to a critical execution mode, stopping lower-criticality tasks at the last instant possible, so as to ensure that it will not miss its deadline (i.e., suffer a criticality violation). This last instant, which is known as the zero-slack instant of task τ_i (Z_i), is calculated using a modified version of the response-time test presented in Section 9.2.1.3 that combines the two modes of execution. Now, to accommodate variations in the worst-case execution time, tasks are given two execution budgets: the nominal execution budget (C_i), which is considered the worst-case execution time when the task does not overrun, and the overloaded execution budget (C_i^o), which is the worst-case execution time when it overruns. As a result, tasks in ZSRM are defined as follows:

$$\tau_i = (C_i,\ C_i^o, T_i, D_i, \zeta_i)$$

Here ζ_i represents the criticality of the task (following the convention that a lower number means higher criticality). With this task model, the response-time test is run for each task τ_i with a timeline that starts with the rate-monotonic execution mode (RM mode) and ends with the critical execution mode (critical mode). In each of the modes, a different set of interfering tasks is considered: In RM mode, the set of all higher-priority tasks (Γ_i^{rm}) is considered; in critical mode, only the set (Γ_i^c) of tasks with higher criticality (including the tasks with lower priority) is considered. Z_i is calculated by first executing the whole job $J_{i,k}$ from task τ_i in critical mode; it is fixed at the time by which the job needs to start so as to finish by its deadline when it suffers preemptions from tasks in Γ_i^c. Then, given Z_i, we calculate the slack (intervals of time where no task in Γ_i^{rm} runs) from that arrival of the job to Z_i. This slack is used to "move" part of the job to execute in the RM mode, moving Z_i toward the end of the period. This step is repeated until no more slack is discovered and Z_i is fixed. A sample response-time timeline and the Z_i are shown in Figure 9.10.

ZSRM provides the following task-based guarantee: A task τ_i is guaranteed to execute for C_i^o before its deadline (D_i) if no task $\tau_j \mid \zeta_j \leq \zeta_i$ executes for more than C_j. This guarantee also provides a graceful degradation guarantee ensuring that if deadlines are missed, they are missed in reversed order of criticality.

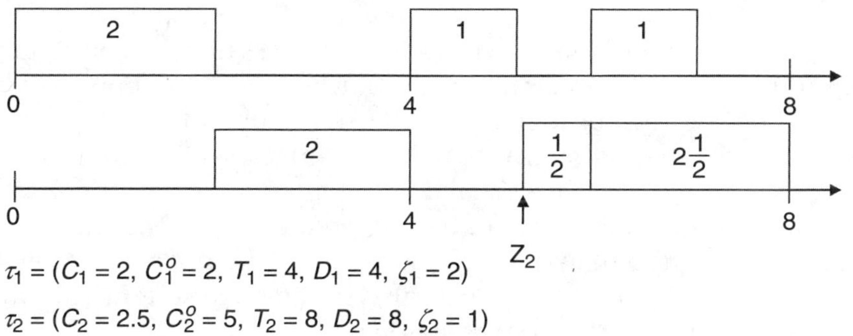

Figure 9.10: *ZSRM response-time timeline*

The ZSRM scheme has been extended to include synchronization protocols [Lakshmanan11], bin-packing algorithms for partitioned multiprocessor scheduling [Lakshmanan10], and utility-based allocation [de Niz13].

Other models of mixed-criticality focus on different levels of assurance for the worst-case execution time as the source of their variation [Baruah08, Vestal07]. In such models, tasks have different WCET for each of the criticality levels of the system and the schedulability of a task is evaluated based on its required level of criticality (which maps to a level of assurance). For such a model, [Baruah11] developed the Own Criticality-Based Priority to assign priorities to a mixed-criticality set of nonrecurrent jobs, later extending this scheme to sporadic tasks.

9.3.3 Managing Other Resources

Real-time scheduling techniques have been generalized for other types of resources so as to provide deterministic guarantees. In this section we discuss two important types: networks and power.

9.3.3.1 Network Scheduling and Bandwidth Allocation

Many systems (e.g., cars) are composed of computer nodes that jointly perform a function. For example, one computer node may take a sensor reading and then send that reading to another computer node; this second computer node may compute a command and actuate this command on the physical environment. The system repeats these sequences on an ongoing basis. For these kinds of systems, it is necessary to ensure that the time from the sensor reading in one node to the actuation in the other node is bounded and less than the so-called end-to-end deadline. To meet this demand, the time from when a message is requested to be transmitted until its transmission is finished must be bounded and known, and we must be able to compute an upper bound on this time. Traditionally, Ethernet with a shared medium (typically a coaxial cable) has been used to enable personal computers to communicate, but this technology has the drawback of having potentially unbounded delays because the arbitration to the medium is randomized. To overcome this problem, engineers of cyber-physical systems have sought communication technologies where the delays can be bounded.

One solution is to synchronize the clocks of all computer nodes and form a static schedule that specifies the time at which a computer node is allowed to transmit a message; this static schedule is then repeated indefinitely. Time-triggered Ethernet (TTP) and the static segment of FlexRay are examples of technologies that rely on this idea. Such technologies offer the advantage that once a schedule has been created, schedulability analysis is not needed, which makes the technology simpler to understand. They also offer fault detection with low latency and can provide very high bit rates. Unfortunately, they also have some drawbacks: They cannot efficiently guarantee timing requirements of sporadic tasks with very tight deadlines but long minimum inter-arrival times, and they tend to be less flexible in the sense that if the message characteristics (e.g., period) of one message stream change, or if a new computer node is attached, it is often necessary to recompute a new schedule and redeploy this schedule on all computer nodes.

Engineers of cyber-physical systems have sought alternative solutions for arbitrating access to the medium, with the controller area network (CAN) bus being one of the most widely deployed options. To date, hundreds of millions of CAN-enabled microcontrollers have been sold.

The CAN bus uses prioritized medium access. It works as follows: Each message is assigned a priority, and a computer node contends for medium access with the highest-priority message of the computer node. The CAN medium access control (MAC) protocol grants access to the computer node with the highest priority, with this message being transmitted. When the transmission is finished, the CAN MAC protocol grants access to the computer node with the highest priority; this message is then transmitted, much like before. This sequence is repeated as long as at least one computer node requests to transmit a message. Hence, one can analyze the timing of messages on a CAN bus by using the scheduling theory of fixed-priority scheduling (discussed earlier in this chapter) while taking the nonpreemptive aspects of communication into account [Davis07, Tindell94].

Because of its popularity, the CAN technology has been extended with new ideas. A wireless version of the CAN bus has been developed [Andersson05]. In addition, a new version of CAN uses much higher bit rate for data payload transmission than for the arbitration [Sheikh10].

9.3.3.2 Power Management

A number of real-time systems operate under power-related restrictions such as limited battery capacity or heat dissipation. In these systems it is important to minimize the power consumption. For this purpose, most processor chips today allow the adjustment of their operating frequency to take advantage of the fact that, in CMOS technology, the power is proportional to the square of the voltage multiplied by the frequency (i.e., $P \propto V^2 \cdot f$). Hence, given that both the frequency and the voltage are adjusted together in these processors, for each unit that the frequency is decreased we have a cubic decrease in their power consumption.

Taking advantage of this fact, the real-time community developed a set of real-time frequency scaling algorithms to minimize the frequency of the processor without missing task deadlines. In particular, [Saewong03] developed a set of algorithms for fixed-priority real-time systems that enables the designer to determine the minimum frequency of a processor to schedule a given taskset. Each algorithm targets different use cases as follows:

- *System clock frequency assignment (Sys-Clock):* The Sys-Clock algorithm is designed for processors for which the overhead of frequency switching is high and it is preferable to fix the frequency at design time.
- *Priority monotonic clock frequency assignment (PM-Clock):* The PM-Clock algorithm is designed for systems with low frequency-switching overhead. In this case, each task is assigned its own frequency.
- *Optimal clock frequency assignment (Opt-Clock):* The Opt-Clock algorithm is based on nonlinear optimization techniques to determine the optimal clock frequency for each task. This algorithm is complex and is recommended for use only at design time.
- *Dynamic PM-clock (DPM-Clock):* The DPM-Clock algorithm adapts to the difference between the average and worst execution times.

These algorithms assume that the tasks are scheduled under the deadline-monotonic scheduler, where a task τ_i is characterized by its period (T_i); its constrained deadline (D_i), which is less than or equal to the period; and its worst-case execution time (C_i). The frequency settings of the processors are normalized to the maximum frequency f_{max}.

The Sys-Clock algorithm finds the minimum frequency that allows all the tasks to meet their deadlines. In particular, this algorithm identifies the earliest completion time of a task and the latest completion time, which indicate the earliest completion at the maximum frequency and the latest instant before a preemption from a higher-priority task would make that task miss its deadline, respectively. The algorithm then calculates the minimum frequency that allows the task to complete exactly at the end of the idle period between the earliest and latest completion times.

Let us illustrate this algorithm with the taskset presented in Table 9.3. The execution timeline for these tasks is presented in Figure 9.11.

Figure 9.11 depicts the execution timeline with the processor running at the higher frequency. Following the Sys-Clock algorithm, we can reduce the frequency to move the execution of the taskset to the end of the idle times—in this example, 20 and 28. This work can be depicted as calculating the frequency scaling factor (α) necessary to extend the total workload of a task and its preempting tasks (β) from the start of the idle interval to the end. In particular, the current workload is calculated as follows:[1]

$$\beta = \frac{C_1 + C_2 + C_3}{f_{max}} = 15$$

$$\alpha = \frac{15}{20} = 0.75$$

The following expression calculates the workload for the second idle interval (from 27 to 28):

$$\beta = \frac{2C_1 + C_2 + C_3}{f_{max}} = 22$$

$$\alpha = \frac{22}{28} = 0.79$$

Table 9.3: *Sample Taskset*

Task	C_i	T_i	D_i
τ_1	7	20	20
τ_2	5	28	28
τ_3	3	30	30

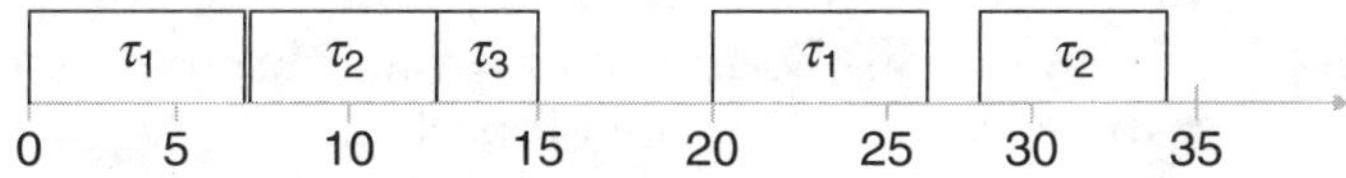

Figure 9.11: *Execution timeline of sample taskset*

1. Note that f_{max} is initially 1.

Selecting the minimum, we obtain 0.75. Once all frequencies for all tasks are calculated, the maximum among them is selected as the fixed frequency.

The PM-Clock algorithm is an adaptation of the Sys-Clock algorithm that starts by calculating the Sys-Clock frequencies for each task. If the frequency of a high-priority task provides more slack to lower-priority tasks, it allows the recalculation of the low-priority frequencies. In this case, every task runs at its own frequency, and the slack of the higher-priority tasks is translated into a lower frequency for the low-priority tasks.

The Opt-Clock algorithm defines an optimization problem over the hyper-period of the taskset to minimize the energy consumption, taking advantage of the convex nature of the power equation ($P(f) = cf^x,\ x \geq 3$).

Finally, the DPM-Clock algorithm is an online voltage scaling algorithm that further decreases the frequency of the processor when tasks complete earlier than anticipated. In particular, it uses the slack for the next ready task with a lower or the same priority than the task with the early completion.

[Aydin06] proposes a system-level energy management approach for real-time systems that take into account not only the frequency of the processor, but also the use of other devices that do not scale with this frequency. Specifically, the authors describe the execution time of a task as follows:

$$C_i(S) = \frac{x_i}{S} + y_i$$

Here x_i is the task work on the processor chip that scales with the selected speed S (corresponding to the frequency) and y_i is the off-chip work that does not scale with this speed. Based on this model, the authors propose an algorithm to derive the optimal static speed to minimize the energy consumption.

More recently, [Moreno12] developed an optimal algorithm for multiprocessors that can independently set their own frequency (uniform multiprocessors). These authors' approach takes advantage of the optimal scheduling algorithm for uniform multiprocessors developed by [Funk09] using a global scheduling approach known as U-LLREF. Using this algorithm, [Moreno12] developed the Growing Minimum Frequency (GMF) algorithm to calculate the minimum frequency at which each processor should run to meet the taskset deadlines.

The GMF algorithm tests the schedulability of a subset of tasks against a subset of the processors. Specifically, tasks are added one by one to the subset of tasks to be scheduled, in a non-increasing order of utilization. Similarly, the set of processors to be used to schedule the tasks are added one by one in a non-increasing order of processor speed (or frequency). The algorithm starts by assigning the minimum frequency to all processors and scheduling a taskset comprising one task on a multiprocessor comprising only one processor. As each task (one by one) is added to the tasks to be scheduled, processors are also added one by one. We then test whether the total utilization of the first i largest tasks is smaller or equal to the sum of the frequencies of the i fastest processors. If that is not the case, we increase the frequency of the slowest processor and repeat the previous step. Once we have considered all the available processors, if we have additional tasks, then we consider all the tasks together and repeat the frequency-increasing step. [Moreno12] proves the optimality of this approach for processors with discrete frequency steps where the steps are equal in size.

9.3.4 Rhythmic Tasks Scheduling

So far, we have seen various methods for proving that all timing requirements are met when tasks have fixed timing parameters. Mature theory has been developed for this scenario. More recently, researchers have explored theories for proving that timing requirements are met when the timing requirements vary as a function of the physical process. We will discuss this work in this section. Given that this area is not yet well developed, we will focus on scheduling on a single processor.

While a large number of algorithms that interact with the physical world, such as control algorithms, can be executed with a fixed period in the steady state, a growing number of algorithms cannot have a fixed period. As an example, consider the powertrain control in a car engine. The execution of a number of tasks is triggered when the crankshaft reaches a specific position. In this case, the interval between two such events varies according to the rpm of the engine. For this kind of system, [Kim12] developed the rhythmic real task model. This model is based on the rate-monotonic scheduling algorithm and is used to allow tasks to continually vary their periods.

For this model, [Kim12] defines tasks whose timing parameters vary according to a state vector $v_s \in R^k$, where k is the number of dimensions used to represent the system status. With this vector, the timing parameters of a task τ_i are defined as $C_i(v_s), T_i(v_s)$, $D_i(v_s)$ and we use $\tau_i^*, C_i^*, D_i^*, T_i^*$ to identify the rhythmic tasks. The variability in the timing parameters of rhythmic tasks is limited by a maximum acceleration parameter derived from period variability as follows:

$$\alpha(v_s) = 1 - \frac{T(v_s, J_{i+1})}{T(v_s, J_i)}$$

This value can be either positive or negative. Similarly, a maximum acceleration interval is defined in terms of the maximum number of job releases in the interval ($n_a(v_s)$).

[Kim12] presents a number of use cases and their analysis. First, these authors evaluate the case of having one rhythmic task and one periodic task. They then use this case to show the basic relationships between these two tasks in terms of schedulability. More specifically, the authors prove that for this case the taskset is schedulable if the following holds in each state:

$$C_1^* \leq \max\left(\frac{T_2 - C_2}{\left\lceil \frac{T_2}{T_1^*} \right\rceil}, T_1^* - \frac{C_2}{\left\lceil \frac{T_2}{T_1^*} \right\rceil} \right)$$

The [Kim12] paper presents a function parameterized with T_1^* to calculate the maximum WCET for a rhythmic task when scheduled with another set of periodic tasks to preserve the schedulability among other analyses. This is defined as follows:

$$f^*_{C_{max}}(T_1^*) = \min_{\forall \tau_i \in \Gamma} \max\left(T_1^* - \frac{\sum_{j=2}^{i} \left\lceil \frac{T_i}{T_j} \right\rceil C_j}{\left\lfloor \frac{T_i}{T_1^*} \right\rfloor}, \frac{T_i - \sum_{j=2}^{i} \left\lceil \frac{T_i}{T_j} \right\rceil C_j}{\left\lceil \frac{T_i}{T_1^*} \right\rceil} \right)$$

The authors also derive the maximum acceleration possible for a taskset with one rhythmic task and multiple periodic tasks for a maximum acceleration interval of $n_\alpha = 1$ as follows:

$$\alpha \leq 1 - \frac{T_1^*}{C_1^*}\left(UB(n) - \sum_{i-2}^{n} \frac{C_i}{T_i} \right)$$

UB(n) is the utilization bound as calculated by [Liu73].

The [Kim12] paper also presents a case study based on the powertrain control system to explore the schedulability conditions of the system.

9.4 Summary and Open Challenges

Ensuring that the cyber part executes in sync with the physical part of a CPS is the new challenge for complex CPS. In this chapter, we discussed the well-established foundations for the predictable execution of real-time systems based on techniques originating from the seminal paper [Liu73] that presented the rate-monotonic scheduling scheme. We presented generalizations including protocols to share resources, multiprocessor scheduling in multiple variants, and applications of these techniques to other types of resources including networks and power consumption.

The new generation of CPS presents a number of challenges that must be addressed. These challenges stem from different sources—specifically, from new hardware architectures that feature multicore processors and an increased variability in the timing parameters due to physical processes with increased dynamism that do not exhibit a traditional steady state.

New hardware architectures demand more complex resource allocation algorithms that can address predictability problems in commercial off-the-shelf (COTS) hardware, given that this hardware is designed to minimize the average-case execution time but makes the worst-case execution time highly unpredictable. Furthermore, hardware accelerators such as GPGPUs, DSP, or even just cores with different speeds create a new level of heterogeneity that must be managed. Finally, as the operating temperature of processors increases, thermal management comes into the picture and will become a more prominent concern in the future.

The increased dynamism of physical processes in new CPS demands new models of resource demands, with smart scheduling mechanisms to bound this dynamism and make it analyzable.

References

[Abel13] A. Abel, F. Benz, J. Doerfert, B. Dorr, S. Hahn, F. Haupenthal, M. Jacobs, A. H. Moin, J. Reineke, B. Schommer, and R. Wilhelm. "Impact of Resource Sharing on Performance and Performance Prediction: A Survey." CONCUR, 2013.

[Altmeyer11] S. Altmeyer, R. Davis, and C. Maiza. "Cache-Related Preemption Delay Aware Response Time Analysis for Fixed Priority Pre-emptive Systems." *Proceedings of the IEEE Real-Time Systems Symposium*, 2011.

[Andersson08] B. Andersson. "Global Static-Priority Preemptive Multiprocessor Scheduling with Utilization Bound 38%." In *Principles of Distributed Systems*, pages 73–88. Springer, 2008.

[Andersson01] B. Andersson, S. K. Baruah, and J. Jonsson. "Static-Priority Scheduling on Multiprocessors." Real-Time Systems Symposium, pages 193–202, 2001.

[Andersson08a] B. Andersson and K. Bletsas. "Sporadic Multiprocessor Scheduling with Few Preemptions." Euromicro Conference on Real-Time Systems, pages 243–252, 2008.

[Andersson12] B. Andersson, S. Chaki, D. de Niz, B. Dougherty, R. Kegley, and J. White. "Non-Preemptive Scheduling with History-Dependent Execution Time." Euromicro Conference on Real-Time Systems, 2012.

[Andersson10] B. Andersson, A. Easwaran, and J. Lee. "Finding an Upper Bound on the Increase in Execution Time Due to Contention on the Memory Bus in COTS-Based Multicore Systems." *SIGBED Review*, vol. 7, no. 1, page 4, 2010.

[Andersson03] B. Andersson and J. Jonsson. "The Utilization Bounds of Partitioned and Pfair Static-Priority Scheduling on Multiprocessors are 50%." Euromicro Conference on Real-Time Systems, pages 33–40, 2003.

[Andersson05] B. Andersson and E. Tovar. "Static-Priority Scheduling of Sporadic Messages on a Wireless Channel." In *Principles of Distributed Systems*, pages 322–333, 2005.

[Andersson06] B. Andersson and E. Tovar. "Multiprocessor Scheduling with Few Preemptions." *Real-Time Computing Systems and Applications*, pages 322–334, 2006.

[Audsley91] N. C. Audsley, A. Burns, M. F. Richardson, and A. J. Wellings. "Hard Real-Time Scheduling: The Deadline Monotonic Approach." *Proceedings of the 8th IEEE Workshop on Real-Time Operating Systems and Software*, 1991.

[Aydin06] H. Aydin, V. Devadas, and D. Zhu. "System-Level Energy Management for Periodic Real-Time Tasks." Real-Time Systems Symposium, 2006.

[Baruah96] S. K. Baruah, N. K. Cohen, C. G. Plaxton, and D. A. Varvel. "Proportionate Progress: A Notion of Fairness in Resource Allocation." *Algorithmica*, vol. 15, no. 6, pages 600–625, 1996.

[Baruah11] S. Baruah, H. Li, and L. Stougie. "Towards the Design of Certifiable Mixed-Criticality Systems." 16th IEEE Real-Time and Embedded Technology and Applications Symposium (RTAS), 2010.

[Baruah08] S. Baruah and S. Vestal. "Schedulability Analysis of Sporadic Tasks with Multiple Criticality Specifications." European Conference on Real-Time Systems, 2008.

[Chattopadhyay11] S. Chattopadhyay and A. Roychoudhury. "Scalable and Precise Refinement of Cache Timing Analysis via Model Checking." *Proceedings of the 32nd IEEE Real-Time Systems Symposium*, 2011.

[Davis11] R. I. Davis and A. Burns. "A Survey of Hard Real-Time Scheduling for Multiprocessor Systems." *ACM Computing Surveys*, vol. 43, no. 4, Article 35, October 2011.

[Davis07] R. I. Davis, A. Burns, R. J. Bril, and J. J. Lukkien. "Controller Area Network (CAN) Schedulability Analysis: Refuted, Revisited and Revised." *Real-Time Systems*, vol. 35, no. 3, pages 239–272, 2007.

[de Niz09] D. de Niz, K. Lakshmanan, and R. Rajkumar. "On the Scheduling of Mixed-Criticality Real-Time Tasksets." *Proceedings of the 30th IEEE Real-Time Systems Symposium*, 2009.

[de Niz13] D. de Niz, L. Wrage, A. Rowe, and R. Rajkumar. "Utility-Based Resource Overbooking for Cyber-Physical Systems." IEEE International Conference on Embedded and Real-Time Computing Systems and Applications, 2013.

[Dhall78] S. K. Dhall, and C. L. Liu. "On a Real-Time Scheduling Problem." *Operations Research*, vol. 26, pages 127–140, 1978.

[Funk09] S. Funk and A. Meka. "U-LLREF: An Optimal Real-Time Scheduling Algorithm for Uniform Multiprocessors." Workshop of Models and Algorithms for Planning and Scheduling Problems, June 2009.

[Hahn02] J. Hahn, R. Ha, S. L. Min, and J. W. Liu. "Analysis of Worst-Case DMA Response Time in a Fixed-Priority Bus Arbitration Protocol." *Real-Time Systems*, vol. 23, no. 3, pages 209–238, 2002.

[Hansen09] J. P. Hansen, S. A. Hissam, and G. A. Moreno. "Statistical-Based WCET Estimation and Validation." Workshop on Worst-Case Execution Time, 2009.

[Huang97] T. Huang. "Worst-Case Timing Analysis of Concurrently Executing DMA I/O and Programs." PhD thesis, University of Illinois at Urbana–Champaign, 1997.

[Joseph86] M. Joseph and P. Pandya. "Finding Response Times in a Real-Time System." *Computer Journal*, vol. 29, no. 5, pages 390–395, 1986.

[Kato07] S. Kato and N. Yamasaki. "Real-Time Scheduling with Task Splitting on Multiprocessors." *Real-Time Computing Systems and Applications*, pages 441–450, 2007.

[Kato09] S. Kato, N. Yamasaki, and Y. Ishikawa. "Semi-Partitioned Scheduling of Sporadic Task Systems on Multiprocessors." Euromicro Conference on Real-Time Systems, pages 249–258, 2009.

[Khemka92] A. Khemka and R. K. Shyamasundar. "Multiprocessor Scheduling of Periodic Tasks in a Hard Real-Time Environment." International Parallel Processing Symposium, 1992.

[Kim14] H. Kim, D. de Niz, B. Andersson, M. Klein, O. Mutlu, and R. Rajkumar. "Bounding Memory Interference Delay in COTS-Based Multi-Core Systems." Real-Time and Embedded Technology and Applications Symposium, 2014.

[Kim12] J. Kim, K. Lakshmanan, and R. Rajkumar. "Rhythmic Tasks: A New Task Model with Continually Varying Periods for Cyber-Physical Systems." IEEE/ACM International Conference on Cyber-Physical Systems, 2012.

[Klein93] M. Klein, T. Ralya, B. Pollak, R. Obenza, and M. G. Harbour. *A Practitioner's Handbook for Real-Time Analysis: Guide to Rate Monotonic Analysis for Real-Time Systems*. Kluwer Academic, 1993.

[Lakshmanan11] K. Lakshmanan, D. de Niz, and R. Rajkumar. "Mixed-Criticality Task Synchronization in Zero-Slack Scheduling." 17th IEEE Real-Time and Embedded Technology and Applications Symposium (RTAS), 2011.

[Lakshmanan10] K. Lakshmanan, D. de Niz, R. Rajkumar, and G. Moreno. "Resource Allocation in Distributed Mixed-Criticality Cyber-Physical Systems." IEEE 30th International Conference on Distributed Computing Systems, 2010.

[Lakshmanan09] K. Lakshmanan, R. Rajkumar, and J. P. Lehoczky. "Partitioned Fixed-Priority Preemptive Scheduling for Multi-Core Processors." Euromicro Conference on Real-Time Systems, pages 239–248, 2009.

[Leung82] J. Y. T. Leung and J. Whitehead. "On the Complexity of Fixed-Priority Scheduling of Periodic, Real-Time Tasks." *Performance Evaluation*, vol. 2, no. 4, pages 237–250, 1982.

[Levin10] G. Levin, S. Funk, C. Sadowski, I. Pye, and S. A. Brandt. "DP-FAIR: A Simple Model for Understanding Optimal Multiprocessor Scheduling." Euromicro Conference on Real-Time Systems, pages 3–13, 2010.

[Liu73] C. L. Liu and J. W. Layland. "Scheduling Algorithms for Multiprogramming in Hard-Real-Time Environment." *Journal of the ACM*, pages 46–61, January 1973.

[Locke90] D. Locke and J. B. Goodenough. *Generic Avionics Software Specification*. SEI Technical Report, CMU/SEI-90-TR-008, December 1990.

[Lundberg02] L. Lundberg. "Analyzing Fixed-Priority Global Multiprocessor Scheduling." IEEE Real Time Technology and Applications Symposium, pages 145–153, 2002.

[Nan10] M. Lv, G. Nan, W. Yi, and G. Yu. "Combining Abstract Interpretation with Model Checking for Timing Analysis of Multicore Software." Real-Time Systems Symposium, 2010.

[Moreno12] G. Moreno and D. de Niz. "An Optimal Real-Time Voltage and Frequency Scaling for Uniform Multiprocessors." IEEE International Conference on Embedded and Real-Time Computing Systems and Applications, 2012.

[Oh98] D. Oh and T. P. Baker. "Utilization Bounds for *N*-Processor Rate Monotone Scheduling with Static Processor Assignment." *Real-Time Systems*, vol. 15, no. 2, pages 183–192, 1998.

[Pellizzoni10] R. Pellizzoni, A. Schranzhofer, J. Chen, M. Caccamo, and L. Thiele. "Worst Case Delay Analysis for Memory Interference in Multicore Systems." Design, Automation, and Test in Europe (DATE), 2010.

[Rajkumar91] R. Rajkumar. *Synchronization in Real-Time Systems: A Priority Inheritance Approach*. Springer, 1991.

[Rajkumar97] R. Rajkumar, C. Lee, J. Lehoczky, and D. Siewiorek. "A Resource Allocation Model for QoS Management." Real-Time Systems Symposium, 1997.

[Regnier11] P. Regnier, G. Lima, E. Massa, G. Levin, and S. A. Brandt. "RUN: Optimal Multiprocessor Real-Time Scheduling via Reduction to Uniprocessor." Real-Time Systems Symposium, pages 104–115, 2011.

[Saewong03] S. Saewong and R. Rajkumar. "Practical Voltage-Scaling for Fixed Priority RT Systems." Real-Time and Embedded Technology and Applications Symposium, 2003.

[Schliecker10] S. Schliecker, M. Negrean, and R. Ernst. "Bounding the Shared Resource Load for the Performance Analysis of Multiprocessor Systems." Design, Automation, and Test in Europe (DATE), 2010.

[Sha90] L. Sha, R. Rajkumar, and J. P. Lehoczky. "Priority Inheritance Protocols: An Approach to Real-Time Synchronization." IEEE Transactions on Computers, vol. 39, no. 9, pages 1175–1185, 1990.

[Sheikh10] I. Sheikh, M. Short, and K. Yahya. "Analysis of Overclocked Controller Area Network." *Proceedings of the 7th IEEE International Conference on Networked Sensing Systems (INSS)*, pages 37–40, Kassel, Germany, June 2010.

[Srinivasan02a] A. Srinivasan and J. H. Anderson. "Optimal Rate-Based Scheduling on Multiprocessors." ACM Symposium on Theory of Computing, pages 189–198, 2002.

[Srinivasan02] A. Srinivasan and S. K. Baruah. "Deadline-Based Scheduling of Periodic Task Systems on Multiprocessors." *Information Processing Letters*, vol. 84, no. 2, pages 93–98, 2002.

[Suzuki13] N. Suzuki, H. Kim, D. de Niz, B. Andersson, L. Wrage, M. Klein, and R. Rajkumar. "Coordinated Bank and Cache Coloring for Temporal Protection of Memory Accesses." IEEE Conference on Embedded Software and Systems, 2013.

[Tindell94] K. Tindell, H. Hansson, and A. J. Wellings. "Analysing Real-Time Communications: Controller Area Network (CAN)." Real-Time Systems Symposium, pages 259–263, 1994.

[Vestal07] S. Vestal. "Preemptive Scheduling of Multi-Criticality Systems with Varying Degrees of Execution Time Assurance." *Proceedings of the IEEE Real-Time Systems Symposium*, 2007.

[Wilhelm08] R. Wilhelm, J. Engblom, A. Ermedahl, N. Holsti, S. Thesing, D. B. Whalley, G. Bernat, C. Ferdinand, R. Heckmann, T. Mitra, F. Mueller, I. Puaut, P. P. Puschner, J. Staschulat, and P. Stenström. "The Worst-Case Execution-Time Problem: Overview of Methods and Survey of Tools." *ACM Transactions on Embedded Computing Systems*, vol. 7, no. 3, 2008.

[Zhu03] D. Zhu, D. Mossé, and R. G. Melhem. "Multiple-Resource Periodic Scheduling Problem: How Much Fairness Is Necessary?" Real-Time Systems Symposium, pages 142–151, 2003.

Chapter 10

Model Integration in Cyber-Physical Systems

Gabor Simko and Janos Sztipanovits

Composition of heterogeneous physical, computational, and communication systems is an important challenge in the engineering of cyber-physical systems (CPS). In model-based design of CPS, heterogeneity is reflected in fundamental differences in the semantics of the models and modeling languages—in causality, time, and physical abstractions. The formalization of CPS modeling language semantics requires understanding the interaction between heterogeneous models. The semantic differences among the integrated models may result in unforeseen problems that surface during system integration, leading to significant cost and time delays. In this chapter, we focus on the composition semantics of heterogeneous models. We discuss the challenges in the model-based design of CPS, and demonstrate the formalization of a complex CPS model integration language. In the case study, we develop both the structural and behavioral semantics for the language using an executable, logic-based specification language.

10.1 Introduction and Motivation

One of the key challenges in the engineering of CPS is the integration of heterogeneous concepts, tools, and languages [Sztipanovits2012]. To address these challenges, a model-integrated development approach for CPS design was introduced by [Karsai2008], which advocates the pervasive use of models throughout the design process, such as application models, platform models, physical system models, environment models, and the interaction models between these models. For embedded systems, a similar approach is discussed in [Sztipanovits1997] and [Karsai2003], in which both the computational processes and the supporting architecture (hardware platform, physical architecture, operating environment) are modeled within a common modeling framework.

It might be argued that modeling the vastly diverse set of CPS concepts using a single generic modeling language is impractical. As an alternative, we can adopt a set of domain-specific modeling languages (DSMLs)—each of which describes a particular domain—and interconnect them through a model-integration language. This strategy is well aligned with the content creation approach: Models are created by domain experts, who are used to their own domain-specific concepts and terminology. To facilitate their work, we need to provide an environment (such as the Generic Modeling Environment [Ledeczi2001]) that allows them to work with these domain-specific languages. In the CPS domain, typical examples of DSMLs include circuit diagrams for electrical engineers and data-flow diagrams for control engineers.

An important question in CPS modeling is how to model the integration of the individual domains, such that the cross-domain interactions are properly accounted for [Sztipanovits2012]. A recently proposed solution [Lattmann2012, Simko2012, Wrenn2012] uses a model-integration DSML for describing model interactions. Such a model-integration language is built on the paradigm of component-based modeling, wherein complex systems are built from interconnected components. From the model integration point of view, the most important part of a component is its interface—that is, the means by which it interacts with its environment. The role of the model-integration language then becomes to unambiguously define the heterogeneous interactions (e.g., message passing, variable sharing, function call or physical interactions) between the interconnected interfaces.

Like any other language, a DSML is defined by its syntax and semantics: The syntax describes the structure of the language (e.g., syntactic elements and their relations), and the semantics describes the meaning of its models. While meta-modeling (and meta-modeling environments) provides a mature methodology for tackling the DSML syntax, efforts to express the semantics of DSML remain in their infancy. Nonetheless, the semantics of a language should not be taken lightly, especially not in the CPS domain. Without unambiguous specifications, different tools may interpret the language in different ways, which could easily lead to situations when the compiler generates code with different behavior than what the verification tool analyzes. This sort of discrepancy might potentially render the results of the formal analysis and verification tools invalid. Furthermore, developing the formal semantics of a language requires the developer to ponder and clearly specify all the details of the language, so as to avoid common design errors in the language. To support the development of a CPS DSML, along with tools operating on that language, it is highly recommended that both the syntax and the semantics of the language be rigorously defined and formalized. Of course, even with formal semantic specifications, a tool may be faulty with respect to these specifications. Nevertheless, as long as the specifications are unambiguous, this problem resides with the tool, rather than being a problem with the language.

The primary focus of this chapter is the generalization of CPS models' behavior, and the demonstration of a logic-based approach for the specification of their semantics. The chapter first describes the challenges related to the model-based design of CPS. These challenges are related to the heterogeneous behavior of CPS, so they directly affect the behavioral semantic specifications. Later in the chapter, we explore modeling languages and their semantics. We then discuss the formalization of the structural and denotational semantics of a CPS modeling language.

10.2 Basic Techniques

For CPS modeling, we need unambiguous semantic specifications for the modeling languages. A key challenge in developing these specifications is found in the heterogeneity of the behaviors represented by the languages; these behaviors range from untimed discrete computations to trajectories over continuous time and space. In this section we present the basic techniques to deal with these issues. In particular, in CPS modeling, we can distinguish four fundamental dimensions of heterogeneity that affect the semantics of the languages:

- Causality considerations (causal versus acausal relations)
- Time semantics (e.g., continuous time, discrete time)
- Physical domains (e.g., electrical, mechanical, thermal, acoustic, hydraulic)
- Interaction models (e.g., model of computation–based, interaction algebra)

After discussing each of these dimensions, we consider the semantics of a domain specific CPS modeling language

10.2.1 Causality

Classical system theory and control theory are traditionally based on input/output signal flows, where causality—the fact that the inputs of a system determine its outputs—plays a key role. For physical systems modeling, however, such a causal model is artificial and inapplicable [Willems2007], because the separation of inputs and outputs is generally unknown at the time of modeling. The problem stems from the mathematical models of physical laws and systems: These models are based on equations, and there are no causal relationships between the involved variables. Indeed, the only causal law in physics is the second law of thermodynamics [Balmer1990], which defines the direction in which the time flows. Recently, acausal[1] physical modeling has gained traction. Indeed, several acausal physical systems modeling languages have been designed, such as bond graph formalism [Karnopp2012], Modelica [Fritzson1998, Modelica2012], Simscape [Simscape2014], EcosimPro [Ecosimpro2014], and others.

Even though the mathematical models for the laws of physics are acausal, we often rely on their causal abstraction. For example, operational amplifiers are often abstracted as input/output systems, such that the output is solely determined by the system's input and has no feedback effects on the input. For real operational amplifiers, these assumptions are only approximations. Nevertheless, they are reasonably accurate in some situations and greatly simplify the design process.

---

1. Acausal relationships between two variables establish a condition but do not force an assignment. For example, the relationship $A = B$ causes an assignment of the value of B into A in a causal model but not in an acausal one, as it validates only that all assignments to A and B make the condition true.

Physical modeling languages, then, usually support both causal and acausal modeling. Clearly, this is the case with all state-of-the-art modelers and languages (e.g., Simulink/Simscape, Modelica, bond graph modeling with modulated elements [Karnopp2012]). Consequently, our behavioral semantic specifications need to be able to describe both acausal and causal models.

10.2.2 Semantic Domains for Time

In software design, one of the most powerful abstraction is the abstraction of time [Lee2009]. Such an abstraction is harmless as long as time is a nonfunctional property of programs. However, in CPS and real-time systems, this premise is often invalid [Lee2008], and timing is a key concept from a functional point of view of the system. For example, in hard real-time systems, results are often worthless or—in the worst case—even catastrophic, if not delivered in time.

Thus, in CPS, we have to properly account for the timely behavior of models. It is known that the behaviors of CPS models are interpreted over a diverse set of semantic domains for time. Some of these semantic domains are summarized here:

- *Logical time:* This time model is used for computational processes, whose behaviors are described by sequences of states. Logical time defines a countable totally ordered time domain, and is typically represented by the natural numbers.
- *Continuous time:* Also known as real time or physical time, this dense time model is often represented by the non-negative real numbers (where zero represents the start of the system) or the non-negative hyper-real numbers. Physical systems are usually considered to have their behavior in the continuous-time regime.
- *Super-dense time:* This time model extends real time with causal ordering for simultaneous events [Maler1992]. Super-dense time is usually used for describing the behavior of discrete-event systems [Lee2007].
- *Discrete time:* This time model denotes a finite number of events on any finite time interval. Therefore, the events can be indexed by the natural numbers. For instance, clock-driven systems have discrete-time semantics.
- *Hyper-dense time:* This time model extends hyper-real time with causal ordering for simultaneous events [Mosterman2013, Mosterman2014]. Hyper-dense time is used for describing discontinuities in physical systems' behavior.

In real-time systems, time is a functional property and we need to explicitly account for the execution times of the software. However, on modern architectures, the exact execution times cannot be accurately predicted because of caching, pipelining, and other advanced techniques. To resolve this problem, several abstractions of time have been proposed [Kirsch2012], such as zero execution time, bounded execution time, and logical execution time:

- *Zero execution time (ZET):* This model abstracts away time by assuming that the execution time is zero—in other words, the computation is infinitely fast. ZET is the basis for synchronous reactive programming.
- *Bounded execution time (BET):* In this model, execution time has an upper bound. A program execution is correct as long as the output is produced within this temporal bound. Note that this model is not really an abstraction, but rather a specification that can be verified for correctness.
- *Logical execution time (LET):* This model abstracts away the real execution time but does not discard it completely, as ZET does [Henzinger2001]. LET denotes the time it takes from reading the input to producing the output, regardless of the real execution time. Unlike BET, the LET abstraction also defines a lower bound equal to the upper bound; that is, LET precisely defines the time when the output is produced. Such an abstraction has a significant impact on the complexity of the behavior. For example, it protects against timing anomalies [Reineke2006] (i.e., when a faster local execution leads to slower global execution).

Timed automata are introduced as a time model for real-time systems by [Abdellatif2010]. Compared to LET semantics, timed automata provide more generic constraints, such as lower time bounds, upper time bounds, and time nondeterminism. To avoid timing anomalies, the authors introduce the notion of time robustness, along with some sufficient conditions to guarantee it.

10.2.3 Interaction Models for Computational Processes

When modeling heterogeneous computational systems, a key question is how to define the interactions between different subsystems. We can distinguish two approaches here.

The interactions may be modeled with models of computation (MoCs)—for example, in Ptolemy II [Goderis2007, Goderis2009]. Ptolemy II is based on the hierarchical composition of various MoCs, such as process networks, dynamic and synchronous dataflows, discrete-event systems, synchronous-reactive systems, finite-state machines, and modal models. Each actor in the hierarchy has a director—a model of computation—that determines the interaction model between its children. The actor abstract semantics [Lee2007] of Ptolemy II is a common abstraction of all MoCs in Ptolemy II, which defines an interface with action methods and method contracts. An actor that conforms to these contracts is said to be domain polymorphic, and the actor can be used by any director operating on actors.

Alternatively, the interactions may be modeled using algebras—for example, in BIP [Basu2006] framework. In BIP, connector and interaction algebras [Bliudze2008] are defined for characterizing different interaction types, such as rendezvous, broadcast, or atomic broadcast. In the BIP framework, the behaviors of components are abstracted as a labeled transition system (LTS), and the interaction algebra establishes the relations between these transition systems.

10.2.4 Semantics of CPS DSMLs

In [Chen2007], a (modeling) language is defined as a tuple of $\langle C, A, S, M_A, M_S \rangle$, where C is the concrete syntax, A is the abstract syntax, S is the semantic domain, and two mappings $M_A : C \rightarrow A$ and $M_S : A \rightarrow S$ map from the concrete syntax to the abstract syntax, and from the abstract syntax to the semantic domain, respectively.

The concrete syntax of a language is the concrete representation that is used for representing programs/models. Traditional programming languages are usually text based, whereas modeling languages often have visual representations. A language describes a set of concepts and relations between these concepts, which is represented by its abstract syntax. The syntactic mapping M_A maps elements of the concrete syntax to corresponding elements of the abstract syntax. Finally, the language's semantics defines the meaning of models by means of semantic mapping(s) M_S from the concepts and relations of the abstract syntax to some semantic domain(s).

To discuss existing work in language semantics, first we need to understand the syntax and semantics of programming languages. The concrete syntax of a programming language is usually described with some context-free grammar, typically in the (extended) Backus-Naur form (BNF) [Backus1959]. Such a grammar describes production rules, which can be used to build parse trees from the source code. The abstract syntax tree of a program is an abstract version of the parse tree, which typically removes some parser-specific details from the parse tree. The static semantics of a programming language describes those properties of its programs that are statically (without executing the program) computable; this is also called the well-formedness rules of the language. Typically, this corresponds to context-sensitive parts of the grammar that are not expressible with a context-free grammar, such as unique naming of variables, static type checking, and scoping. Finally, the dynamic semantics of a language describes its dynamic aspects, meaning the sequences of computations described by its programs.

In contrast, the syntaxes of domain-specific modeling languages are typically described with meta-models. Generally, meta-models describe graph structures, and models are represented using abstract syntax graphs. Furthermore, instead of static semantics, models have structural semantics [Jackson2009]. In [Chen2005], structural semantics is defined as the meaning of a model in terms of the structure of the model instances. Similar to static semantics, structural semantics describes the well-formedness rules of a language; however, structural semantics is not necessarily static. In model-based design, models may represent dynamic structures that evolve through model transformations, in which case the structural semantics describes invariants for these transformations. The dynamic behavior of a model is described by its behavioral semantics. Note that the behaviors represented by modeling languages are generally interpreted on different semantic domains than the domains for programming languages. For example, models can represent physical systems, for which the behaviors are trajectories over continuous time and space.

Throughout the rest of this chapter, we will concentrate on the denotational behavioral semantics of modeling languages. Denotational semantics describes the language semantics by mapping its phrases to mathematical objects, such as numbers, tuples, functions, and so on. An advantage of denotational semantics is that it provides mathematically rigorous specifications for programs without specifying any computational procedures for calculating the results. This approach results in abstract specifications that describe what programs (models) do, instead of describing how they do it.

10.3 Advanced Techniques

In this section, we discuss the formalization of the Cyber-Physical Systems Modeling Language (CyPhyML), a model integration language for heterogeneous CPS components. CyPhyML encompasses several sub-languages, such as a language for describing the composition of CPS components, a language for describing design spaces with multiple choices, and so on. In the following discussion, we consider only the composition sub-language; when we use the term "CyPhyML," we mean to refer to this language.

10.3.1 ForSpec

For the specifications, we use the ForSpec specification language, which is an extension of the Microsoft FORMULA [Jackson2010, Jackson2011] language that includes constructs for the specification of behavioral semantics. ForSpec is a constraint logic language over algebraic data types for writing formal specifications.

In this section, we provide a short overview of the language. For more information, see the language's website at https://bitbucket.org/gsimko/forspec as well as the documentation of FORMULA.

The *domain* keyword specifies a domain (analogous to a meta-model) that is composed of type definitions, data constructors, and rules. A model of the domain consists of a set of facts (also called initial knowledge) that are defined using the data constructors of the domain. The well-formed models of the domain are defined with the *conforms* rules. Given a model, if the constraints of the conforms rules are satisfied by the least fixed-point model obtained from the initial set of facts by applying the domain rules until a fixed point is reached, the model is said to conform to its domain.

ForSpec has a complex type system based on built-in types (`Natural`, `Integer`, `Real`, `String`, and `Bool`), enumerations, data constructors, and union types. Enumerations are sets of constants defined by enumerating all their elements. For example, `bool ::= {true,false}` denotes the usual two-valued Boolean type.

Union types are unions of types in the set-theoretical sense; that is, the elements of a union type are defined by the union of the elements of the constituent types. Union types are defined using the notation `T ::= Natural + Integer`, which defines type `T` as the union of natural and integer numbers; that is, type `T` denotes the integer numbers.

Data constructors can be used for constructing algebraic data types. Such terms can represent sets, relations, partial and total functions, injections, surjections, and bijections. Consider the following type definitions:

```
A::= new (x: Integer, y:String).
B::= fun (x: Integer, -> y:String).
C::= fun (x: Integer => y:A).
D::= inj (x: Integer -> y: String).
E::= bij (x:A => y: B).
F:: = Integer + G.
G::= new (Integer, any F).
H::= (x: Integer, y: String).
```

Data type `A` defines A-terms by pairing integers and strings (and `A` also stands for the data constructor for this type; for example, `A(5,"f")` is an A-term), where the optional `x` and `y` are the accessors for the respective values. Data type `B` defines a partial function (functional relation) from the domain of `Integers` to the codomain of `Strings`. Similarly, `C` defines a total function from strings to A-terms, `D` defines a partial injective function, and `E` defines a bijective function between A-terms and B-terms. Type `F` is the union of the integers and `G`, where `G` is a data type composed of an integer and an `F` term. Note that `F` and `G` are mutually dependent, which would cause an error message during static type checking. To avoid the error message, we use the `any` keyword in the definition of `G`.

While the previous data types (and constructors) are used for defining the initial facts in models, derived data types are used for representing facts derived from the initial knowledge by means of rules. For example, the derived data type `H` defines a term over pairs of integers and strings.

Set comprehensions are defined using the form `{head|body}`, which denotes the set of elements formed by `head` that satisfies `body`. Set comprehensions are used by built-in operators such as `count` or `toList`. For instance, given a relation `Pair ::= new (State,State)`, the expression `State(X), n = count(Y|Pair(X,Y))` counts the number of states paired with state `X`.

Rules are used for deducing constants and derived data types. They have the form (where $X' \subseteq X$):

$$A_0(X') \text{ :- } A_1(X)\,, \cdots, A_n(X)\,, \text{no } B_1(X), \cdots, \text{no } \; B_m(X)$$

Whenever there is a substitution for X, where all the ground terms $A_1(X), \cdots, A_n(X)$ are derivable and none of $B_1(X), \cdots, B_m(X)$ is derivable, then $A_0(X')$ is derivable. The use of negation (no) is stratified, which implies that rules generate a unique minimal set of derivations; that is, there is a least fixed-point model.

For example, we can calculate all the paths between nodes in a graph with the following specifications:

```
// node is a type that consists of an integer
node :: new (Integer).
// edges are formed from pairs of nodes
edge ::= new (node, node).
// path is a derived data type generated by the following rules:
path ::= (node, node).
path(X,Y) :- edge(X,Y).
path(X,Y) := path(X,Z), edge(Z,Y).
```

To help write multiple rules with the same left-hand term, the semi-colon operator is used: its meaning is logical disjunction. For instance, in *A*(*X*) :- *S*(*X*); *T*(*X*), any substitution for *X*, such that *S*(*X*) or *T*(*X*) is derivable, makes *A*(*X*) derivable.

Type constraint `x:A` is true if and only if variable `x` is of type `A`, while `x` is `A` is satisfied for all derivations of type `A`. Similarly, `A(x,"a")` is a type constraint, which is satisfied by all substitutions for variable *x* such that the resulting ground term is a member of the knowledge set (note that the second subterm is already a ground term in this example). Besides type constraints, ForSpec supports relational constraints, such as equality of ground terms, and arithmetic predicates (e.g., less than, greater equal) over reals and integers. The special symbol _ denotes an anonymous variable, which cannot be referred to elsewhere.

The well-formed models of a domain conform to the domain specifications. The conforms rules of a domain determine its well-formed models.

The language elements discussed so far are borrowed from FORMULA. In addition to these concepts, ForSpec extends the language with structures for expressing denotational semantic specifications. The most important extensions are the definition of semantic functions and semantic equations, as shown in the following example:

```
add ::= new (lhs:expr, rhs:expr).
expr ::= add + …
S : expr -> Integer.
S [[add]] = summa
where summa = S ([[add.lhs]] + S [[add.rhs]].
```

This defines a semantic function `S` that maps expressions to integers. The semantic equation for the `add` operator says that given an add term in the model, its semantics is an integer `summa` that is the summation of the semantics of the left-hand side (`lhs`) and the right-hand side (`rhs`) of the add term.

Under the hood, ForSpec generates auxiliary data types and rules for the execution of such specifications, which can be converted to the equivalent FORMULA specifications. For the exact semantics of semantic functions and equations, we refer the reader to the website of ForSpec.

Note that ForSpec has a printing tool that will convert the previous example to the following form:

```
add ::= new ( lhs:expr, rhs: expr).
expr ::= add+ …
S : expr -> Integer.
S⟦add⟧ = summa
where summa = S⟦add.lhs⟧ + S⟦add.rhs⟧.
```

10.3.2 The Syntax of CyPhyML

The GME meta-model [Ledeczi2001] of CyPhyML is shown in Figure 10.1. The main building blocks of CyPhyML are the components, which represent physical or computational elements with ports on their interfaces. Component assemblies are used for building composite structures by composing components and other component assemblies. Component assemblies also facilitate encapsulation and port hiding. Two types of ports are used in CyPhyML: acausal power ports, for representing physical interaction points, and causal signal ports, for representing information flow between components. Both the physical and information flows

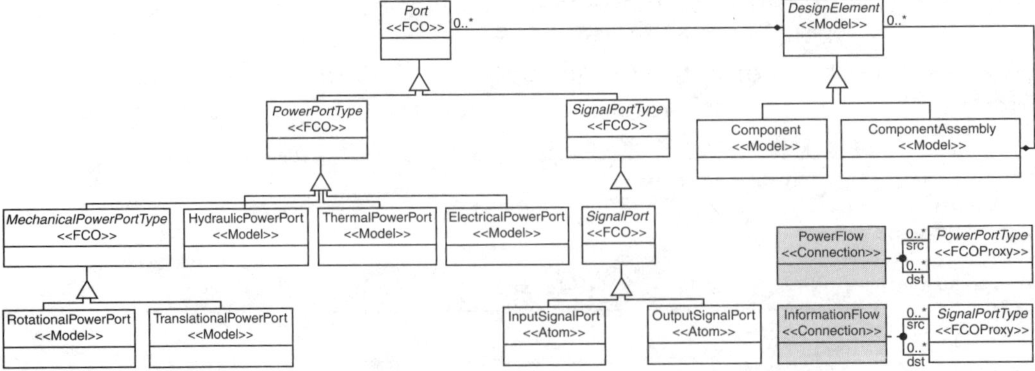

Figure 10.1: *GME meta-model for the composition sub-language of CyPhyML*

are interpreted over the continuous-time domain. CyPhyML distinguishes power ports by type, such as electrical power ports, mechanical power ports, hydraulic power ports, and thermal power ports.

Formally, a CyPhyML model M is a tuple $M = \langle C, A, P, contain, portOf, E_P, E_S \rangle$ with the following interpretation:

- C is a set of components.
- A is a set of component assemblies.
- $D = C \cup A$ is the set of design elements.
- P is the union of the following sets of ports: $P_{rotMech}$ is a set of rotational mechanical power ports, $P_{transMech}$ is a set of translational mechanical power ports, $P_{multibody}$ is a set of multibody power ports, $P_{hydraulic}$ is a set of hydraulic power ports, $P_{thermal}$ is a set of thermal power ports, $P_{electrical}$ is a set of electrical power ports, P_{in} is a set of continuous-time input signal ports, and P_{out} is a set of continuous-time output signal ports. Furthermore, P_P is the union of all the power ports and P_S is the union of all the signal ports.
- contain: $D \to A^*$ is a containment function, whose range is $A^* = A \cup \{\text{root}\}$, the set of design elements extended with a special root element root.
- portOf: $P \to D$ is a port containment function, which uniquely determines the container of any port.
- $E_P \subseteq P_P \times P_P$ is the set of power flow connections between power ports.
- $E_S \subseteq P_S \times P_S$ is the set of information flow connections between signal ports.

We can formalize this language using the following algebraic data types:

```
// Components, component assemblies, and design elements
Component ::= new (name: String, …, id:Integer).
ComponentAssembly ::= new (name: String, …, id:Integer).
DesignElement ::= Component + ComponentAssembly.
// Components of a component assembly
ComponentAssemblyToCompositionContainment ::=
   (src:ComponentAssembly, dst:DesignElement).
// Power ports
TranslationalPowerPort ::= new (…, id:Integer).
RotationalPowerPort ::= new (…, id:Integer).
ThermalPowerPort ::= new (…, id:Integer).
HydraulicPowerPort ::= new (…, id:Integer).
ElectricalPowerPort ::= new (…, id:Integer).
```

continued

```
// Signal ports
InputSignalPort ::= new (…, id:Integer).
OutputSignalPort ::= new (…, id:Integer).
// Ports of a design element
DesignElementToPortContainment ::= new (src:DesignElement, dst:Port).
// Union types for ports
Port ::= PowerPortType + SignalPortType.
MechanicalPowerPortType ::= TranslationalPowerPort +
     RotationalPowerPort. PowerPortType ::= MechanicalPowerPortType +
     ThermalPowerPort +
     HydraulicPowerPort + ElectricalPowerPort.
SignalPortType ::= InputSignalPort + OutputSignalPort.
// Connections of power and signal ports
PowerFlow ::=
      new (name:String,src:PowerPortType,
           dst:PowerPortType,…).InformationFlow ::=
            new (name:String,src:SignalPortType,dst:SignalPortType,…).
```

10.3.3 Formalization of Semantics

In this section we provide the formal semantics of the language in four parts: structural semantics, denotational semantics, denotational semantics of power port connections, and semantics of signal port connections.

10.3.3.1 Structural Semantics

A CyPhyML model is considered to be well formed if it does not contain any dangling ports, distant connections, or invalid port connections. In such a case, it conforms to the following domain:

```
conforms
  no dangling(_),
  no distant(_),
  no invalidPowerFlow(_),
  no invalidInformationFlow(_).
```

To meet this requirement, we need to define a set of auxiliary rules. Dangling ports are ports that are not connected to any other ports:

```
dangling ::= (Port).
dangling(X) :- X is PowerPortType,
  no {P | P is PowerFlow, P.src = X} ,
  no { P | P is PowerFlow, P.dst = X }.
dangling(X) :- X is SignalPortType,
  no { I|I is InformationFlow, I.src = X},
  no { I | I is InformationFlow, I.dst = X }.
```

A distant connection connects two ports belonging to different components, such that the components have different parents, and neither component is a parent of the other one:

```
distant ::= (PowerFlow+InformationFlow).
distant(E) :-
  E is PowerFlow+InformationFlow,
  DesignElementToPortContainment(PX,E.src),
  DesignElementToPortContainment(PY,E.dst),
  PX != PY,
  ComponentAssemblyToCompositionContainment(PX,PPX),
  ComponentAssemblyToCompositionContainment(PY,PPY),
  PPX != PPY, PPX != PY, PX != PPY.
```

A power flow is valid if it connects power ports of the same type:

```
validPowerFlow ::= (PowerFlow).
validPowerFlow(E) :- E is PowerFlow,
  X=E.src, X:TranslationalPowerPort,
  Y=E.dst, Y:TranslationalPowerPort.
validPowerFlow(E) :- E is PowerFlow,
  X=E.src, X:RotationalPowerPort,
  Y=E.dst, Y:RotationalPowerPort.
validPowerFlow(E) :- E is PowerFlow,
  X=E.src, X:ThermalPowerPort,
  Y=E.dst, Y:ThermalPowerPort.
validPowerFlow(E) :- E is PowerFlow,
  X=E.src, X:HydraulicPowerPort,
  Y=E.dst, Y:HydraulicPowerPort.
validPowerFlow(E) :- E is PowerFlow,
  X=E.src, X:ElectricalPowerPort,
  Y=E.dst, Y:ElectricalPowerPort.
```

If a power flow is not valid, it is invalid:

```
invalidPowerFlow ::= (PowerFlow).
invalidPowerFlow(E) :- E is PowerFlow, no validPowerFlow(E).
```

An information flow is invalid if a signal port receives signals from multiple sources, or if an input port is the source of an output port:

```
invalidInformationFlow ::= (InformationFlow).
invalidInformationFlow(X) :-
  X is InformationFlow,
  Y is InformationFlow,
  X.dst = Y.dst, X.src != Y.src.
```

continued

```
invalidInformationFlow(E) :-
  E is InformationFlow,
  X = E.src, X:InputSignalPort,
  Y = E.dst, Y:OutputSignalPort.
```

Note that output ports can be connected to output ports.

10.3.3.2 Denotational Semantics

The denotational semantics of a language is described by a semantic domain and a mapping that maps the syntactic elements of the language to this semantic domain. In this section, we specify a semantic mapping from CyPhyML to a hybrid differential-difference equations semantic unit [Simko2013a]. We use the semantic anchoring framework [Chen2005] for the denotational semantic specification of CyPhyML, as shown in Figure 10.2.

Acausal CPS modeling languages distinguish between acausal power ports and causal signal ports. In CyPhyML, each power port contributes two variables to the equations, and the denotational

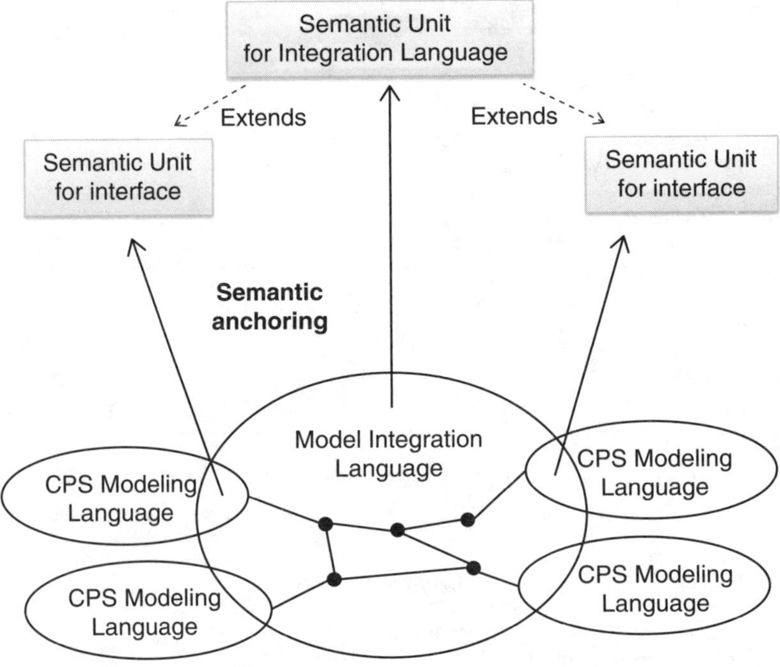

Figure 10.2: *Denotational semantic specification of CyPhyML*

semantics of CyPhyML is defined in terms of equations over these variables. Signal ports transmit signals with strict causality. Consequently, if we associate a signal variable with each signal port, the variable of a destination port is enforced to denote the same value as the variable of the corresponding source port. This relationship goes in only one way: The value of the variable at the destination port cannot affect the source variable along the connection in question.

The semantic function for power ports is mapping power ports to pairs of continuous-time variables:

```
𝒫𝒫 : PowerPort → cvar, cvar.
𝒫𝒫 ⟦CyPhyPowerPort⟧ =
  (cvar("CyPhyML_effort",CyPhyPowerPort.id),
  cvar("CyPhyML_flow",CyPhyPowerPort.id)).
```

The semantic function of signal ports is mapping signal ports to continuous-time variables:

```
𝒮𝒫 : SignalPort → cvar+dvar.
𝒮𝒫⟦CyPhySignalPort⟧ = cvar("CyPhyML_signal",CyPhySignalPort.id).
```

10.3.3.3 Denotational Semantics of Power Port Connections

The semantics of power port connections is defined through their transitive closure. Using fixed-point logic, we can easily express the transitive closure of connections as the least-fixed-point solution for `ConnectedPower`. Informally, `ConnectedPower(x,y)` expresses that power ports x and y are interconnected through one or more power port connections:

```
ConnectedPower ::= (src:CyPhyPowerPort, dst:CyPhyPowerPort).
ConnectedPower(x,y) :-
  PowerFlow(_,x,y,_,_), x:CyPhyPowerPort, y:CyPhyPowerPort;
  PowerFlow(_,y,x,_,_), x:CyPhyPowerPort, y:CyPhyPowerPort;
  ConnectedPower(x,z), PowerFlow(_,z,y,_,_), y:CyPhyPowerPort;
  ConnectedPower(x,z), PowerFlow(_,y,z,_,_), y:CyPhyPowerPort.
```

More precisely, $P_x = \{y \mid ConnectedPower(x, y)\}$ is the set of power ports reachable from power port x.

The behavioral semantics of CyPhyML power port connections is defined by a pair of equations generalizing the Kirchoff equations. Their form is as follows:

$$\forall x \in CyPhyPowerPort.\left(\sum_{y \in \{y|ConnectedPower(x,y)\}} e_y = 0\right)$$

$$\forall x, y(ConnectedPower(x,y) \rightarrow e_x = e_y)$$

We can formalize this form in the following way:

```
𝒫: ConnectedPower ⟶ eq+addend.
𝒫⟦ConnectedPower⟧ =
  eq(sum("CyPhyML_powerflow",flow1.id),0)
  addend(sum("CyPhyML_powerflow",flow1.id),flow1)
  addend(sum("CyPhyML_powerflow",flow1.id), flow2)
  eq(effort1, effort2)
where
  x = ConnectedPower.src, y = ConnectedPower.dst, x != y,
  DesignElementToPortContainment(cx,x),cx:Component,
  DesignElementToPortContainment(cy,y),cy:Component,
  𝒫𝒫⟦x⟧ = (effort1,flow1),
  𝒫𝒫⟦y⟧ = (effort2,flow2).
```

The explanation of why such a pair of power variables (effort and flow) are used for describing physical connections is beyond the scope in this chapter. The interested reader can find a great introduction to this topic in [Willems2007].

10.3.3.4 Semantics of Signal Port Connections

A signal connection path (`ConnectedSignal`) is a directed path along signal connections. We can use fixed-point logic to find the transitive closure by solving for the least fixed point of `ConnectedSignal`. Informally, `ConnectedSignal(x,y)` expresses that there is a signal path (chain of connections) from signal port `x` to signal port `y`.

```
ConnectedSignal ::= (CyPhySignalPort,CyPhySignalPort).
ConnectedSignal(x,y) :-
  InformationFlow(_,x,y,_,_),
  x:CyPhySignalPort,
  y:CyPhySignalPort.
```

```
ConnectedSignal(x,y) :-
  ConnectedSignal(x,z),
  InformationFlow(_,z,y,_,_),
  y:CyPhySignalPort.
```

More precisely, $P_x = \{y \mid ConnectedSignal(x, y)\}$ is the set of signal ports reachable from signal port x.

A signal connection (`SignalConnection`) is a `ConnectedSignal` such that its endpoints are the signal ports of components (therefore leaving out any signal ports that are ports of component assemblies):

```
SignalConnection ::= (src:CyPhySignalPort,dst:CyPhySignalPort).
SignalConnection(x,y) :-
  ConnectedSignal(x,y),
  DesignElementToPortContainment(cx,x), cx:Component,
  DesignElementToPortContainment(cy,y), cy:Component.
```

The behavioral semantics of CyPhyML signal connections is defined as variable assignment. The value of the variables associated with the source and the destination of a signal connection are equal:

$$\forall x, y\left(SignalConnection(x, y) \rightarrow s_y = s_x\right)$$

```
𝒮: SignalConnection ⟶ eq.
𝒮 ⟦SignalConnection⟧ =
  eq(𝒮𝒫 ⟦SignalConnection.dst⟧, 𝒮𝒫⟦SignalConnection.src⟧).
```

10.3.4 Formalization of Language Integration

So far, we have formally defined the semantics of the compositional elements of CyPhyML, but we have not specified how components are integrated into CyPhyML. In this section, we develop the semantics of the integration of external languages—that is, a bond graph language, the Modelica language, and the ESMoL language. In the future, we could easily add other languages to this list by following the same steps presented here.

Bond graphs are multi-domain graphical representations for physical systems that describe the structure of energy flows [Karnopp2012]. [Simko2013] introduced a bond graph language, along with its formal semantics. Here, we consider an extended bond graph language that defines also power ports: A bond graph component interacts with its environment through these ports. Each power port is adjacent to exactly one bond; therefore, a power port represents a pair of power variables—the power variables of its unique bond. The bond graph language we consider here also contains output signal ports for measuring efforts and flows at bond graph junctions, and modulated bond graph elements that are controlled by input signals fed to the bond graph through input signal ports. Note that the effort and flow variables of the bond graph language are different from the effort and flow variables of CyPhyML; they denote different entities in different physical domains. The semantics of the languages formalize these differences precisely.

Modelica is an equation-based object-oriented language [Fritzson 1998] used for systems modeling and simulation. Modelica supports component-based development through its model and connector concepts. Models are components with internal behavior and a set of ports called connectors. Models are interconnected by connecting their connector interfaces. A connector is a set of variables (e.g., input, output, acausal flow or potential), and the connection of different connectors defines relations over their variables. We discuss the integration of a restricted set of Modelica models in CyPhyML—namely, models that contain connectors consisting of either exactly one input/output variable or a pair of potential and flow variables.

The Embedded Systems Modeling Language (ESMoL) [Porter2010] is a language and tool suite for modeling and implementing computational systems and hardware platforms. ESMoL consists of several sub-languages for defining platform and software architectures, describing the deployment of software on hardware, and specifying the scheduling of execution. In this chapter, by ESMoL we refer to the statechart variant sublanguage of ESMoL, which is used to model software controllers. Statechart is based on periodic time-triggered execution semantics, and its components expose periodic discrete-time signal ports on their interface.

10.3.4.1 Integration of Structure

The role of CyPhyML in the integration process is to establish meaningful and valid connections between heterogeneous models. Component integration is an error-prone task because of the slight differences between different languages. For instance, during the formalization we found the following discrepancies:

- Power ports have different meanings in different modeling languages.
- Even if the semantics is the same, there are differences in the naming conventions.

- Connecting the signals of ESMoL to the signals of CyPhyML requires a conversion between discrete-time and continuous-time signals.

To formalize the integration of external languages, we extend CyPhyML with the semantic interfaces of these languages. Hence, we need language elements for representing models of these heterogeneous languages, their port structures, and the port mapping between the ports and the corresponding CyPhyML ports.

We formalize the models and their containment in CyPhyML as follows:

```
BondGraphModel ::= new (URI:String, id:Integer).
ModelicaModel ::= new (URI:String, id:Integer).
ESMoLModel ::= new (URI:String, id:Integer, sampleTime:Real).
Model ::= BondGraphModel + ModelicaModel + ESMoLModel.
// A relation describing the containment of bond graph models in CyPhyML
   components
ComponentToBondGraphContainment ::= new (Component => BondGraphModel).
...
```

Note the fields of `ESMoLModel:`. Since ESMoL models are periodic discrete-time systems, we need real values describing their period and initial phase in the continuous-time world. The interface ports and port mappings follow:

```
// Bond graph power ports (and similarly for the other languages)
BGPowerPort ::= MechanicalDPort + MechanicalRPort + ...
...
// Port mappings for bond graph power ports (and similarly for other
languages) BGPowerPortMap ::= (src:BGPowerPort,dst:CyPhyPowerPort).
...
// All the power ports in CyPhyML and the integrated languages:
PowerPort ::= CyPhyPowerPort + BGPowerPort + ModelicaPowerPort.
// All the signal ports in CyPhyML and the integrated languages:
SignalPort ::= ElectricalSignalPort +
           BGSignalPort +
           ModelicaSignalPort +
           ESMoLSignalPort.
// List of all ports:
AllPort ::= PowerPort + SignalPort.

// Mapping from model ports to CyPhyML ports
PortMap ::= BGPowerPortMap +
           BGSignalPortMap +
           ModelicaPowerPortMap +
           ModelicaSignalPortMap +
           SignalFlowSignalPortMap.
```

An integrated model (i.e., a CyPhyML model integrated with other models) is considered well formed if it conforms to the original CyPhyML domain, and if its port mappings are valid:

```
conforms no invalidPortMapping.
```

A port mapping is invalid if it connects incompatible ports, or if the interconnected ports are not part of the same CyPhyML component:

```
invalidPortMapping :- M is PortMap, no compatible(M).
invalidPortMapping :-
  M is BGPowerPortMap,
  BondGraphToPortContainment(BondGraph,M.src),
  DesignElementToPortContainment(CyPhyComponent,M.dst),
    no ComponentToBondGraphContainment(CyPhyComponent,BondGraph).
  …
// Compatible denotes that port mapping M is valid (i.e., the
corresponding ports
  are compatible)
compatible ::= (PortMap).
compatible(M) :- M is BGPowerPortMap(X,Y), X:MechanicalRPort,
  Y:RotationalPowerPort.
…
```

10.3.4.2 Bond Graph Integration

The semantics of bond graph power ports is explained by mapping to pairs of continuous-time variables:

```
𝓑𝓖𝓟𝓟 : BGPowerPort ⟶ cvar, cvar.
𝓑𝓖𝓟𝓟 ⟦BGPowerPort⟧ =
  (cvar("BondGraph_effort",BGPowerPort.id),
   cvar("BondGraph_flow",BGPowerPort.id)).
```

The semantics of bond graph signal ports is explained by mapping to continuous-time variables:

```
𝓑𝓖𝓢𝓟: BGSignalPort ⟶ cvar.
𝓑𝓖𝓢𝓟 ⟦BGSignalPort⟧ = cvar("BondGraph_signal",port.id).
```

The behavioral semantics of bond graph power port mappings for the hydraulic and thermal domains is the equality of the associated port variables. We can formalize it with the following rules:

```
𝓑𝓖𝓟: BGPowerPortMap ⟶ eq+diffEq.
𝓑𝓖𝓟 ⟦BGPowerPortMap⟧ =
  eq(cyphyEffort, bgEffort)
  eq(cyphyFlow, bgFlow)
where
  bgPort = BGPowerPortMap.src,
  cyphyPort = BGPowerPortMap.dst,
  bgPort : HydraulicPort + ThermalPort,
  𝓟𝓟⟦cyphyPort⟧ = (cyphyEffort, cyphyFlow),
  𝓑𝓖𝓟𝓟 ⟦bgPort⟧ = (bgEffort, bgFlow).
```

In the mechanical translational domain, the effort of CyPhyML power ports denotes absolute position and the flow denotes force, whereas for bond graphs, the effort is force and the flow is velocity. In the mechanical rotational domain, the effort of CyPhyML power ports denotes absolute rotation angle and the flow denotes torque, whereas for bond graphs the effort is torque and the flow is angular velocity. Their interconnection in CyPhyML is formalized by the following equations:

```
𝓑𝓖𝓟 ⟦BGPowerPortMap⟧ =
  diffEq(cyphyEffort, bgFlow)
  eq(bgEffort, cyphyFlow)
where
  bgPort = BGPowerPortMap.src,
  cyphyPort = BGPowerPortMap.dst,
  bgPort : MechanicalDPort + MechanicalRPort,
𝓟𝓟⟦cyphyPort⟧ = (cyphyEffort, cyphyFlow),
𝓑𝓖𝓟𝓟 ⟦bgPort⟧ = (bgEffort, bgFlow).
```

For the electrical domain, the bond graph electrical power ports denote a pair of physical terminals (electrical pins); in the CyPhyML language, they denote single electrical pins. In both cases, the flow (the current) through the pins is the same, but the interpretation of voltage differs. In the bond graph case, the effort variable belonging to the electrical power port denotes the difference of the voltages between the two electrical pins. In the CyPhyML case, the effort variable denotes the absolute voltage (with respect to an arbitrary ground). The semantics of electrical power port mapping is the equality of the flows and efforts, which means that the negative terminal of the bond graph electrical power port is automatically grounded to the CyPhyML ground:

```
𝓑𝓖𝓟 ⟦BGPowerPortMap⟧ =
  eq(bgFlow, cyphyFlow)
  eq(bgEffort, cyphyEffort)
```

continued

```
where
  bgPort = BGPowerPortMap.src,
  cyphyPort = BGPowerPortMap.dst,
  bgPort : ElectricalPort,
  𝒫𝒫⟦cyphyPort⟧ = (cyphyEffort, cyphyFlow),
  ℬ𝒢𝒫𝒫⟦bgPort⟧ = (bgEffort, bgFlow).
```

Finally, the denotation of the bond graph and CyPhyML signal port mappings is equality of the interconnected port variables:

```
ℬ𝒢𝒮 : BGSignalPortMap ⟶ eq.
ℬ𝒢𝒮⟦BGSignalPortMap⟧ =
    eq(ℬ𝒢𝒮𝒫⟦BGSignalPortMap.src⟧, 𝒮𝒫⟦BGSignalPortMap.dst⟧).
```

10.3.4.3 Modelica Integration

The semantics of Modelica power ports is explained by mapping to pairs of continuous-time variables:

```
ℳ𝒫𝒫 : ModelicaPowerPort ⟶ cvar,cvar.
ℳ𝒫𝒫 ⟦ModelicaPowerPort⟧ =
  (cvar("Modelica_potentia",ModelicaPowerPort.id),
   cvar("Modelica_flow",ModelicaPowerPort.id)).
```

The semantics of Modelica signal ports is explained by mapping to continuous-time variables:

```
ℳ𝒮𝒫: ModelicaSignalPort ⟶ cvar.
ℳ𝒮𝒫 ForSpecDenotationModelicaSignalPort    =
   cvar("Modelica_signal",ModelicaSignalPort.id).
```

The semantics of the Modelica and CyPhyML power port mappings is equality of the power variables:

```
ℳ𝒫: ModelicaPowerPortMap ⟶ eq.
ℳ𝒫⟦ModelicaPowerPortMap⟧ =
  eq(cyphyEffort, modelicaEffort)
  eq(cyphyFlow, modelicaFlow)
where
  modelicaPort = ModelicaPowerPortMap.src,
  cyphyPort = ModelicaPowerPortMap.dst,
𝒫𝒫⟦cyphyPort⟧ = (cyphyEffort, cyphyFlow),
ℳ𝒫𝒫⟦modelicaPort⟧ = (modelicaEffort, modelicaFlow).
```

The semantics of the Modelica and CyPhyML signal port mappings is equality of the signal variables:

```
𝓜𝓢 : ModelicaSignalPortMap ⟶ eq.
𝓜𝓢 ⟦ModelicaSignalPortMap⟧ =
  eq(𝓜𝓢𝓟 ⟦ModelicaSignalPortMap.src⟧,
   𝓢𝓟 ⟦ModelicaSignalPortMap.dst⟧).
```

10.3.4.4 Signal-Flow Integration

The semantics of the ESMoL signal ports is explained by mapping to discrete-time variables, and the periodicity of the discrete variable is determined by the sample time of its container block:

```
𝓔𝓢𝓟 : ESMoLSignalPort ⟶ dvar, timing.
𝓔𝓢𝓟 ⟦ESMoLSignalPort⟧ = (Dvar, timing(Dvar, container.sampleTime, 0))
where
  Dvar = dvar("ESMoL_signal", ESMoLSignalPort.id),
  BlockToSF_PortContainment(container,ESMoLSignalPort).
```

While signal ports in signal flow have discrete-time semantics, signal ports in CyPhyML are continuous-time. Thus, signal-flow output signals are integrated into CyPhyML by means of the `hold` operator:

$$\forall x, y(SignalFlowSignalPortMap(x, y) \rightarrow e_y = hold(e_x))$$

```
𝓔𝓢 : SignalFlowSignalPortMap ⟶ eq+timing.
𝓔𝓢 ⟦SignalFlowSignalPortMap⟧ = eq(cyphySignal, hold(signalflowSignal))
where
  signalflowPort = SignalFlowSignalPortMap.src,cyphyPort =
  SignalFlowSignalPortMap.dst,signalflowPort : OutSignal,
  𝓢𝓟 ⟦cyphyPort⟧ = cyphySignal,
  𝓔𝓢𝓟 ⟦signalflowPort⟧ = (signalflowSignal,_).
```

For the opposite direction, we can use the sample operator. The sample rate of the sampling function is defined by the signal-flow block containing the port:

$$\forall x, y(SignalFlorSignalPortMap(x, y) \rightarrow s_x = sample_r(s_y))$$

```
ES ⟦SignalFlowSignalPortMap⟧ = eq(signalflowSignal,
  sample(cyphySignal,samp.period,samp.phase))
where
  signalflowPort = SignalFlowSignalPortMap.src,
  cyphyPort = SignalFlowSignalPortMap.dst,
  signalflowPort : InSignal,
  SP ⟦cyphyPort⟧ = cyphySignal,
  ESP ⟦signalflowPort⟧ = (signalflowSignal,samp).
```

10.4 Summary and Open Challenges

In this chapter, we provided an overview of the theoretical and practical considerations for composing heterogeneous dynamic behavior in CPS. Using CyPhyML (a model integration language) as an example, we demonstrated how ForSpec (an executable logic-based specification language) can be used for the formalization of the integration language and the composition.

In our example, the structural semantics describes the well-formedness rules in the language, and the denotational semantics is given by mapping to the domain of differential algebraic equations. This mapping provides a mathematically rigorous and unambiguous description of the behavior.

Finally, we described the integration of three heterogeneous languages—that is, the integration of bond graphs, Modelica models, and ESMoL components into CyPhyML. The integration of these heterogeneous components led to the incremental definition of the structural and denotational semantics for CyPhyML. Since new features are increasingly being introduced into the still-evolving language of CyPhyML, the language and its semantic definitions are open for further extensions.

Our approach has two advantages: We used an executable formal specification language, which lends itself to model conformance checking, model checking, and model synthesis; and both the structural and behavioral specifications were written using the same logic-based language, so both can be used for deductive reasoning. In particular, structure-based proofs about behaviors become feasible. It remains to future work to leverage the specifications for performing such symbolic formal analysis.

The results described in this chapter have been used for creating the Semantic Backplane for OpenMETA, the integrated component- and model-based tool suite developed for the Defense Advanced Research Project Agency (DARPA) Adaptive Vehicle Make (AVM) program [Simko2012]. OpenMETA provides a manufacturing-aware design flow, which covers both cyber and physical design aspects [Sztipanovits2014]. The Semantic Backplane includes the formal specification of all semantic interfaces and the CyPhyML model integration language, the formal specification of all model transformation tools used for tool integration, and the specification of data models used in the overall information architecture. It had major significance in terms of keeping the model and tool integration elements of a very large design tool suite consistent, and by providing a reference documentation for developers and users of the tool suite. At this point, the connection between the production tools of OpenMETA and the formal specifications in the Semantic Backplane is just partially automated; their tight integration is a future challenge.

References

[Abdellatif2010] T. Abdellatif, J. Combaz, and J. Sifakis. "Model-Based Implementation of Real-Time Applications." *Proceedings of the 10th ACM International Conference on Embedded Software*, pages 229–238, 2010.

[Backus1959] J. W. Backus. "The Syntax and Semantics of the Proposed International Algebraic Language of the Zurich ACM-GAMM Conference." *Proceedings of the International Conference on Information Processing*, 1959.

[Balmer1990] R. T. Balmer. *Thermodynamics*. West Group Publishing, St. Paul, MN, 1990.

[Basu2006] A. Basu, M. Bozga, and J. Sifakis. "Modeling Heterogeneous Real-Time Components in BIP." *Proceedings of the 4th IEEE International Conference on Software Engineering and Formal Methods*, pages 3–12, 2006.

[Bliudze2008] S. Bliudze and J. Sifakis. "The Algebra of Connectors: Structuring Interaction in BIP." *IEEE Transactions on Computers*, vol. 57, no. 10, pages 1315–1330, October 2008.

[Chen2005] K. Chen, J. Sztipanovits, S. Abdelwalhed, and E. Jackson. "Semantic Anchoring with Model Transformations." In *Model Driven Architecture: Foundations and Applications*, vol. 3748 of *Lecture Notes in Computer Science*, pages 115–129. Springer, Berlin/Heidelberg, 2005.

[Chen2007] K. Chen, J. Sztipanovits, and S. Neema. "Compositional Specification of Behavioral Semantics." *Proceedings of the Conference on Design, Automation and Test in Europe (DATE)*, pages 906–911, 2007.

[Ecosimpro2014] EcosimPro. www.ecosimpro.com.

[Fritzson1998] P. Fritzson and V. Engelson. "Modelica: A Unified Object-Oriented Language for System Modeling and Simulation." In *ECOOP'98: Object-Oriented Programming*, vol. 1445 of *Lecture Notes in Computer Science*, pages 67–90. Springer, Berlin/Heidelberg, 1998.

[Goderis2007] A. Goderis, C. Brooks, I. Altintas, E. A. Lee, and C. Goble. "Composing Different Models of Computation in Kepler and Ptolemy II." In *Computational Science: ICCS 2007*, vol. 4489 of *Lecture Notes in Computer Science*, pages 182–190. Springer, Berlin/Heidelberg, 2007.

[Goderis2009] A. Goderis, C. Brooks, I. Altintas, E. A. Lee, and C. Goble. "Heterogeneous Composition of Models of Computation." *Future Generation Computer Systems*, vol. 25, no. 5, pages 552–560, May 2009.

[Henzinger2001] T. Henzinger, B. Horowitz, and C. Kirsch. "Giotto: A Time-Triggered Language for Embedded Programming." In *Embedded Software*, vol. 2211 of *Lecture Notes in Computer Science*, pages 166–184. Springer, Berlin/Heidelberg, 2001.

[Jackson2011] E. Jackson, N. Bjørner, and W. Schulte. "Canonical Regular Types." *ICLP (Technical Communications)*, pages 73–83, 2011.

[Jackson2010] E. Jackson, E. Kang, M. Dahlweid, D. Seifert, and T. Santen. "Components, Platforms and Possibilities: Towards Generic Automation for MDA." *Proceedings of the 10th ACM International Conference on Embedded Software (EMSOFT)*, pages 39–48, 2010.

[Jackson2009] E. Jackson and J. Sztipanovits. "Formalizing the Structural Semantics of Domain-Specific Modeling Languages." *Software and Systems Modeling*, vol. 8, no. 4, pages 451–478, 2009.

[Karnopp2012] D. Karnopp, D. L. Margolis, and R. C. Rosenberg. *System Dynamics Modeling, Simulation, and Control of Mechatronic Systems.* John Wiley and Sons, 2012.

[Karsai2008] G. Karsai and J. Sztipanovits. "Model-Integrated Development of Cyber-Physical Systems." In *Software Technologies for Embedded and Ubiquitous Systems*, pages 46–54. Springer, 2008.

[Karsai2003] G. Karsai, J. Sztipanovits, A. Ledeczi, and T. Bapty. "Model-Integrated Development of Embedded Software." *Proceedings of the IEEE*, vol. 91, no. 1, pages 145–164, 2003.

[Kirsch2012] C. M. Kirsch and A. Sokolova. "The Logical Execution Time Paradigm." In *Advances in Real-Time Systems*, pages 103–120. Springer, 2012.

[Lattmann2012] Z. Lattmann, A. Nagel, J. Scott, K. Smyth, J. Ceisel, C. vanBuskirk, J. Porter, T. Bapty, S. Neema, D. Mavris, and J. Sztipanovits. "Towards Automated Evaluation of Vehicle Dynamics in System-Level Designs." ASME 32nd Computers and Information in Engineering Conference (IDETC/CIE), 2012.

[Ledeczi2001] A. Ledeczi, M. Maroti, A. Bakay, G. Karsai, J. Garrett, C. Thomason, G. Nordstrom, J. Sprinkle, and P. Volgyesi. "The Generic Modeling Environment." Workshop on Intelligent Signal Processing, vol. 17, Budapest, Hungary, 2001.

[Lee2008] E. A. Lee. "Cyber Physical Systems: Design Challenges." 11th IEEE International Symposium on Object Oriented Real-Time Distributed Computing (ISORC), pages 363–369, 2008.

[Lee2009] E. A Lee. "Computing Needs Time." *Communications of the ACM*, vol. 52, no. 5, pages 70–79, 2009.

[Lee2007] E. A. Lee and H. Zheng. "Leveraging Synchronous Language Principles for Heterogeneous Modeling and Design of Embedded Systems." *Proceedings of the 7th ACM/IEEE International Conference on Embedded Software (EMSOFT)*, pages 114–123, 2007.

[Maler1992] O. Maler, Z. Manna, and A. Pnueli. "From Timed to Hybrid Systems." In *Real-Time: Theory in Practice, Lecture Notes in Computer Science*, pages 447–484. Springer, 1992.

[Modelica2012] Modelica Association. "Modelica: A Unified Object-Oriented Language for Physical System Modeling." Language Specification, Version 3.3, 2012.

[Mosterman2013] P. J. Mosterman, G. Simko, and J. Zander. "A Hyperdense Semantic Domain for Discontinuous Behavior in Physical System Modeling." Compositional Multi-Paradigm Models for Software Development, October 2013.

[Mosterman2014] P. J. Mosterman, G. Simko, J. Zander, and Z. Han. "A Hyperdense Semantic Domain for Hybrid Dynamic Systems to Model with Impact." 17th International Conference on Hybrid Systems: Computation and Control (HSCC), 2014.

[Porter2010] J. Porter, G. Hemingway, H. Nine, C. vanBuskirk, N. Kottenstette, G. Karsai, and J. Sztipanovits. *The ESMoL Language and Tools for High-Confidence Distributed Control Systems Design. Part 1: Design Language, Modeling Framework, and Analysis.* Technical Report ISIS-10-109, Vanderbilt University, Nashville, TN, 2010.

[Reineke2006] J. Reineke, B. Wachter, S. Thesing, R. Wilhelm, I. Polian, J. Eisinger, and B. Becker. "A Definition and Classification of Timing Anomalies." *Proceedings of 6th International Workshop on Worst-Case Execution Time (WCET) Analysis*, 2006.

[Simko2012] G. Simko, T. Levendovszky, S. Neema, E. Jackson, T. Bapty, J. Porter, and J. Sztipanovits. "Foundation for Model Integration: Semantic Backplane." ASME 32nd Computers and Information in Engineering Conference (IDETC/CIE), 2012.

[Simko2013] G. Simko, D. Lindecker, T. Levendovszky, E. K. Jackson, S. Neema, and J. Sztipanovits. "A Framework for Unambiguous and Extensible Specification of DSMLs for Cyber-Physical Systems." IEEE 20th International Conference and Workshops on the Engineering of Computer Based Systems (ECBS), 2013.

[Simko2013a] G. Simko, D. Lindecker, T. Levendovszky, S. Neema, and J. Sztipanovits. "Specification of Cyber-Physical Components with Formal Semantics: Integration and Composition." ACM/IEEE 16th International Conference on Model Driven Engineering Languages and Systems (MODELS), 2013.

[Simscape2014] Simscape. MathWorks. http://www.mathworks.com/products/simscape.

[Sztipanovits2014] J. Sztipanovits, T. Bapty, S. Neema, L. Howard, and E. Jackson. "OpenMETA: A Model and Component-Based Design Tool Chain for Cyber-Physical Systems." In *From Programs to Systems: The Systems Perspective in Computing*, vol. 8415 of *Lecture Notes in Computer Science*, pages 235–249. Springer, 2014.

[Sztipanovits1997] J. Sztipanovits and G. Karsai. "Model-Integrated Computing." *Computer*, vol. 30, no. 4, pages 110–111, 1997.

[Sztipanovits2012] J. Sztipanovits, X. Koutsoukos, G. Karsai, N. Kottenstette, P. Antsaklis, V. Gupta, B. Goodwine, J. Baras, and S. Wang. "Toward a Science of Cyber-Physical System Integration." *Proceedings of the IEEE*, vol. 100, no. 1, pages 29–44, 2012.

[Willems2007] J. C. Willems. "The Behavioral Approach to Open and Interconnected Systems." *IEEE Control Systems*, vol. 27, no. 6, pages 46–99, 2007.

[Wrenn2012] R. Wrenn, A. Nagel, R. Owens, H. Neema, F. Shi, K. Smyth, D. Yao, J. Ceisel, J. Porter, C. vanBuskirk, S. Neema, T. Bapty, D. Mavris, and J. Sztipanovits. "Towards Automated Exploration and Assembly of Vehicle Design Models." ASME 32nd Computers and Information in Engineering Conference (IDETC/CIE), 2012.

About the Authors

Ragunathan (Raj) Rajkumar is the George Westinghouse Professor in Electrical and Computer Engineering at Carnegie Mellon University. Among other companies such as TimeSys, he founded Ottomatika, Inc., which focused on software for self-driving vehicles and was acquired by Delphi. He has chaired several international conferences, has three patents, has authored one book and co-edited another, and has published more than 170 refereed papers in conferences and journals. Eight of his papers have received Best Paper awards. Dr. Rajkumar received his B.E. (Hons.) degree from the University of Madras, India, in 1984, and his M.S. and Ph.D. degrees from Carnegie Mellon University, Pittsburgh, Pennsylvania, in 1986 and 1989, respectively. His research interests include all aspects of cyber-physical systems.

Dionisio de Niz is a Principal Researcher at the Software Engineering Institute at Carnegie Mellon University. He received a master of science in information networking from the Information Networking Institute and a Ph.D. in electrical and computer engineering from Carnegie Mellon University. His research interests include cyber-physical systems, real-time systems, and model-based engineering. In the real-time arena, he has recently focused on multicore processors and mixed-criticality scheduling, and has led a number of projects on both fundamental research and applied research for the private industry and government organizations. Dr. de Niz worked on the reference implementation and a commercial version of the Real-Time Java Specification.

Mark Klein is Senior Member of the Technical Staff at the Software Engineering Institute and Technical Director of its Critical System Capabilities Directorate, which conducts research in cyber-physical systems and advanced mobile systems. His research has spanned various facets of software engineering, dependable real-time systems and numerical methods. Klein's most recent work focuses on design and analysis principles for systems at scale, including cyber-physical systems. Previously, as Technical Lead of the Architecture-Centric Engineering Initiative, he led research on the analysis of software architectures, architecture evolution, economics-driven architecting, architecture competence, architecture tradeoff analysis, attribute-driven architectural design, scheduling theory, and applied mechanism design. His work in real-time systems involved the development of rate-monotonic analysis (RMA), the extension of the theoretical basis for RMA, and its application to realistic systems. Klein's earliest work involved research in high-order finite-element methods for solving fluid flow equations arising in oil reservoir simulation. He is co-author of many papers and three books: *A Practitioner's Handbook for Real-Time Analysis: Guide to Rate Monotonic Analysis for Real-Time Systems*, *Evaluating Software Architecture: Methods and Case Studies*, and *Ultra-Large-Scale Systems: The Software Challenge of the Future*.

About the Contributing Authors

Abdullah Al-Nayeem received his Ph.D. from University of Illinois at Urbana–Champaign in 2013. His Ph.D. dissertation title was "Physically-Asynchronous Logically-Synchronous (PALS) System Design and Development." Currently, he is working as a Software Engineer at Google's Pittsburgh office.

Björn Andersson has been Senior Member of Technical Staff at the Software Engineering Institute at Carnegie Mellon University since March 2011. He received his M.Sc. degree in electrical engineering in 1999 and his Ph.D. degree in computer engineering in 2003, both from Chalmers University, Sweden. His main research interests are presently in the use of multicore processors in real-time systems and principles for cyber-physical systems.

Karl-Erik Årzén received his M.Sc. degree in electrical engineering in 1981 and his Ph.D. degree in automatic control in 1987, both from Lund University, Sweden. He is currently Professor at the Department of Automatic Control, Lund University. His research interests include embedded real-time systems, feedback computing, cloud control, and cyber-physical systems.

Anaheed Ayoub is a Principal Engineer at Mathworks. Before joining Mathworks, she was a postdoctoral researcher in computer science at the University of Pennsylvania. Her research interests include model-based design workflows involving formal modeling, verification, and code generation and validation of safety-critical real-time systems. She received her M.S. and Ph.D. degrees in computer engineering from Ain Shams University, Cairo, Egypt.

Anton Cervin received his M.Sc. in computer science and engineering in 1998 and his Ph.D. degree in automatic control in 2003, both from Lund University, Sweden. He is currently an Associate Professor at the Department of Automatic Control, Lund University. His research interests include embedded and networked control systems, event-based control, real-time systems, and computer tools for analysis and simulation of controller timing.

Sagar Chaki is a Principal Researcher at the Software Engineering Institute at Carnegie Mellon University. He received a B.Tech. in computer science and engineering from the Indian Institute of Technology, Kharagpur, in 1999, and a Ph.D. in computer science from Carnegie Mellon University in 2005. These days, he works mainly on model-checking software for real-time and cyber-physical systems, but he is generally interested in rigorous and automated approaches for improving software quality. Dr. Chaki has developed several automated software verification tools, including two model checkers for C programs, MAGIC and Copper. He has co-authored more than 70 peer-reviewed publications. More details about Dr. Chaki and his current work can be found at http://www.contrib.andrew.cmu.edu/~schaki/.

Sanjian Chen is a Ph.D. candidate in computer and information science at the University of Pennsylvania. His research interests include data-driven modeling, machine learning, formal analysis, and system engineering with applications to cyber-physical systems and human-software interaction. He received the Best Paper award at the 2012 IEEE Real-Time Systems Symposium (RTSS).

Edmund M. Clarke is now Professor Emeritus in the Computer Science Department at Carnegie Mellon University. He was the first recipient of the FORE Systems Endowed Professorship in 1995 and became a University Professor in 2008. He received a B.A. degree from the University of Virginia, an M.S. degree from Duke University, and a Ph.D. from Cornell University, and taught at Duke and Harvard Universities before joining Carnegie Mellon University in 1982. His research interests include hardware and software verification and automatic theorem proving. In particular, his research group developed symbolic model checking using BDDs, bounded model checking using fast CNF satisfiability solvers, and pioneered the use of counterexample-guided abstraction refinement (CEGAR). He was a co-founder of the conference on computer-aided verification (CAV). He has received numerous awards for his contributions to formal verification of hardware and software correctness, including the IEEE Goode Award, ACM Kanellakis Award, ACM Turing Award, CADE Herbrand Award, and CAV Award. He received the 2014 Franklin Institute Bower Award and Prize for Achievement in Science for verification of computer systems. He received an Einstein Professorship from the Chinese Academy of

Sciences and honorary doctorates from the Vienna University of Technology and the University of Crete. Dr. Clarke is a member of the National Academy of Engineering and the American Academy of Arts and Sciences, a Fellow of the ACM and IEEE, and a member of Sigma Xi and Phi Beta Kappa.

Antoine Girard is a Senior Researcher (Directeur de Recherche) at CNRS. He received his Ph.D. in applied mathematics from the Institut National Polytechnique de Grenoble in 2004. He was a postdoctoral researcher at University of Pennsylvania in 2004–2005 and at the Verimag Laboratory in 2006. From 2006 to 2015, he was an Associate Professor at the Université Joseph Fourier. His research interests mainly deal with analysis and control of hybrid systems, with an emphasis on computational approaches, approximation, abstraction, and applications to cyber-physical systems. In 2009, he received the George S. Axelby Outstanding Paper Award from the IEEE Control Systems Society. In 2014, he was awarded the CNRS Bronze Medal. In 2015, he was appointed as a junior member of the Institut Universitaire de France (IUF).

Arie Gurfinkel received a Ph.D. in computer science from the Computer Science Department of University of Toronto in 2007. He is a Principal Researcher at the Carnegie Mellon Software Engineering Institute. His research interests lie in the intersection of formal methods and software engineering, with an emphasis on automated reasoning about software systems. He has co-led development of a number of automated verification tools, including the first multivalued model-checker XChek, the software verification frameworks UFO and SeaHorn, and the hardware model-checker Avy.

John J. Hudak is a senior member of the technical staff at the Software Engineering Institute (SEI) in the Architecture Practices initiative. He has a MSEE from Carnegie Mellon University and has also been in the Ph.D. program in the university's Electrical and Computer Engineering Department. He is responsible for developing and applying model-based engineering methodologies based on the AADL for real-time embedded system development. In addition, he is an instructor for the SEI Model-Based Engineering with AADL course, which has been delivered to academia and industry, and has led and participated in a number of independent technical assessment teams for government projects. His interests include dependable real-time systems, computer hardware and software architecture, model-based verification, software reliability, and control engineering. Before joining the SEI, he was a member of Carnegie Mellon Research Institute, an applied R&D division within Carnegie Mellon University. He served in various technical and managerial capacities in R&D projects addressing the needs of industry. Hudak is a Senior Member of the IEEE, adjunct faculty member of University of Pittsburgh (Johnstown), and has a PE license in the state of Pennsylvania.

Marija Ilić holds appointments in the departments of ECE and EPP at Carnegie Mellon University, where she has been a tenured faculty member since October 2002. Dr. Ilić received her M.Sc. and D.Sc. degrees in Systems Science and Mathematics from Washington University in St. Louis and earned her MEE and Dip. Ing. at the University of Belgrade. She is an IEEE Fellow and an IEEE distinguished lecturer, as well as a recipient of the First Presidential Young Investigator Award for Power Systems. In addition to her academic work, Dr. Ilić is a consultant for the electric power industry and the founder of New Electricity Transmission Software Solutions, Inc. From September 1999 until March 2001, Dr. Ilić was a Program Director for Control, Networks, and Computational Intelligence at the National Science Foundation. Dr. Ilić has co-authored several books on the subject of large-scale electric power systems, and has co-organized an annual multidisciplinary Electricity Industry conference series at Carnegie Mellon (http://www.ece.cmu.edu/~electriconf) with participants from academia, government, and industry. Dr. Ilić is the founder and co-director of the Electric Energy Systems Group at Carnegie Mellon University (http://www.eesg.ece.cmu.edu).

BaekGyu Kim received his Ph.D. in computer science from the University of Pennsylvania. His research interests include modeling and verification of safety-critical medical devices and automotive systems, and the automated implementation of such systems through formal methods.

Cheolgi Kim received his Ph.D. from KAIST in 2005. He worked at the University of Illinois at Urbana–Champaign from 2006 to 2012 as a postdoctoral student and a Visiting Research Scientist, researching on cyber-physical systems and safety-critical system frameworks. Currently, he is an Associate Professor of Software Department with Korea Aerospace University.

Tiffany Hyun-Jin Kim is a Research Scientist at HRL Laboratories. She obtained her B.A. in computer science at the University of California at

Berkeley, her M.S. in computer science at Yale University, and her Ph.D. in electrical and computer engineering at Carnegie Mellon University. Her research interests include user-centric security and privacy, network security, trust management, and applied cryptography.

Andrew King received his Ph.D. in computer science from the University of Pennsylvania. His research interests include the modeling, verification, and certification of distributed systems and software—in particular, systems that can be integrated and reconfigured at runtime. He received his B.S. and M.S. degrees in computer science from Kansas State University.

Insup Lee is the Cecilia Fitler Moore Professor of Computer and Information Science and serves as the Director of the PRECISE Center at the University of Pennsylvania. He holds a secondary appointment in the Department of Electrical and Systems Engineering. His research interests include cyber-physical systems, real-time and embedded systems, runtime assurance and verification, trust management, and high-confidence medical systems. He received a Ph.D. in computer science from the University of Wisconsin, Madison. He is an IEEE fellow. He received the IEEE TC-RTS Outstanding Technical Achievement and Leadership Award in 2008.

John Lehoczky is the Thomas Lord university professor of statistics and mathematical sciences at Carnegie Mellon University. He works in the area of real-time systems and is most known for his work on the development of rate-monotonic scheduling and real-time queueing theory. He is the co-recipient of the 2016 IEEE Simon Ramo Medal awarded for "technical leadership and contributions to fundamental theory, practice, and standardization for engineering real-time systems." He is a fellow of the ASA, IMS, INFORMS, and AAAS, and is an elected member of the ISI.

Yilin Mo is an Assistant Professor in the School of Electrical and Electronic Engineering at Nanyang Technological University. He received his Ph.D. in electrical and computer engineering from Carnegie Mellon University in 2012 and his bachelor of engineering degree from the Department of Automation, Tsinghua University in 2007. Prior to his current position, he was a postdoctoral scholar at Carnegie Mellon University in 2013 and at California Institute of Technology from 2013 to 2015. His research interests include secure control systems and networked control systems, with applications in sensor networks and power grids.

Adrian Perrig is a Professor at the Department of Computer Science at ETH Zürich, Switzerland, where he leads the network security group. He is also a Distinguished Fellow at CyLab, and an Adjunct Professor of Electrical and Computer Engineering, and Engineering and Public Policy, at Carnegie Mellon University. Dr. Perrig's research revolves around building secure systems; in particular, he is working on the SCION secure future Internet architecture.

Alexander Roederer is a Ph.D. candidate in computer and information science at the University of Pennsylvania. His research interests include the application of machine learning to high-frequency, multisource data streams—in particular, to develop clinical decision support systems. He received his B.S. in Computer science and mathematics from the University of Miami and his M.S.E. in computer and information science from the University of Pennsylvania.

Matthias Rungger is a postdoctoral researcher at the Technical University of Munich. He is affiliated with the Hybrid Control Systems Group in the Department of Electrical and Computer Engineering. His research interests lie in the broad area of formal methods in control, including analysis and control of cyber-physical systems and abstraction-based controller design. Matthias spent two years, from 2012 to 2014, as postdoctoral researcher at the Electrical Engineering Department of the University of California, Los Angeles. He received his Dr.-Ing. (Ph.D.) in 2011 from the University of Kassel and his Dipl.-Ing. (M.Sc.) in electrical engineering in 2007 from the Technical University of Munich.

Lui Sha graduated with Ph.D. from Carnegie Mellon University in 1985. Currently, he is Donald B. Gillies Chair Professor at University of Illinois at Urbana–Champaign. His team's work on safety-critical real-time systems has impacted many large-scale high-technology programs, including GPS, Space Station, and Mars Pathfinder. Currently, his team is developing the technologies for Certifiable Multicore Avionics and Medical Best Practice Guidance Systems (Medical GPS). He is a co-recipient 2016 of IEEE Simon Ramo Medal, a Fellow of IEEE and ACM, and a member of NASA Advisory Council.

Gabor Simko is a Senior Software Engineer at Google Inc. He received his M.Sc. in technical informatics from Budapest University of Technology and Economics in 2008, and his M.Sc. in biomedical engineering from the same university in 2010. He graduated with a Ph.D. in computer science from Vanderbilt University in 2014, where his thesis topic was the formal semantic specification of domain-specific modeling languages for cyber-physical systems. His interests include speech recognition, voice-activity detection, formal verification of hybrid systems, and formal specification of modeling languages.

Bruno Sinopoli received his Dr. Eng. degree from the University of Padova in 1998 and his M.S. and Ph.D. in electrical engineering from the University of California at Berkeley in 2003 and 2005, respectively. Dr. Sinopoli joined the faculty at Carnegie Mellon University, where he is an Associate Professor in the Department of Electrical and Computer Engineering with courtesy appointments in Mechanical Engineering and in the Robotics Institute; he is also co-director of the Smart Infrastructure Institute. His research interests include the modeling, analysis and design of secure-by-design cyber-physical systems with application to interdependent infrastructures, Internet of Things, and data-driven networking.

Oleg Sokolsky is a Research Associate Professor of Computer and Information Science at the University of Pennsylvania. His research interests include the application of formal methods to the development of cyber-physical systems, architecture modeling and analysis, and specification-based monitoring, as well as software safety certification. He received his Ph.D. in computer science from Stony Brook University.

John A. Stankovic is the BP America Professor in the Computer Science Department at the University of Virginia. He served as chair of the department for 8 years. He is a Fellow of both the IEEE and the ACM. He has been awarded an Honorary Doctorate from the University of York. Dr. Stankovic won the IEEE Real-Time Systems Technical Committee's Award for Outstanding Technical Contributions and Leadership. He also won the IEEE Technical Committee on Distributed Processing's Distinguished Achievement Award (inaugural winner). He has seven Best Paper awards, including one for ACM SenSys 2006. He also has two Best Paper runner-up awards, including one for IPSN 2013. He has also been a finalist for four other Best Paper awards. Dr. Stankovic has an h-index of 107 and more than 41,000 citations. In 2015, he was awarded the University of Virginia Distinguished Scientist Award, and in 2010, he received the School of Engineering's Distinguished Faculty Award. Dr. Stankovic also received a Distinguished Faculty Award from the University of Massachusetts. He has given more than 35 keynote talks at conferences and many distinguished lectures at major universities. Currently, he serves on the National Academy's Computer Science Telecommunications Board. He was the Editor-in-Chief for *IEEE Transactions on Distributed and Parallel Systems* and was founder and co-editor-in-chief for *Real-Time Systems Journal*. His research interests are in real-time systems, wireless sensor networks, wireless health, cyber-physical systems, and the Internet of Things. Dr. Stankovic received his Ph.D. from Brown University.

Janos Sztipanovits is currently the E. Bronson Ingram Distinguished Professor of Engineering at Vanderbilt University and founding director of the Vanderbilt Institute for Software Integrated Systems. Between 1999 and 2002, he worked as program manager and acting deputy director of DARPA Information Technology Office. He leads the CPS Virtual Organization and is co-chair of the CPS Reference Architecture and Definition public working group established by NIST in 2014. In 2014–2015, he served as academic member of the Steering Committee of the Industrial Internet Consortium. Dr. Sztipanovits was elected Fellow of the IEEE in 2000 and external member of the Hungarian Academy of Sciences in 2010.

Paulo Tabuada was born in Lisbon, Portugal, one year after the Carnation Revolution. He received his "Licenciatura" degree in aerospace engineering from Instituto Superior Tecnico, Lisbon, in 1998 and his Ph.D. degree in electrical and computer engineering in 2002 from the Institute for Systems and Robotics, a private research institute associated with Instituto Superior Tecnico. Between January 2002 and July 2003, he was a postdoctoral researcher at the University of Pennsylvania. After spending three years at the University of Notre Dame as an Assistant Professor, he joined the Electrical Engineering Department at the University of California, Los Angeles, where he established and directs the Cyber-Physical Systems Laboratory. Dr. Tabuada's contributions to cyber-physical systems have been recognized by multiple awards, including the NSF Career Award in 2005, the Donald P. Eckman Award in 2009, the George S. Axelby Award in 2011, and the Antonio Ruberti Prize in 2015. In 2009, he co-chaired the International Conference Hybrid Systems: Computation and Control (HSCC'09) and joined its steering committee in 2015. In 2012, he was program co-chair for the Third IFAC Workshop on Distributed Estimation and Control in Networked Systems (NecSys'12), and in 2015, he was program co-chair for the IFAC Conference on Analysis and Design of Hybrid Systems. He also served on the editorial board of *IEEE Embedded Systems Letters* and *IEEE Transactions on Automatic Control*. His latest book, on verification and control of hybrid systems, was published by Springer in 2009.

Index

A

B

D

F

G

H

J

K

L

M

Q

R

S

T

Z